ACCELERATED STUDIES IN PHYSICS AND CHEMISTRY

A Mastery-Oriented Introductory Curriculum

JOHN D. MAYS

AUSTIN, TEXAS
2012

© 2012 John D. Mays

All rights reserved. No part of this book may be reproduced or transmitted in any form or by any means, electronic or mechanical, including photocopying, recording, or by information storage and retrieval systems, without the written permission of the publisher, except by a reviewer who may quote brief passages in a review.

Scriptural quotations are from The Holy Bible, English Standard Version, copyright © 2001 by Crossway Bibles, a publishing ministry of Good News Publishers. Used by permission. All rights reserved.

Published by
Novare Science and Math
P. O. Box 92934
Austin, Texas 78709-2934

novarescienceandmath.com

Printed in the United States of America

ISBN: 978-0-615-60643-9

Other titles by John D. Mays, published by Novare Science and Math:

Teaching Science so that Students Learn Science
 A Paradigm for Christian Schools

The Student Lab Report Handbook
 A Guide to Content, Style and Formatting for Effective Science Lab Reports

These may all be found at novarescienceandmath.com.

To all of my students, present and past.

ACKNOWLEDGEMENTS

As I developed the course for which this book is the new text, I have had the great privilege of working with many excellent people. I wish to express my thanks to my colleagues in the Math-Science Department at Regents School of Austin, particularly Chris Corley, Cathy Waldo and Dr. Christina Swan. Each of these has contributed in many, many ways, both great and small, to helping *Accelerated Studies in Physics and Chemistry* (*ASPC*) become what it is. Thanks also to many of the other excellent faculty at Regents, people who are not only colleagues but good friends.

I have also benefitted tremendously from the hundreds of conversations I have had with my dear friend Dr. Chris Mack, who has also helped me in my teaching in ways too numerous to name.

Finally, I wish to express my deep appreciation to Rod Gilbert, Head of School at Regents School. Without his repeated challenges to me to keep improving things until they were finally right, I would have been tarred and feathered long ago, and *ASPC* would have gone the way of all the earth.

Pax Christi to all of these.

ASPC

CONTENTS

Preface	**i**
Learning vs. Not Learning	i
Skills and Prerequisites	ii
Assignments, Homework and the Weekly Workload	iii
Companion Resources	iii
Preface to Students	v
Chapter I The Nature of Scientific Knowledge	**1**
Truth and Facts	2
Science	2
Theories	2
Hypotheses	5
Experiments	5
The Scientific Method, Experimental Variables and Controls	6
Chapter II Motion and the Medieval Model of the Heavens	**11**
Computations in Physics	11
MKS Units	12
Dimensional Analysis	12
Accuracy and Precision	13
Significant Digits	13
Scientific Notation	16
Problem Solving Methods	16
Motion, Velocity and Acceleration	16
Velocity, Acceleration and Slope	21
Graphical Analysis of Motion	22
The Medieval Model of the Heavens	26
Key Historical Events and Scientists	30
Chapter III Newton's Laws of Motion	**46**
Matter, Inertia, Mass and Force	46
Newton's Laws of Motion	47
Showing Units of Measure in Computations	51
Weight	52
Applying Newton's Laws of Motion	55
How a Rocket Works	56
Chapter IV Variation and Proportion	**61**
The Language of Nature	61
Independent and Dependent Variables	62
Common Types of Variation	63
Normalizing Equations	64
Chapter V Energy	**78**
What is Energy?	79
The Law of Conservation of Energy	79
Forms of Energy	79
The Energy Trail	82
Calculations With Energy	83

Work	87
Applying Conservation of Energy	88
Conservation of Energy Problems	90
The Effect of Friction on a Mechanical System	95
Energy in the Pendulum	95

Chapter VI Heat and Temperature — **102**

Temperature Scales	102
Temperature Unit Conversions	103
Definitions of Common Terms	104
Heat Transfer Processes	106
Kinetic Theory of Gases	109
Specific Heat Capacity	109
Thermal Conductivity	110
Distinguishing Between Specific Heat Capacity and Thermal Conductivity	110

Chapter VII Waves, Sound and Light — **117**

The Anatomy of Waves	118
Categorizing Waves	119
Wave Calculations	120
Wave Phenomena	124
Sound Waves	130
Frequencies of Sound Waves	130
Loudness of Sound	131
Connections Between Scientific and Musical Terms	131
Electromagnetic Waves and Light	132
Harmonics and Timbre	134

Chapter VIII Electricity and DC Circuits — **140**

The Amazing History of Electricity	141
Charge and Static Electricity	146
How Static Electricity Can Form	147
Electric Current	151
The Water Analogy	151
The Basic DC Electric Circuit	154
Two Secrets	155
Electrical Variables and Units	156
Ohm's Law	158
What Exactly Are Resistors and Why Do We Have Them?	158
Through? Across? In?	159
Voltages Are Relative	160
Power in Electrical Circuits	161
Two-Resistor Networks	163
Equivalent Resistance	165
Significant Digits in Circuit Calculations	167
Larger Resistor Networks	168
Kirchhoff's Laws	172
Putting it All Together to Solve DC Circuits	175

Chapter IX Fields and Magnetism — **191**
  Types of Fields — 191
  Laws of Magnetism — 194
  The Right-Hand Rule — 195
  Solenoids, Generators and Transformers — 197
Chapter X Substances — **209**
  Review of Some Basics — 210
  Types of Substances — 210
  More on Solutions — 218
  Physical and Chemical Properties and Changes — 220
Chapter XI Atomic Models — **225**
  The History of Atomic Models — 225
  Density — 234
Chapter XII The Bohr and Quantum Models of the Atom — **242**
  Atomic Spectra and the Bohr Model — 244
  The Quantum Model — 248
  Electron Configuration Notation — 253
  Atomic Masses, Mass Numbers and Isotopes — 254
Chapter XIII Atomic Bonding — **259**
  Atomic Bonds — 260
  Valence Electrons and Energy Levels — 261
  Goals Atoms Seek to Fulfill — 264
  Metallic Bonds — 265
  Ionic Bonds — 266
  Valence Numbers and Ionic Compound Binary Formulas — 269
  Covalent Bonds — 270
  Diatomic Gases — 273
  Hydrogen — 274
  Polyatomic Ions — 274
Chapter XIV Chemical Reactions — **281**
  Four Types of Chemical Reactions — 281
  Balancing Chemical Equations — 284
  Energy in Chemical Reactions — 287
  Reaction Rates and Collision Theory — 289
Appendix A Conversion Factors, Prefixes and Physical Constants — 294
Appendix B Chapter Objective Lists — 295
Appendix C Laboratory Experiments — 302
Appendix D Scientists Required for Cumulative Review — 324
Appendix E Unit Conversions Tutorial — 325
Appendix F References — 329
Index — 331

PREFACE

At the school where I teach we always refer to the Accelerated Studies in Physics and Chemistry course as ASPC. ASPC is an introductory course designed for students who are capable of and motivated toward honors-level study in science and mathematics. I teach the course to high school freshmen, but the course could occur in a different year, depending on the school's science curriculum programming and the ability levels of the students. I developed the philosophy and teaching methods in the course over a ten-year period, and they are described in detail in my book, *Teaching Science so that Students Learn Science*[1]. The reigning paradigm for this course is *mastery* of the basic tools of physics and chemistry. These tools include fundamental concepts and applications, computational skills, analytical skills and historical information.

LEARNING VS. NOT LEARNING

Most high school physical science texts are 400, 500 or 600 pages long, or even longer. I suppose that is because the authors believe that they can't leave anything out, and that an introduction to physics and chemistry should cover everything. The trouble is, there is no way the student can learn everything. On their part, teachers feel the obligation to "cover" what is in the text, or at least most of it, trusting that the author of the text must be a good judge of what course content for a year of science should look like. As a result, teachers often end up teaching through the text superficially, perpetuating what I call the Cram-Pass-Forget cycle. Teaching this way is common in our time, and in classes that operate this way, which is most classes, students cram for tests, pass them, and then forget most of the material in about three weeks.

This disastrous cycle, standard operating procedure in nearly every school in our nation, has brought us to the point of total exasperation. It is a very rare student in our day who masters what you teach him in a math or science class, remembers it, and can use it weeks or months down the road when it is needed. As a teacher, I consider this circumstance unacceptable. And thus the cornerstone of my career as a science and math teacher has been to develop pedagogical methods and curriculum materials that lead students *en masse* to mastery and retention.

ASPC, and this new text, are different from most science courses and most science texts. Over the course of some ten to twelve years teaching ninth grade science I focused on the process of mastery, constantly studying and tweaking the curriculum to discover how much material students should reasonably master and retain in one year. I discovered that in a demanding academic environment, such as the school where I teach, a mastery-based course philosophy requires leaving out many conventional topics, including some that many would assume were sacrosanct for introductory physics, such as simple machines, momentum and mechanical power. But most high school freshmen, including the really bright ones, cannot master a 400+ page text in one year.

---

1 *Teaching Science so that Students Learn Science*, by John D. Mays (2010). Published by Novare Science and Math and available at novarescienceandmath.com.

If mastery is the teaching goal, as it is mine, hard choices must be made so that the course demands remain reasonable within the framework of other course programming.

But while carving the content down to size, I also discovered that mastery of fewer basics is infinitely more powerful and satisfying than covering more topics superficially and forgetting most of them. In fact, a large percentage of my senior physics students still remember a great deal of what they learned from me in ASPC when they were freshmen. As a result, we can move through the senior physics course – which includes all of the standard topics – at a rapid pace. In short, the methods work.

SKILLS AND PREREQUISITES

As you will see when you look inside, the problems in the examples and problem sets almost amount to a celebration of basic math skills that are used ubiquitously in science. These skills include using scientific notation, performing unit conversions, using the metric prefixes, determining significant digits, using memorized formulas for area and volume, and solving equations for unknowns. Every one of these skills is basic and essential. Students study every one of them in their pre-algebra courses. However, science classes in which the use of these skills is continuous so that students truly master them are few. The way to address this problem is to use these skills every day, week after week.

Time and again I have brought classes of ordinary, grade-level students to mastery in the use of these skills. I would emphasize them just as heavily if this text were designed for them. But the content and pace of the course for which this text is designed is calibrated for accelerated students. How much more then should they be expected to handle the use of these skills with near automaticity!

Students in ASPC should have already completed one year of high school algebra. At our school, the students in ASPC are typically in ninth grade, completed Algebra in eighth grade, and are taking Geometry concurrently with ASPC. When I teach the course I assume the students have mastered the emphasized math skills mentioned above. Naturally, some of them haven't. But they usually get up to speed during the first few weeks of class, and since the course quizzes require them to stay there they generally do. Except for the tutorial in Appendix E, this text does not review methods for performing unit conversions or using scientific notation. However, I do review these in detail in class, and nearly all of the computational exercises are designed to give students in-depth practice at both of these skills. It is also important at that time to make sure students know how to use the EE or EXP buttons on their calculators so that values in scientific notation are entered into the calculator correctly. My motto in class is, if I am pushing buttons on my calculator, the students are, too.

At our school we have learned that math placement in and of itself is not adequate to ensure that placement in ASPC is appropriate. We use a variety of other criteria to identify those students who are well organized, study diligently, are responsible to submit and follow up on assignments, have a track record of above average performance on standardized tests, and earn solid grades of B or higher in their eighth grade science and math classes. These are the students most suited for placement in ASPC.

ASSIGNMENTS, HOMEWORK AND THE WEEKLY WORKLOAD

I do not assign very much homework in ASPC. Instead, I usually give students a fair amount of time to work on exercises in class. The major exception to this is the six lab reports, which are completed entirely outside of class.

The reason outside assignments are kept to a minimum is that mastery requires regular review and practice. Since my goal for students is mastery, I help them achieve that goal by encouraging them to spend their study time at home rehearsing the material we covered in class to get it firmly in their memories, and working through review exercises to keep older material fresh. If they keep up with these tasks they will spend two to three hours per week outside of class studying and rehearsing course material. Since I do not wish for the student workload to be any higher than this I give students time in class to work on new assignments. Good students complete virtually all of the exercises in class. Less efficient students will inevitably end up completing some of their work at home.

Memory work is a significant part of every science course, and ASPC is no different. It is impossible to think and converse about physics unless one not only understands the concepts, but also knows the major laws, equations, conversion factors, metric prefixes, and a few physical constants by heart. Likewise, no one can participate effectively in a study or discussion in chemistry unless many of the common chemical symbols and formulae are known by heart. This is the reason I require students to memorize specific sets of information, as indicated in the Objectives Lists found at the beginning of each chapter. General cultural and scientific literacy also requires that students have a modest amount of historical information in their heads, so I have built in requirements to this effect in the course. Parents sometimes disagree with me about this, sometimes rather strongly. But having spent a significant portion of my adult life in graduate school, 14 years as a professional in the engineering world, and longer than that as an educator, I am firmly persuaded that my point of view on this is sound.

But having established the need for some memorization, I will also hasten to add that there is no point in having students memorize reference data such as particle masses, element densities, or information from the Periodic Table of the Elements. Instead, students need to learn how to use resources like the Periodic Table or data tables and whenever possible I provide these for students to use on their quizzes and exercises. As a specific case in point, I require students to know the major conversion factors for working within the metric system, and others for working within the U.S. Customary System of units (which most of them have known since they were children). But the only conversion factor I require them to memorize for converting between these two systems is 1 inch = 2.54 centimeters. This one factor is used a lot and has the beauty of being exact. One can also get by with it if one has to even without any of the other cross-system factors. Other factors for converting miles to meters or gallons to liters can always be looked up when needed.

COMPANION RESOURCES

There are four important companion resources designed to be used alongside this text. These are all available from Novare Science and Math at novarescienceandmath.com.

Companion Volume: *Teaching Science so that Students Learn Science*

Accelerated or not, to achieve the mastery and automaticity described above, with the basic math skills as well as with all of the physics and chemistry content, students need to use what they learn every week. The weekly cumulative quiz regimen I use in ASPC is a very important part of this process, and is one of the hallmarks of the course. The exercises, Weekly Review Guides, "daily questions," and other activities are all oriented toward enabling students to perform well on these quizzes, which account for the large majority of each student's grade. As I mentioned above, all of these teaching methods are described in *Teaching Science so that Students Learn Science*. This book should be regarded as an essential companion volume to this text for anyone teaching the course.

Companion Volume: *Favorite Experiments for Physics and Physical Science*

Favorite Experiments describes in detail the background, apparatus and practical considerations for the six experiments listed in the present volume, the Charles' Law demonstration described in Chapter IV, and other demonstrations developed specifically for ASPC. *Favorite Experiments* also describes all of the experiments I use in my upper-level Physics course. Each presentation includes illustrative photographs. The background material students need for each experiment is included in the present volume, in Appendix C. But for the full details to assure that each experiment is a success, teachers will want to avail themselves of the detailed descriptions available in *Favorite Experiments*.

Teacher Resource CD

To facilitate the teaching methodology that teachers should use with *ASPC*, the following resources are available on CD.

- *Quiz Bank* This is a set of 29 quizzes for the weekly quiz. The files are in Microsoft Word to facilitate editing. PDF files of handwritten keys for the computations are included.

- *Weekly Review Guide Bank* ASPC students are given a Weekly Review Guide each week beginning with week three. The Weekly Review Guides focus on rehearsal and review of previously-covered material so that it stays fresh. This set of 25 review guides is also formatted in Microsoft Word.

- *Course Schedule* This lesson schedule is based on class meetings four days per week, and covers the entire text in one regular school year.

Companion Student Text: *The Student Lab Report Handbook*

Copies of this book should be supplied to high school freshmen so they can use the book as a resource for science lab report writing year after year throughout high school and on into college. The *Handbook* presents virtually everything students need to know in order to write excellent lab reports.

PREFACE TO STUDENTS

This course is designed to challenge students while bringing them to a solid level of mastery in introductory physics and chemistry. For you to succeed in ASPC there are a few essential things you, as the student, must bring to the table.

First, this course is taught at an accelerated, or honors, level. Although equal before God, we are not all created with equal abilities in every area. To succeed in ASPC you should have above average abilities in mathematics (although hard work can compensate quite a bit).

The second thing you must have is a sincere desire to learn. This is not the same as liking science. Nearly everyone likes science. We are all fascinated by octopuses and lasers, explosions and rockets. But since ASPC is specifically designed to enable the student to learn, master and retain core knowledge and skills, you must desire to learn. And learning takes work, as does achievement in anything. If you play on a sports team, or if you play a musical instrument, you probably really like doing it. But you also know that it takes a lot of practice to get good at it. Some of this practice is painful and a lot of it is exhausting. But because you love your sport or your instrument you don't mind the pain, the fatigue, and the hours spent.

The same things apply to academics in any course designed the way this one is. If you are really to master any subject of study, real effort is required. You need to desire to learn, knowing that this means hard work is involved. If you are not prepared to work diligently, either because of indifference, or laziness, or lack of motivation, then this course is not for you. You need to sign up for the regular science course available at your school rather than ASPC.

Third, you need to ask questions. Having taught this course for nearly ten years, I can attest that students who are engaged in learning ask questions. If you never ask any, it is a sign to your teacher that there is trouble ahead for you. When reviewing new topics questions will always arise. Ask them! When working through problem sets, there will be times when you do not understand something. Ask your instructor about it. This is an important two-way street. If you are engaged, questions will arise. You will get the answers you need by asking your instructor. And when you do, the instructor knows that since you are asking questions you are engaged with the material. The instructor will also know from your questions how well you are progressing, and this will enable him or her to bring things into the classroom that will help you and other students along.

Fourth, you need to be organized. If you are studying well, your notebook will be well organized. If it is not well organized, it is a sign that you are not doing the work that is required. When students claim that they are studying hard while their notebooks are messy and disorganized, it is clear that their claims are exaggerated. Effective study in ASPC requires you to have separate sections in your course binder for notes, quizzes, review guides, homework papers, lab reports,

Not a good sign.

and practice problems. These papers should all be filed in your binder, in order, so you can easily find and use them as you review.

Finally, you must apply yourself to all of the exercises. In addition to the regular exercises assigned in each chapter, you will be given a Weekly Review Guide containing a list of exercises and review activities. The review guides will tell you to make and rehearse flash cards, work problems, recite memory work, and other tasks. You must be diligent about accomplishing these tasks if you desire to succeed in ASPC. I have outlined the essential components for a complete study strategy in the box below.

For years students have said that ASPC was one of the classes in which they learned the most. If you have the aptitude and the interest, and you are ready to work hard, you will learn a lot and have a great year.

A SOLID STUDY STRATEGY

Your grade in ASPC will be very strongly based on your performance on the weekly quizzes that will occur throughout the year. These quizzes are cumulative, which means that once new material is covered in class you are responsible for it on quizzes all year long.

To be prepared for the weekly cumulative quizzes, successful students always establish a weekly study regimen encompassing each of the tasks listed below. You should spread out your review work for ASPC so that you spend time with the material at least two or three separate days each week. Most students find that an hour spent two to three times per week is adequate for solid performance in the course.

These are the documents you must pay attention to and use in your weekly studies:

- Chapter Objectives Lists (14 total for the year)
- Scientists List (see Appendix D)
- Conversion Factors and Constants (see Appendix A)
- Weekly Review Guides

1. Study the Objectives List for each new chapter carefully. Make it your policy that you will be able to do everything on the list (that is, for the objectives that have been covered so far in class) before quiz day each week.

2. Look over Objectives Lists from previous chapters regularly. Identify any item that you cannot do or cannot remember how to do and follow up on it.

3. Develop, maintain and practice flash cards for each major category of information that you need to know. I recommend these four separate stacks

of flash cards: 1) Technical terms, laws and equations; 2) Scientists and experiments; 3) Special lists to memorize (as indicated by the Objectives Lists); and 4) Conversion factors, prefixes and constants. Also, on cards for equations, put down the units of measure for the variables involved and make saying those units part of your flash card practice routine.

4. Read every chapter in this text at least once. Ideally, every time your instructor covers new material you should read the sections in this book corresponding to that material within 24 hours.

5. Go through the exercises described in the Weekly Review Guide every week. Work each of the four review computations. The Review Guide will prompt you to rehearse your flash cards, review older topics, and so on. Take the Weekly Review Guide seriously and do what it says.

6. Raise questions in class as often as you can. Asking questions and interacting with the instructor and the rest of the class is a very effective way to help your brain engage, focus and remember.

7. Go back and read the chapters in this book again when you are a month or two down the road. You will be amazed at how much easier it is to remember things when you have reread a chapter. (Besides, reading is more fun than flash cards.)

8. When you are working on exercises involving computations, check your answers against the answer key. Every time you get an incorrect answer, dig in and stay with the problem until you identify your mistake and obtain the correct answer. If you can't figure out a problem after 10 or 15 minutes, raise the question in class.

9. Every time you lose significant points on a quiz, follow up and fill in the gaps in your learning. If you didn't understand something, raise the question with your instructor. If you forgot something, rehearse it more thoroughly until you have it down. If you failed to commit something to memory, or didn't have it in your flash cards, then add it to the cards and commit it to memory. If you were not proficient enough at one or more of the computations, look up some similar problems from the exercises or from previous quizzes and practice them thoroughly. Always follow up *before the next quiz*. Remember, the quizzes are cumulative and the same questions are going to come up again and again.

If you have the aptitude for studying science and math at an accelerated level and you study for the course according to this study plan, you cannot help but be successful in the course.

(By the way, you should apply this same strategy to many of your other classes. It will work there, too!)

ASPC

CHAPTER I
THE NATURE OF SCIENTIFIC KNOWLEDGE

> **OBJECTIVES**
>
> After studying this chapter and completing the exercises, students will be able to do each of the following tasks, using supporting terms and principles as necessary:
>
> 1. Define science, theory, hypothesis, and scientific fact.
> 2. Explain the difference between truth and scientific facts, and describe how we obtain knowledge of each.
> 3. Describe the "Cycle of Scientific Enterprise," including the relationships between facts, theories, hypotheses, and experiments.
> 4. Explain what a theory is and describe the two main characteristics of a theory.
> 5. Explain what is meant by the statement, "a theory is a model."
> 6. Explain the role and importance of theories in scientific research.
> 7. State and describe the steps of the "scientific method."
> 8. Define explanatory, response and lurking variables in the context of an experiment.
> 9. Explain why an experiment must be designed to test only one explanatory variable at a time. Use the procedures the class followed in the Pendulum Lab as a case in point.
> 10. Explain the purpose of the control group in an experiment.
> 11. Describe the possible implications of a negative experimental result. In other words, if the hypothesis is not confirmed, explain what this might imply about the experiment, the hypothesis or the theory itself.

There are many different kinds of knowledge. As Christians we are very concerned about truth, which is knowledge that is somehow revealed to us by God. As students of the natural sciences, we are also concerned about scientific facts, which represent a different kind of knowledge.

Some people handle this distinction by calling religious teachings one kind of truth, and scientific teaching a different kind of truth. But scientific knowledge is not static. It is always changing as new discoveries are made. On the other hand, the core teachings of Christianity do not change. This indicates to me that the knowledge we have from the Scriptures is a different kind of knowledge than what we learn from scientific investigations. To handle this distinction I have developed a model of knowledge that will help us understand the differences between what the Scriptures teach us and what scientific investigations teach us. This model is not perfect, nor is it exhaustive, but it is very useful, as all good models are.

TRUTH AND FACTS

Christians believe in truths that have been revealed to us, that are absolute and unchanging. Scientific facts, by their very nature, are not like this. So our definitions for truth and for scientific facts are going to take this into account. **Truth** is knowledge that is revealed to us by God, either by **Special Revelation** (the Bible) or by **General Revelation** (nature). This means that truth is not discovered the same way scientific facts are discovered. True propositions are true for all people, all times and all places. Truth never changes. Some examples of revealed truths are:

- Jesus is the divine Son of God.
- All have sinned and fall short of what God requires.
- All people must die once and then face judgment.
- God is the creator of all that is.
- God loves us.

Scientific **facts** (which we will simply call facts) are propositions that are supported by a great deal of evidence. They are discovered by observation and experiment, and by making inferences from what we observe or the results of our experiments. Facts are *correct as far as we know*, but they can change as more information becomes known.

The distinction between truth and facts is crucial. As believers, we embrace the absolute truths we find in Scripture. Facts can change; truth does not. If we get these confused we may have a hard time discussing and defending the faith and distinguishing between the precious truths we know absolutely, such as that Jesus is our redeemer and rose from the dead, and things that may turn out not to be correct after all.

SCIENCE

Science is the process of using experiment, observation and logical thinking to build "mental models" of the natural world. These mental models are called theories. We do not and cannot know the natural world perfectly or completely, so we construct mental models of how it works. These are our scientific theories. Theories never explain the world to us perfectly. To know the world perfectly we would have to know it as God knows it, which we cannot (at least not in the present age). So theories always have their limits, but we hope they get better and better and more complete over time, explaining more and more physical phenomena (facts).

Scientific knowledge is continuously changing and advancing through a cyclic process that I call **The Cycle of Scientific Enterprise**, represented in Figure 1-1. In the next few paragraphs we will examine how this cycle works.

THEORIES

Theories are the grandest thing in science. In fact, it is fair to say that theories are the *glory* of science, and developing good theories is what science is all about. Electromagnetic field theory, atomic theory, quantum theory, the general theory of relativity – these are all theories in physics that have had a profound effect on scientific progress and on the way we all live. Now, even though many people do not realize it,

Chapter I   The Nature of Scientific Knowledge

Figure 1-1. The Cycle of Scientific Enterprise.

all scientific knowledge is theoretically based. Let me explain. A **theory** is a mental model or explanatory system that explains and relates together most or all of the facts (the data) in a certain sphere of knowledge. A theory is not a hunch or a guess or a wild idea. Theories are the mental structures we use to make sense of the data we have. We cannot understand any scientific data without a theory to organize it and explain it. This is why I wrote that all scientific knowledge is theoretically based. And for this reason, it is inappropriate and scientifically incorrect to scorn these explanatory systems as "merely a theory" or "just a theory." Theories are explanations that account for a lot of different facts. If a theory has stood the test of time that means it has wide support within the scientific community.

It is popular in some circles to speak dismissively of certain scientific theories as if they represented some kind of untested speculation. It is simply incorrect – and very unhelpful – to speak this way. As students in high school science, one of the important things you need to understand is the nature of scientific knowledge, the purpose of theories, and the way scientific knowledge progresses. These are the issues this chapter is about.

All useful scientific theories must possess several characteristics. The two most important ones are:

- The theory accounts for and explains most or all of the related facts.
- The theory enables new hypotheses to be formed and tested.

Theories take decades or even centuries to form. If a theory gets replaced by a new, better theory, this also usually takes decades or even centuries to happen. No theory is ever "proved" or "disproved" and we do not speak of them in this way. We also do not speak of them as being "true," because, as we already learned, we do not use the word "truth" when speaking of scientific knowledge. Instead we speak of facts being correct as far as we know, or of current theories as representing our best understanding, or of theories being good (i.e., useful) models that lead to accurate predictions.

An experiment in which the hypothesis is confirmed is said to support the theory. After such an experiment the theory is stronger, but it is not proved. If a hypothesis is not confirmed by an experiment, the theory might be weakened, but it is not disproved. Scientists require a great deal of experimental evidence before a new theory can be established as the best explanation for a body of data. This is why it takes so long for theories to develop. And since no theory ever explains everything perfectly, there are always phenomena we know about that our best theories do not adequately explain. Of course, scientists continue their work in a certain field hoping eventually to have a theory that does explain all of the facts. But since no theory can explain everything perfectly, it is impossible for one experimental failure to bring down a theory. Just as it takes a lot of evidence to establish a theory, so it would take a large and growing body of conflicting evidence before scientists would abandon an established theory.

Earlier I wrote that theories are mental **models**. This statement needs a bit more explanation. A model is a representation of something, and models are designed for a purpose. You have probably seen a model of the organs in the human body in a science classroom or textbook. A model like this is a physical model, and its purpose is to help people understand how the human body is put together. A mental model is not physical; it is an intellectual understanding, although we often use illustrations or physical models to help communicate to one another our mental ideas. But as in the example of the model of the human body, a theory is also a model. That is, a theory is a representation of how part of the world works. Frequently our models take the form of mathematical equations that allow us to make numerical predictions and calculate the results of experiments. The more accurately a theory represents the way the world works, which we judge by forming new hypotheses and testing them with experiments, the better and more successful the theory is.

To summarize, a good theory represents the natural world accurately. This means the model will be useful, because if a theory is an accurate representation, then it will lead to accurate predictions about nature. When a theory repeatedly leads to predictions that are confirmed in scientific experiments, it is a good theory.

HYPOTHESES

A **hypothesis** is a positively stated, informed prediction about what will happen in certain circumstances. We say a hypothesis is an *informed* prediction because when we form hypotheses we are not just speculating out of the blue. We are applying a certain theoretical understanding of the subject to the new situation before us and predicting what will happen or what we expect to find in the new situation based on the theory the hypothesis is coming from. Every scientific hypothesis is based upon a particular theory.

Often hypotheses are worded as IF-THEN statements, such as, "If various forces are applied to a pickup truck, then the truck will accelerate at a rate that is in direct proportion to the force." Every scientific hypothesis is based on a theory, and it is the hypothesis that is directly tested by an experiment. If the experiment turns out the way the hypothesis predicts, the hypothesis has been confirmed, and the theory it came from is strengthened. Of course, the hypothesis may not be confirmed by the experiment. I will describe how scientists respond to this situation in the next section.

The terms *theory* and *hypothesis* are often used interchangeably in common speech, but in science they mean different things. For this reason you should make note of the distinction.

One more point about hypotheses. A hypothesis that cannot be tested is not a scientific hypothesis. For example, horoscopes purport to predict the future with statements like, "You will meet someone important to your career in the coming weeks." Statements like this are so vague they are untestable, and do not qualify as scientific hypotheses.

EXPERIMENTS

Effective **experiments** are difficult to perform. Thus, for any experimental outcome to become regarded as a "fact" it must be replicated by several different experimental teams, often working in different labs around the world. Once confirmed, the result of an experiment gives rise to new facts. This is the case regardless of whether the hypothesis is confirmed or not. But if the outcome of an experiment does not confirm the hypothesis we have to consider all of the possibilities for why this might have happened. Why didn't our theory, which is our best model of the natural world, enable us to form a correct prediction? There are a number of possibilities.

- The experiment may have been flawed. Scientists will double check everything about the experiment, making sure all equipment is working properly, double checking the calculations, looking for lurking variables that may have inadvertently influenced the outcome, verifying that the measurement instruments are good enough to do the job, and so on. They will also wait for other experimental teams to try the experiment to see if they get the same results or different results, and then compare. (Although, naturally, every scientific team would like to be the first one to complete an important new experiment.)
- The hypothesis may not have been based on a correct understanding of the theory. Maybe the experimenters did not understand the theory well enough, and maybe the hypothesis is not a correct statement of what the theory says will happen.
- The values used in the calculation of the hypothesis' predictions may not have been

accurate or precise enough, throwing off the hypothesis' predictions.
- Finally, if all else fails, and the hypothesis still cannot be confirmed by experiment, it is time to look again at the theory. Maybe the theory can be altered to account for this new fact. If the theory simply cannot account for the new fact, then the theory has a weakness, namely, there are facts it doesn't account for adequately. If enough of these weaknesses accumulate, then over a long period of time (like decades) the theory might eventually need to be replaced with a different theory, that is, another, better theory that does a better job of explaining all the facts we know. Of course, for this to happen someone would have to conceive of a new theory, which usually takes a great deal of scientific insight. And remember, it is also possible that the facts themselves can change.

THE SCIENTIFIC METHOD, EXPERIMENTAL VARIABLES AND CONTROLS

The so-called **scientific method** that you have been studying ever since about fourth grade is simply a way of conducting reliable experiments. Experiments are an important part of the Cycle of Scientific Enterprise, and so the scientific method is important to know. You probably remember studying the steps in the scientific method from prior courses, so I have listed them in Table 1-1 without further comment.

1.	State the problem.	5.	Collect data.
2.	Research the problem.	6.	Analyze the data.
3.	Form a hypothesis.	7.	Form a conclusion.
4.	Conduct an experiment.	8.	Repeat the work.

Table 1-1. Steps in the scientific method.

We will be discussing variables and measurements a lot in this course, so we should take the opportunity here to identify some of the language researchers use during the experimental process. In a scientific experiment the researchers have a question they are trying to answer (from the State the Problem step in the scientific method), and typically it is some kind of question about the way one physical quantity affects another one. So the researchers design an experiment in which one quantity can be manipulated (that is, deliberately varied in a controlled fashion) while the value of another quantity is monitored.

A simple example of this in everyday life that you can easily relate to would be to vary the amount of time you spend each week studying for your math class in order to see what effect the time spent has on the grades you earn. If you reduce the time you spend, will your grades go down? If you increase the time, will they go up? A precise answer depends on a lot of things, of course, including the person involved, but in general we would all agree that if a student varies the study time enough we would expect to see the grades vary as well. And in particular, we would expect more study time to result in higher grades.

Now let us consider this same concept in the context of scientific experiments. An experiment typically involves some kind of complex system that the scientists are modeling. The system could be virtually anything in the natural world – a galaxy, an atom, a mixture of chemicals, a tomato or a badger. The variables in the scientists' mathematical models of the system correspond to the physical quantities that can be manipulated or measured in the system.

When performing an experiment the variable that is deliberately manipulated by the researchers is called the **explanatory variable**. As the explanatory variable is manipulated, the researchers monitor the effect this variation has on the **response variable**. In the example of study time versus math grade, the study time is the explanatory variable and the grade earned is the response variable.

Usually, a good experimental design will allow only one explanatory variable to be manipulated at a time so that the researchers can tell definitively what its effect is on the response variable. If more than one explanatory variable were changing during the course of the experiment, researchers may not be able to tell which one was causing the effect on the response variable.

A third kind of variable that plays a role in experiments is the **lurking variable**. A lurking variable is a variable that is affecting the response variable without the researchers being aware of it. This is bad, of course, because the researchers will not be able to make a correct conclusion about the effect of the known explanatory variables on the response variable under study. So researchers have to study their experimental projects very carefully to minimize the possibility of lurking variables affecting their results.

Often experiments are designed to test some new kind of treatment by comparing the results of the new treatment to those obtained using a conventional treatment or no treatment at all. This is the situation in medical research all the time for experiments testing new therapies, medications, or procedures. In an experiment like this the subjects under study will be divided into two groups, the **control group** and the **experimental group**. The control group receives no treatment or some kind of standard treatment. The experimental group receives the new treatment being tested. The results of the experimental group are assessed by comparing them to the control group.

Another example will help to clarify all of these terms. Let's say researchers have identified a gene that relates to a plant's ability to resist drought. According to their *theoretical understanding* of how genes work in the biological systems of the plant, they *hypothesize* that if they modify the gene in a certain way the plant will be able to bear better fruit during drought conditions. To test this hypothesis by *experiment*, the scientists develop a group of the plants with the modified gene. They then place the plants in a test plot, along with other plants that do not have the modified gene, and see how they perform. The plants with the modification are in the *experimental group*, and the plants without the modified gene are in the *control group*.

The *response variable* is the quality of the plant's fruit. Researchers expect that under drought conditions the fruit of the modified plants will be better than the fruit of the plants that have not been modified. The *explanatory variable* is the presence or absence of the modified gene. The plants are exposed to drought conditions in the experiment. If the modified plants produce higher quality fruit than the control group, then the hypothesis is confirmed, and the theory that led to the hypothesis has gained credibility. One can imagine many different *lurking variables* that could affect the outcome of this experiment without the scientists' awareness. For example, the modified plants could be planted in locations that receive different amounts of moisture or sun than the locations where the control group plants are. In a good experimental design researchers seek to identify such factors and take measures to assure that they do not affect the outcome of the experiment.

CHAPTER I EXERCISES

Here is a tip that applies to all of your written responses in this course: Avoid pronouns! Pronouns almost always make your responses vague or ambiguous. If you want to receive full credit for written responses, avoid them. (Oops. I mean, avoid pronouns!)

Study Questions

Answer the following questions with a few complete sentences.

1. Distinguish between theories and hypotheses.

2. Explain why a single experiment can never prove or disprove a theory.

3. Explain how an experiment can still provide very valuable data, even if the hypothesis under test is not confirmed.

4. Explain the difference between truth and facts, and describe the sources of each.

5. State the two primary characteristics of a theory.

6. Does a theory need to account for *all* known facts? Why or why not?

7. It is common to hear people say, "I don't accept that; it's just a theory." What is the error in a comment like this?

8. Distinguish between facts and theories.

9. Distinguish between explanatory variables, response variables, and lurking variables.

10. Why do good experiments that seek to test some kind of new treatment or therapy include a control group?

11. Explain specifically how the procedure you followed in the Pendulum Lab satisfied every principle of the "scientific method."

12. This chapter argues that scientific facts should not be regarded as *true*. Someone might question this and ask, If they aren't true, then what are they good for? How would you answer this question?

13. Explain what a model is, and why theories are often described as models.

14. Consider an experiment that did not deliver the result the experimenters had expected. In other words, the result was negative because the hypothesis was not confirmed. There are many reasons why this might have

happened. Consider each of the following elements of the Cycle of Scientific Enterprise. For each one, describe how it might have been the driving factor that resulted in the experiment's failure to confirm the hypothesis.

a. the experiment
b. the hypothesis
c. the theory

15. Identify the explanatory and response variables in the Pendulum Lab, and identify two realistic possibilities for ways the results may have been influenced by lurking variables.

CHAPTER II
MOTION AND THE MEDIEVAL MODEL OF THE HEAVENS

> **OBJECTIVES**
>
> Memorize and learn how to use these equations:
>
> $$d = vt \qquad a = \frac{v_f - v_i}{t}$$
>
> After studying this chapter and completing the exercises, students will be able to do each of the following tasks, using supporting terms and principles as necessary:
>
> 1. Define and distinguish between velocity and acceleration.
> 2. Use scientific notation correctly with a scientific calculator.
> 3. Calculate distance, velocity and acceleration using the correct equations, MKS units, and correct dimensional analysis.
> 4. Use from memory the conversion factors, metric prefixes, and physical constants listed in Appendix A.
> 5. Explain the difference between accuracy and precision, and apply these terms to questions about measurement.
> 6. Demonstrate correct understanding of precision by using the correct number of significant digits in calculations and rounding.
> 7. Draw and interpret graphs of distance, velocity, and acceleration vs. time, and describe an object's motion from the graphs.
> 8. Describe the key features of the Ptolemaic model of the heavens, including all of the spheres and regions in the model.
> 9. State several additional features of the medieval model of the heavens and relate them to the theological views of the medieval Church.
> 10. Briefly describe the roles and major discoveries of Copernicus, Tycho, Kepler, and Galileo in the Copernican Revolution. Also, describe the significant later contributions of Isaac Newton and Albert Einstein to our theories of motion and gravity.
> 11. Describe the theoretical shift that occurred in the Copernican Revolution and how the Christian Church was involved.
> 12. State Kepler's three Laws of Planetary Motion.
> 13. Describe how the gravitational theories of Kepler, Newton and Einstein illustrate the way the Cycle of Scientific Enterprise works.

COMPUTATIONS IN PHYSICS

In this chapter you will begin mastering the skill of applying mathematics to the study of physics. To do this well you need to know a number of things about the way measurements are handled in scientific work. You also need to have a solid problem solving strategy that you can depend on to help you solve problems correctly without becoming confused. These are the topics of the next few sections.

MKS UNITS

Units of measure are crucial in science. Science is about making measurements and a measurement without units of measure is a meaningless number. For this reason, your answers to computations in this course must *always* show the units of measure.

The two major systems you should know about for units of measure are the SI (from the French *Système international d'unités*), typically known in the United States as the metric system, and the USCS (U.S. Customary System). You have studied these systems before and should be already familiar with some of the SI units and prefixes.

You are probably also already familiar with how cumbersome the USCS is. One problem is that there are many different units of measure for every kind of physical quantity. Just for measuring distance, for example, we have the inch, foot, yard, and mile. The USCS is also full of random numbers like 3, 12 and 5,280, and there is no inherent connection between units for different types of quantities. By contrast, the SI system is simple. There is only one basic unit for each kind of quantity. Prefixes based on powers of ten are used on all of the units to accommodate different sizes of measurements. And units for different types of quantities relate to one another in some way. Unlike the gallon and the foot, which have nothing to do with each other, the liter is 1,000 cubic centimeters. For all of these reasons the USCS is not used much in scientific work.

A subset of the SI system is the **MKS system**. The MKS system uses the *meter*, the *kilogram*, and the *second* (hence, "MKS") as primary units. There are also four other primary units, some of which we will encounter later on. There are also many derived units that are combinations of these three primary units. Examples of derived units that you will learn about and use in this course are the newton (N) for measuring force, the joule (J) for measuring energy, and the watt (W) for measuring power.

Dealing with different systems of units can become very confusing. But the wonderful thing about sticking to the MKS system is that any calculation performed with MKS units will give a result in MKS units. This is why the MKS system is so handy and why we will use it almost exclusively.

To convert the units of measure given in problems into MKS units you will need to know the conversion factors listed in Appendix A. Appendix A also lists several common unit conversions that you are not required to memorize, but should have handy when working problem assignments.

DIMENSIONAL ANALYSIS

Dimensional analysis is a term that refers to all of the work of dealing with units of measure in computations. This work includes converting units from one set of units to another and using units consistently in equations. You are probably already familiar with methods for performing unit conversions. I will provide you with a lot of practice (coming up soon!), but if you need a refresher on using unit conversion factors to convert from one set of dimensions to another, please refer to the tutorial in Appendix E.

ACCURACY AND PRECISION

The terms **accuracy** and **precision** refer to the limitations inherent in making measurements. Science is all about investigating nature and to do that we must make measurements. Accuracy relates to error, which is the difference between a measured value and the true value. The lower the error is in a measurement, the better the accuracy. Error can be caused by human mistakes, malfunctioning equipment, incorrectly calibrated instruments, or unknown factors that are influencing a measurement without the knowledge of the experimenter. All measurements contain error, because (alas!) perfection is simply not a thing we have access to in this world.

Precision refers to the resolution or degree of "fine-ness" in a measurement. The limit to the precision that can be obtained in a measurement is ultimately dependent on the instrument being used to make the measurement. If you want greater precision, you must use a more precise instrument. The precision of a measurement is indicated by the number of **significant digits** (or significant figures) included when the measurement is written down (see below).

It is important that you are able to distinguish between accuracy and precision. Here is an example to illustrate the difference. Let's say you have a bathroom weight scale that is very accurate. If you measure your weight on it you can be assured that the weight reading is correct. But the scale only reads your weight to the nearest pound. This rounding to the nearest pound is a limit on the precision of the scale. Now imagine that you have a friend that has a different type of scale that reads weights to the nearest 1/100th of a pound. This scale is much more precise, and can give 5-digit weight readings instead of the 3-digit readings the other scale gives. However, the more precise scale might be quite inaccurate. A poor mechanical design might mean that the scale gives you 5-digit readings that are always a few pounds too high or too low.

SIGNIFICANT DIGITS

The precision in a measurement is indicated by the number of significant digits it contains. Thus, the number of digits we write in any value we deal with in science is very important. The number of digits is meaningful, because it shows the precision that was present in the instrument used to make the measurement. So when you perform a calculation with physical quantities (measurements) you can't simply write down all the digits shown by your calculator. There is a correct number of digits to write down. You need to know how to determine what that correct number of digits is, and you must use the correct number of digits in all of your calculations.

Here is the rule for determining how many significant digits there are in a given measurement value:

> *The number of significant digits (or figures) in a number is found by counting all the digits from left to right beginning with the first nonzero digit on the left. When no decimal is present, trailing zeros are not considered significant.*

Let's apply this rule to the following numbers to see how it works.

15,679   This value has 5 significant digits.

21.0005   This value has 6 significant digits.

37,000   This value has only 2 significant digits, because when there is no decimal trailing zeros are not significant. Notice that the word *significant* here is a reference to the *precision* of the measurement, which in this case is rounded to the nearest thousand. The zeros in this value are certainly *important*, but they are not *significant* in the context of precision.

0.0105   This value has 3 significant digits, because we start counting with the first nonzero digit on the left.

0.001350   This value has 4 significant digits. Trailing zeros count when there is a decimal.

The significant digit rules enable us to tell the difference between two measurements like 13.05 m and 13.0500 m. Mathematically, of course, these values are equivalent. But they are different in what they tell us about the process of how the measurements were made. The first measurement has 4 significant digits. The second measurement is more precise. It has 6 significant digits, and was made with a more precise instrument.

Now, just in case you are bothered by the zeros at the end of 37,000 that were not significant, here is one more way to think about significant digits that may help. The precision in a measurement depends on the instrument used to make the measurement. If we express the measurement in different units, this should not change the precision. A measurement of 37,000 grams is equivalent to 37 kilograms. Whether we express this value in grams or kilograms, it still has 2 significant digits.

When you make measurements yourself, as you will do in the laboratory experiments in this course, you need to know the rules for which digits are significant in the reading you are making on the measurement instrument. Life can get tricky when using digital instruments, but for digital instruments commonly found in high school science labs all of the digits are significant except the leading zeros. For analog instruments, here is the rule for determining how many significant digits there are in a measurement you are making:

The significant digits in a measurement are all of the digits known with certainty, plus one digit at the end that must be estimated between the finest marks on the scale of your instrument.

Figure 2-1 illustrates this point with two different types of measurements. The photograph on the left side of this figure shows a rule being used to measure the length of a brass block in millimeters (mm). We know the first two digits of the length with certainty; the block is clearly between 31 mm and 32 mm long. We have to estimate the third significant digit. The marks on the rule are in 0.5 mm increments. Comparing the edge of the block with these marks I would estimate the next digit to be a 6, giving

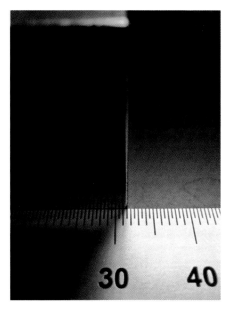 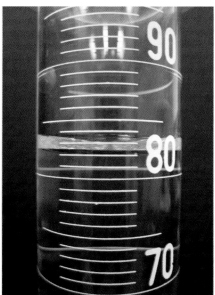

Figure 2-1. Reading the significant digits in a measurement.

a measurement of 31.6 mm. Two digits of this measurement are known with certainty, the third one was estimated, and the measurement has three significant digits.

The photograph on the right side of this figure shows a liquid volume measurement in milliliters (mL) being made with a piece of apparatus called a graduated cylinder. Notice in this figure that when measuring liquid volume the surface of the liquid curls up at the edge of the cylinder. This curved surface is called a meniscus, and in the photo the entire meniscus appears reddish in color, making it easy to see. The liquid measurement must be made at the bottom of the meniscus for most liquids. (You will use graduated cylinders in experiments we perform later on in this course.) For the liquid shown in the figure we know the first two digits of the volume measurement with certainty, because the volume is clearly between 82 mL and 83 mL at the bottom of the meniscus. We have to estimate the third digit, and I would estimate the line to be at 40% of the distance between 82 and 83, giving a reading of 82.4 mL.

It is important for you to keep the significant digits rules in mind when you are making measurements and entering data for your lab reports. The data in your lab journal and the values you use in your calculations and report should correctly reflect the use of the significant digits rules. Note also the helpful fact that when a measurement is written in scientific notation the digits that are written down in front of the power of 10 (the stem, also called the mantissa) *are* the significant digits.

Now that you know the rules for making measurements, you will also need to know the rules for combining the measured values into calculated values, including any unit conversions that must be performed. Here are the two rules for using significant digits in our calculations in this course:

Rule 1   Count the significant digits in each of the values you will use in a calculation, including the conversion factors you will need to use. (Conversion factors that are exact are not considered.) Determine how many significant digits there are in the least precise of all of these values. The result of your calculation must have this same number of significant digits. (This is the rule for multiplying and dividing, which is what most of our calculations entail. There is another rule for adding and subtracting that you will learn when you take chemistry.)

Rule 2   When performing a multi-step calculation you must keep at least one extra digit during intermediate calculations, and round off to the final number of significant digits you need at the very end. This practice will make sure that small round-off errors don't add up during the calculation. This extra digit rule also applies to unit conversions performed as part of the computation.

As I present example problems in the coming chapters I will frequently refer to these rules and show how they apply to the example at hand.

SCIENTIFIC NOTATION

I this course we are assuming you already know how to use scientific notation in computations. However, you must also make sure you are correctly using the EE or EXP feature on your scientific calculator for executing computations that involve values in scientific notation. It is very common for students to do this incorrectly and obtain results that are off by a factor of ten. Since every calculator is a little different in the way its keys are labeled and in the way it presents results in the display, you should be sure to ask your instructor any time you are in doubt about how to enter or read numbers in scientific notation. There are some brief comments on this at the end of Appendix E.

PROBLEM SOLVING METHODS

Organizing problems on your paper in a reliable and orderly fashion is an essential practice. Physics problems can get very complex, and good solution practices can often make the difference between getting most or all of the points for a problem, or getting few or none. Each time you start a new problem you must set it up and follow the steps according to the outline presented in the box on the next page, entitled, "ASPC Problem Solving Method."

MOTION, VELOCITY AND ACCELERATION

In this course we will study two types of **motion**: motion at a constant **velocity**, when an object is not accelerating, and motion with a **uniform acceleration**. Defining these terms is a lot simpler if we stick to motion in one dimension, that is, motion in a straight line. So in this course, this is what we will do and we will leave more complicated kinds of motion to future physics courses. With this simplification we can say that an object's velocity is simply the rate at which the distance to where it started is changing. A girl walking at a velocity of three miles per hour is increasing the distance between herself and where she started at a rate of three miles every hour.

ASPC PROBLEM SOLVING METHOD
Solid Steps to Reliable Problem Solving

In ASPC you are learning how to use math to solve scientific problems. Developing a sound and reliable method for approaching problems is one of the most important elements in the course. You should adhere closely to the following steps:

1. Write down the given quantities at the left side of the page. Include the variable quantities given in the problem statement and the variable you must solve for. Make a mental note of the precision in each given quantity.
2. For each given quantity that is not already in MKS units, work immediately to the right of it to convert the units of measure into MKS units. To help prevent mistakes, always use horizontal fraction bars in your units and unit conversion factors. Write the results of these unit conversions with one extra digit of precision over what you will need in your final result.
3. Write the standard form of the equation you will use to solve the problem.
4. If necessary, use algebra to get the variable you are solving for alone on the left side of the equation. Never put values into the equation until this step is done.
5. Write the equation again with the values in it, using only MKS units, and compute the result.
6. If you were asked to state the answer in non-MKS units, perform the final unit conversion now.
7. Write the result with the correct number of significant digits and the correct units of measure.

EXAMPLE PROBLEM
If you want a complete and happy life, do 'em just like this!

A car is traveling at 35.0 mph. The driver then accelerates uniformly at a rate of 0.15 m/s² for 2 minutes and 10.0 seconds. Determine the final velocity of the car in mph.

Step 1   Write down the given information in a column down the left side of your page, using horizontal lines for the fraction bars in the units of measure.

$v_i = 35.0 \ \dfrac{\text{mi}}{\text{hr}}$

$a = 0.15 \ \dfrac{\text{m}}{\text{s}^2}$

$t = 2 \text{ min } 10.0 \text{ s}$

$v_f = ?$

(This example is continued on the next page.)

Step 2 Perform the needed unit conversions, writing the conversion factors to the right of the given quantities, just where you wrote them in the previous step.

$$v_i = 35.0 \, \frac{\text{mi}}{\text{hr}} \cdot \frac{1609 \text{ m}}{\text{mi}} \cdot \frac{1 \text{ hr}}{3600 \text{ s}} = 15.6 \, \frac{\text{m}}{\text{s}}$$

$$a = 0.15 \, \frac{\text{m}}{\text{s}^2}$$

$$t = 2 \text{ min } 10.0 \text{ s} = 130.0 \text{ s}$$

$$v_f = ?$$

Step 3 Write the equation you will use in its standard form.

$$a = \frac{v_f - v_i}{t}$$

Step 4 Perform the algebra necessary to get the unknown you are solving for alone on the right side of the equation.

$$a = \frac{v_f - v_i}{t}$$

$$at = v_f - v_i$$

$$v_f = v_i + at$$

Step 5 Insert the values (using only values in MKS units) and compute the result.

$$v_f = v_i + at = 15.6 \, \frac{\text{m}}{\text{s}} + 0.15 \, \frac{\text{m}}{\text{s}^2} \cdot 130.0 \text{ s} = 35.1 \, \frac{\text{m}}{\text{s}}$$

Step 6 Convert to non-MKS units, if required in the problem.

$$v_f = 35.1 \, \frac{\text{m}}{\text{s}} \cdot \frac{1 \text{ mi}}{1609 \text{ m}} \cdot \frac{3600 \text{ s}}{1 \text{ hr}} = 78.5 \, \frac{\text{mi}}{\text{hr}}$$

Step 7 Write the result with correct significant digits and units of measure.

$$v_f = 79 \text{ mph}$$

(Optional Step 8: Revel in the satisfaction of knowing that once you get this down you can work physics problems perfectly nearly every time!)

Note that an object's velocity is a measure of how fast it is going, not whether its velocity is changing or not. When the velocity of an object is not constant the object is accelerating, and the value of the acceleration is the rate at which the velocity is increasing or decreasing. If the velocity is increasing or decreasing at a constant rate, as with, say, a falling object, we say the acceleration is *uniform*. In this course, all problems involving acceleration will involve only uniform acceleration. (Calculus is required to solve problems in which the acceleration is not uniform. I doubt you are ready for that!)

Now we are ready to present the first two equations you need to know for solving problems in physics. One equation is for motion at a constant velocity, and the other is for motion with uniform acceleration. For motion with a constant velocity, the equation is

$$d = vt$$

where d is distance in meters (m), v is velocity in meters per second (m/s) and t is time in seconds (s). Note here that the MKS units for velocity are meters per second (m/s).

In this text I am using the colored line and box that follows to indicate an example computation. So here is our first one. In this example I will be using the Problem Solving Method as described in the accompanying box. And remember, all of the unit conversion factors you need are listed in Appendix A.

Sound travels 1,120 ft/s in air. How much time will it take to hear the crack of a gun fired 1,695.5 m away?

First write down the given information and perform the required unit conversions so that all given information is in MKS units. Check to see how many significant digits your result must have, and do the unit conversions with one extra digit. The given speed of sound has three significant digits, so we will do our unit conversions with four digits.

$$v = 1{,}120 \ \frac{\text{ft}}{\text{s}} \cdot \frac{0.3048 \ \text{m}}{\text{ft}} = 341.4 \ \frac{\text{m}}{\text{s}}$$

$$d = 1{,}695.5 \ \text{m}$$

$$t = ?$$

Then write the appropriate equation to use.

$$d = vt$$

Perform any necessary algebra, insert the values in MKS units, and compute the result.

$$t = \frac{d}{v} = \frac{1695.5 \ \text{m}}{341.4 \ \frac{\text{m}}{\text{s}}} = 4.966 \ \text{s}$$

Finally, round the result so that it has the correct number of significant digits. In the velocity unit conversion, and in the calculated result I used 4 significant digits. The

19

given velocity has 3 significant digits and the given distance has 5 significant digits. Thus, our result must be reported with 3 significant digits, but all intermediate calculations must use one extra digit. This is why I used 4 digits. But now we are finished, and our result must be rounded to 3 significant digits, because the least precise measurement in the problem has 3 significant digits. Rounding our result accordingly we have

$t = 4.97$ s

Earlier I defined **acceleration** as changing velocity. If an object's velocity is changing, the object is accelerating, and the value of the acceleration is the rate at which the velocity is changing. The equation we use to calculate uniform acceleration in terms of an initial velocity v_i and a final velocity v_f is

$$a = \frac{v_f - v_i}{t}$$

where a is the acceleration (m/s²), t is the time spent accelerating (s), and v_i and v_f are the initial and final velocities, respectively, (m/s). I will discuss the units for acceleration in more detail in the next section, but here you should note that the MKS units for acceleration are meters per second squared (m/s²).

Because this equation results in negative accelerations when the initial velocity is greater than the final velocity, you can see that a negative value for acceleration means the object is slowing down. In future physics courses you may learn more sophisticated interpretations for what a negative acceleration means, but in this course you are safe associating negative accelerations with decreasing velocity. In common speech people sometimes use the term "deceleration" when an object is slowing down, but mathematically we just say the acceleration is negative.

We must be very careful to distinguish between velocity (m/s) and acceleration (m/s²). Acceleration is a measure of how fast an object's velocity is changing. To see the difference, note that an object can be at rest ($v = 0$) and accelerating at the same instant. This happens every time an object starts from rest, because if an object at rest is to begin moving it must accelerate. It also happens when a ball thrown straight up reaches its highest point and stops for an instant before coming back down. At its highest point the ball is simultaneously at rest and accelerating.

A truck is moving with a velocity of 42 mph (miles per hour) when the driver hits the brakes and brings the truck to a stop. The total time required to stop the truck is 8.75 s. Determine the acceleration of the truck, assuming the acceleration is uniform.

Begin by writing the givens and performing the unit conversions.

$$v_i = 42 \ \frac{mi}{hr} \cdot \frac{1609 \ m}{mi} \cdot \frac{1 \ hr}{3600 \ s} = 18.8 \ \frac{m}{s}$$

$$v_f = 0$$

$$t = 8.75 \ s$$

$$a = ?$$

Now write the equation and complete the problem.

$$a = \frac{v_f - v_i}{t} = \frac{0 - 18.8 \ \frac{m}{s}}{8.75 \ s} = -2.15 \ \frac{m}{s^2}$$

The initial velocity has 2 significant digits, so I did the calculations with 3 significant digits until the end. Now we round off to 2 digits giving

$$a = -2.2 \ \frac{m}{s^2}$$

If you kept all of the digits in your calculator throughout the calculation and rounded to 2 digits at the end you would have $-2.1 \ m/s^2$. This answer is fine, too. Remember, the last digit of a measurement or computation always contains some uncertainty, so it is reasonable to expect small variations in the last significant digit.

If you haven't yet read the example problem in the green Problem Solving Method Box you should read it now to see a slightly more difficult example using this same equation.

VELOCITY, ACCELERATION AND SLOPE

You know that the generic equation for a line in slope-intercept form is

$$y = mx + b$$

If the line goes through the origin, the y-intercept is zero and this equation becomes

$$y = mx$$

In order to understand velocity and acceleration better, and to prepare for some of the analysis we will undertake next, let's take a minute to compare this equation to the two equations for velocity and acceleration introduced above.

The equation relating distance and velocity is $d = vt$. This equation is a linear equation in the same form as the equation for a line. The dependent variable (y) in this equation is distance, d, the independent variable (x) is time, t, and the slope (m) is the velocity v. From this comparison you should be able to see that if an object is moving at a constant velocity, the velocity is the *rate* at which the distance is changing. This means that on a graph of distance versus time the slope of the line is the velocity of the moving object. The steeper the slope, the faster the object is going. A horizontal line has a slope of zero, meaning the velocity is zero and the object is at rest.

Let's look at the acceleration equation the same way. The equation we use for uniform acceleration is

$$a = \frac{v_f - v_i}{t}$$

Assume for a moment that the initial velocity, v_i, is zero, and the final velocity, v_f, is just v, the velocity at any particular time. This gives us

$$a = \frac{v}{t}$$

Now if we then solve this equation for v we have

$$v = at$$

As before, this equation is a linear equation in which the dependent variable is the velocity, v, the independent variable is time, t, and the slope is the acceleration, a. Written this way we can see that the acceleration is the *rate* at which the velocity is changing.

If you think about this for a bit it will help you make sense of the units for acceleration, m/s^2. These units are often a mystery to students. But since acceleration is the rate at which the velocity is changing, the acceleration simply means that the velocity is increasing by so many meters per second each second. Put another way, the slope of a line is "rise over run." If velocity with units of m/s is on the vertical axis of a graph, and time with units of s is on the horizontal axis, then the units of the slope of a line on the graph (rise over run) will be the velocity units over the time units, or

$$\frac{\frac{m}{s}}{s} = \frac{\frac{m}{s}}{\frac{s}{1}} = \frac{m}{s} \cdot \frac{1}{s} = \frac{m}{s^2}$$

In summary, velocity is simply the rate an object's distance (from wherever it started) is changing. Graphically, distance (rise) is in meters and time (run) is in seconds, so the velocity, which is the slope on a graph of $d = vt$, has units of m/s. Likewise, the acceleration is the rate an object's velocity is changing. Graphically, velocity (rise) is in meters per second (m/s) and time (run) is in seconds, so the acceleration, which is the slope of a line on a graph of $v = at$, has units of m/s^2.

GRAPHICAL ANALYSIS OF MOTION

Analyzing motion graphically is a powerful tool. When you understand the graphical analysis in this section you will be off to a solid start in being able to think conceptually and quantitatively about motion the way a student of physics should be able to do.

The graphs in Figure 2-2 show representative curves for three different motion states an object can be in: at rest (no motion), moving at a constant velocity, and accelerating uniformly. Each vertical group of curves depicts distance, velocity, and acceleration as functions of time. In the first group the object is at rest, which means the distance from the object to the "starting line" (from which we measure how far it has gone) is a constant. The only way this can happen is for the object to be at rest. (Well, yes, technically the object could be moving in a circle, but we are not going to consider that in this course!)

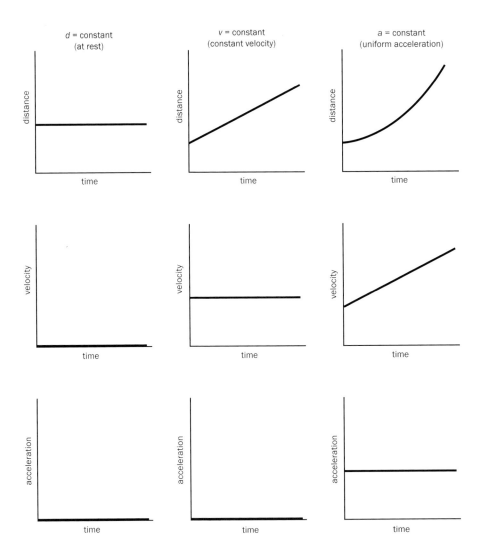

Figure 2-2. Graphical depictions of states of motion.

In the second group the velocity is a constant. This means the distance to the starting line is changing at a constant rate, and the object is not accelerating. In the third group the object is accelerating uniformly, which means its velocity is changing at a constant rate. You will notice that I always started the distance graphs at an arbitrary point. This is just to make the curves as generic as possible. When drawing your own curves if you wish to start the distance curves at the origin that is fine. However, the origin on a velocity graph represents a velocity of zero, meaning the object is at rest. Thus, you can only start a velocity graph at the origin if you are depicting an object that is starting from rest.

Interestingly, when an object is accelerating the distance graph will no longer be linear. Instead, it is a type of curve called a quadratic. We will learn a bit more about this type of curve in Chapter IV, but essentially it is the type of curve that occurs when the relationship between y and x, that is, between the distance and time in this case, can be modeled by an equation such as $y = kx^2$, or, in our specific case, $d = kt^2$. In an equation like this k is simply a constant that depends on the circumstances.

In these time diagrams the distance graph is the only one that will ever be curved. All the others are linear. Also make note that in this course if an object is accelerating the acceleration will always be uniform. On graphs like these this means the acceleration will always be a horizontal line. This horizontal line will be at a positive value when the object is speeding up, and at a negative value when the object is slowing down.

As I mentioned above, when an object is accelerating the graph of the object's distance vs. time will have a quadratic curvature. This curvature makes this graph more complex than the others, so we should look more closely at this type of graph. There are four ways this graph can curve, depending on what the object is doing. These are shown in Figure 2-3.

The first thing to notice is that if the object is going forward the distance must be increasing, so the curve has to be going up. If it is getting steeper, then the object is speeding up. If it is getting less steep, the object is slowing down. The only way the curve can slope downward is if the object is going backwards, so that the distance to the starting line is decreasing. Just as before, if the curve is getting steeper, the object is going faster. If the curve is getting less steep (more horizontal) the object is slowing down.

Figure 2-4 highlights additional details of distance and velocity graphs. When a distance graph curves all the way over to horizontal it means the object has come

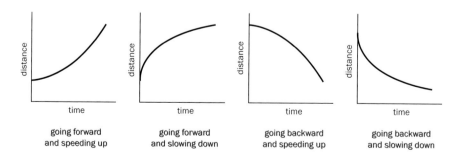

Figure 2-3. Curvature possibilities for distance vs. time graphs.

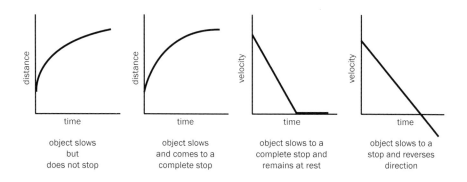

Figure 2-4. Details for distance and velocity graphs.

to a stop. If it stays stopped, then the distance graph will become a horizontal line. A downward sloping velocity graph means the object is slowing, but if the velocity curve actually goes below the horizontal axis that means the velocity is negative and the object is going in reverse.

When given a description of an object's motion you need to be able to piece together segments from different representative curves to represent motion in different time intervals. For example, a vehicle could be traveling at one velocity, accelerate for a while, and then travel at a new velocity. There would be three distinct time intervals associated with this motion, one for the constant speed at the beginning, one for the acceleration in the middle, and one for the new constant speed motion at the end.

Consider a car driving down the road at a constant velocity. The driver then accelerates uniformly to a higher velocity, and then continues at this new, higher velocity. Draw diagrams of d vs. t, v vs. t, and a vs. t depicting this scenario. Show the time intervals distinctly in your diagrams and align your time intervals vertically.

There are three distinct time intervals in this scenario. First, there is a period of time at the initial constant velocity. Then there is an interval when the car is accelerating. Finally, the car continues at the new velocity.

The graphical depiction of this sequence of events is shown in Figure 2-5. I have drawn the three graphs (distance, velocity and acceleration vs. time) above one another so that the time intervals can be aligned in each graph. The three time intervals are separated by the dashed lines. Notice some key details. First, on the distance graph, the slope is higher in the third interval than in the first, because the car's velocity is higher. Second, the two linear sections on the distance graph are smoothly connected to the curved (quadratic) section in the middle. There should never be kinks (sharp corners) in a distance graph. The quadratic curvature only occurs in the distance graph, and only when the car is accelerating. Finally, the acceleration graph is zero everywhere except in the middle when the car is accelerating, and there the curve is a horizontal line representing uniform acceleration.

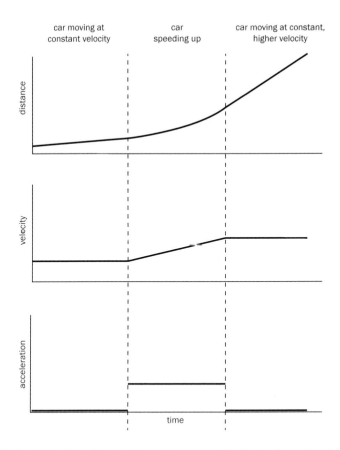

Figure 2-5. Combining time intervals to make a complete graph of an object's motion.

THE MEDIEVAL MODEL OF THE HEAVENS

As outlined in the historical survey in the next section, the view people had of the planets and stars in the medieval period began back with the Greek philosophers Plato and Aristotle in the 5th and 4th centuries BC. The famous Alexandrian astronomer Ptolemy worked out a detailed mathematical system for this model in the 2nd century AD. Over the next thousand years this model of the heavens was adopted by everyone in the West, including the Christian Church. Then the Copernican Revolution happened and over the course of 150 years this model completely collapsed. The reason we study this model in detail in this course is that the history of how it developed and how it crashed is a world-class example of how science works through theories to model nature and how those theories evolve and change over time, just as we discussed in the first chapter. (The Copernican Revolution also happens to be very interesting history!)

We are going to consider many of the mechanical features of Ptolemy's model, but before we do you need to consider a few things about the way the motion of the planets in the night sky appears to observers on earth. If you go out and look at, say,

Mars each night and make a note of its location against the stars, you will see that it is in a slightly different place each night. The planet is gradually working its way along in a pathway against the starry background night after night. If you track the planet for several months or a year it will move quite far. Moreover, there will be periods of time lasting several weeks when the nightly progress of the planet reverses course. Mars appears to be backing up! This apparent backing up is called **retrograde motion**.

Nowadays we easily explain the movement of the planets in the sky, as well as retrograde motion, by looking at the geometry of where earth is and where the planet is as we all orbit around the sun. None of the planets actually reverses course in its orbit, but depending on where earth is and where the planet is (on the same side of the sun, on opposite sides of the sun, and so on) a given planet will appear to be moving one direction or another relative to the stars. But as I will describe below, in the Ptolemaic model all of the planets, and the sun as well, orbit around the earth, not the sun. Additionally, the heavenly bodies all rotate around the earth once each day. This makes the smaller motion of the planets, and their retrograde motions, harder to explain.

Now, to you and me, who all grew up in a time when we know for a fact that the planets and the earth orbit the sun, it seems obvious to us that day and night are caused by the earth's rotation on its axis. We have heard about this all of our lives. But stop and consider that if all we had to go on was our simple observations, it does *appear* that everything is orbiting around the earth. Don't the sun and moon rise each day and track across the sky and set? Don't the planets and stars all do the same thing? Also, it doesn't feel at all like earth is rotating. We all know that anytime we spin in a circle, like people on a merry-go-round, we have to hold on to keep from falling off. We can also feel the wind in our hair. Again, if we had something with us on the merry-go-round that was tall and flexible, like a sapling, it would not stay vertical when it is moving in a circular fashion like this. Instead, it would bend over because of the acceleration pulling it in its circular motion.

These principles seemed *obvious* to *everyone* before 1500, and to everyone except a few cutting-edge astronomers right up to 1642 when Galileo died. Only a crazy person would imagine that the earth was spinning. They all knew that the earth was huge – Eratosthenes had made a very accurate estimate of the earth's circumference as far back as 240 BC. So if something that big were spinning in a circle once a day the people on its surface would be moving very fast (1,000 miles per hour on the equator, actually!) and thus we would have to be hanging on for dear life! The trees would be laying down, and we would be constantly feeling winds that would make a hurricane feel like a calm summer day! People used these arguments all the way up until the time of Galileo to prove that there was no way the earth was orbiting the sun and spinning around once a day. These were very persuasive arguments.

So the prevailing model of the heavens continued to be the Ptolemaic model for a very long time. Now, let's go back to how the Ptolemaic model deals with the planetary motions. In Ptolemy's model the planets did not move in simple circular orbits around the earth. Instead, they moved in **epicycles**. An epicycle is a circular path around a center point, and the center point itself travels on a circular path around the earth. Figure 2-6 depicts a planet moving in a path defined by an epicycle. The motion of a planet moving on an epicycle would be like that of a person in the "tea cup" ride at an amusement park.

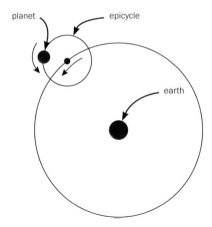

Figure 2-6. A planet moving in a path defined by an epicycle around the earth.

With this background, we are ready now to list some of the basic mechanical features of the Ptolemaic model. Here are a few:

- There are seven heavenly bodies.
- All heavenly bodies are perfectly spherical.
- All heavenly bodies move in circular orbits called **spheres**.
- All of the spheres are centered on the earth, so this system is a **geocentric** system.
- Corruption and change only exist on earth. All other places in the universe, including all the heavenly bodies and stars, are perfect and unchanging.
- All of the spheres containing the heavenly bodies and all the stars rotate completely around the earth every 24 hours.
- Epicycles are used to explain retrograde motion.
- The heavenly bodies inhabit spheres around the earth where their orbits are. In the model there are nine spheres plus the region beyond the spheres. The first seven spheres contain the heavenly bodies. The arrangement of the spheres is as follows:

Sphere 1	Moon
Sphere 2	Mercury
Sphere 3	Venus
Sphere 4	Sun
Sphere 5	Mars
Sphere 6	Jupiter
Sphere 7	Saturn
Sphere 8	The Firmament. This region consists of the stars arranged in their constellations according to the zodiac.

| Sphere 9 | The *Primum Mobile*. This Latin name means "first mover." This sphere rotates around the earth every 24 hours and drags all the other spheres with it, making them all move. |
| Beyond | The Empyrean. This is the region beyond the spheres. The Empyrean is the abode of God, or the gods. |

Among the different astronomers of the ancient world there were those who held to variations on this basic model. For example, some astronomers reckoned that Mercury and Venus orbited the sun while the other heavenly bodies orbited the earth. But the basic Ptolemaic model was as described here. Figure 2-7 depicts this arrangement graphically.

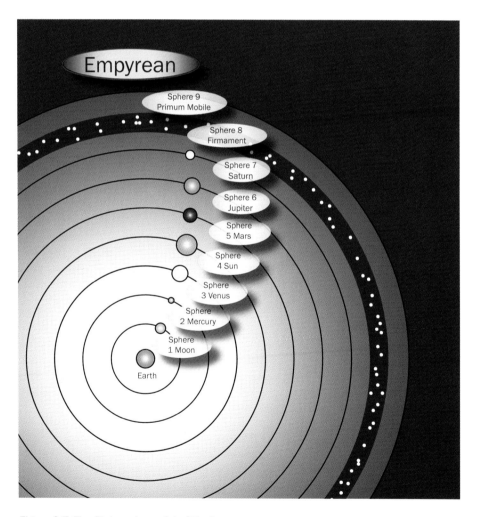

Figure 2-7. The Ptolemaic model of the heavens.

KEY HISTORICAL EVENTS AND SCIENTISTS

Figure 2-8. Ptolemy.

The model outlined above was first suggested by Plato and Aristotle in the 5th and 4th centuries BC, but was worked out into full mathematical detail by **Ptolemy**, an Alexandrian astronomer, in the second century AD (Figure 2-8). By the medieval period 1,000 years later, Church theology had become closely intertwined with the geocentric Ptolemaic model.

The Ptolemaic model made sense to everyone at the time. To the authorities in the Church, the Ptolemaic model seemed to line up with the Bible in many ways. A few of these correlations may be described as follows:

- Many biblical passages seemed to describe the sun and stars going around the earth:

    Psalm 104:5 *He set the earth on its foundations, so that it should never be moved.*
    Psalm 104:19 *He made the moon to mark the seasons; the sun knows its time for setting.*
    Psalm 19:6 *Its rising is from the end of the heavens, and its circuit to the end of them.* (This verse refers to the sun.)
    Ecclesiastes 1:5 *The sun rises and the sun goes down, and hastens to the place where it rises.*
    Judges 5:20 *From heaven the stars fought, from their courses they fought.*

- Seven is the biblical number of perfection, so it made sense that God's creation contains seven heavenly bodies.
- Circles are the most perfect shape, regarded as divine from the times of the ancient Greeks, so the spherical bodies inhabiting spheres in which they move seemed to reflect the perfection of their Creator.
- Corruption was thought to exist only on earth, and it seemed this was obviously because of the curse that resulted from the Fall of man.

Because of these seemingly close linkages between the Bible and the Ptolemaic model, the Church in the medieval period made the category error of believing the model to be the truth. As you recall from Chapter I, this confusion of scientific theory and truth was a mistake. In fact, it was a very bad mistake, with horribly tragic consequences. Scientific theories are not truth, they are models we use to understand the natural world. But Church doctrine in this period was closely linked with the geocentric view of the heavens, so to reject this view was to reject the teaching of the Church and the Bible itself. Such rejection was regarded as heresy, and was punishable by excommunication, banishment, or death. The woeful episode that followed was not finally over until 2000, when Pope John Paul II issued a formal apology for the trial of Galileo.

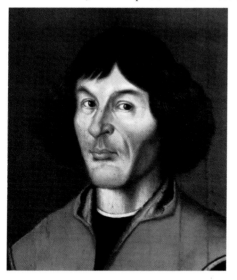

Nicolaus Copernicus (1473-1543), a Polish astronomer, first proposed a detailed heliocentric model, with the earth rotating on its axis, all the planets moving in circular orbits around the sun, and the moon orbiting the earth (Figure 2-9). Copernicus' system was about as accurate, and about as complex, as the Ptolemaic system. Copernicus' model still used circular orbits, and because of this he still had to use epicycles to make the model work. Still, the model was an arrangement that was a lot closer to today's understanding than the Ptolemaic model was. Copernicus didn't want to offend the Church with his ideas, so he published his work privately to his friends in 1514. His work only became public at his death in 1543.

Figure 2-9. Nicolaus Copernicus.

Tycho Brahe (1546-1601), a Danish nobleman and astronomer, built an observatory in Copenhagen (Figure 2-10). (In the literature everyone calls him Tycho; no one

---

Nicolaus Copernicus gave us a beautiful description of our Creator, one that is often quoted. In the preface to his book *On the Revolutions of Heavenly Spheres* he dedicated the book to Pope Paul III. Copernicus wrote:

"I can reckon easily enough, Most Holy Father, that as soon as certain people learn that in these books of mine which I have written about the revolutions of the spheres of the world I attribute certain motions to the terrestrial globe, they will immediately shout to have me and my opinions hooted off the stage."

Copernicus went on to review the shortcomings of the work of other astronomers, and then justified his own work:

"Accordingly, when I had meditated upon this lack of certitude in the traditional mathematics concerning the composition of movements of the spheres of the world, I began to be annoyed that the philosophers, who in other respects had made a very careful scrutiny of the least details of the world, had discovered no sure scheme for the movements of the machinery of the world, which has been built for us by the Best and Most Orderly Workman of all."

– from Nicolaus Copernicus, *On the Revolutions of Heavenly Spheres* (1543)

calls him Brahe. I love the name Tycho, so I will call him that, too.) In 1597 Tycho moved to Prague and became Imperial Mathematician for Rudolph II, who was King of Bohemia and Holy Roman Emperor there. Tycho spent his life cataloging astronomical data for over 1,000 stars (doing this without a telescope), and he published a catalog identifying the positions of 777 stars with unprecedented accuracy.

Tycho witnessed two key astronomical events. First, in 1563 he observed a conjunction between Jupiter and Saturn. A conjunction is when two planets are in a straight line with the earth, so that from earth they appear to be in the same place in the sky. Tycho predicted the date for this conjunction using Copernicus' new model. The prediction was close (this is good) but was still off by a few days (not so good) indicating that there was still something lacking in Copernicus' model. (There was. The orbits are not circular as Copernicus thought.) Second, in 1572 Tycho observed what he called a "nova" (which is Latin for *new*; today we would call it a supernova) and proved that it was a new star. This discovery rocked the Renaissance world because it was very strong evidence that the stars were not perfect and unchanging as Aristotle had thought and as medieval cosmology and Church doctrine declared.

Figure 2-10. Tycho Brahe.

Johannes Kepler (1571-1630), a German astronomer, was invited in 1600 to join the research staff at Tycho's observatory in Prague, and became the Imperial Mathematician there the very next year after Tycho's death (Figure 2-11). He had access to Tycho's massive body of research data and Kepler used it to develop his famous three **Laws of Planetary Motion**, the first two of which were published in 1609. He discovered the third law a few years later and published it in 1619.

Kepler was an amazing scientist. In addition to his astronomical discoveries, he made important discoveries in geometry and optics. He figured out some of the major principles of gravity later synthesized

Figure 2-11. Johannes Kepler.

by Isaac Newton, and he was the first to hypothesize that the sun exerted a force on the earth.

Kepler's Laws of Planetary Motion are as follows:

First Law	Each of the planetary orbits is an ellipse, with the sun at one focus.

You probably haven't studied **ellipses** yet in math. An ellipse is a geometric figure similar to a circle, except instead of having a single point locating the center, an ellipse has two points called *foci* (plural, and pronounced FOH-sigh; the singular is *focus*) that define the shape of the ellipse. Out in space the planets travel on paths defined by a geometrical ellipse. The planetary orbits all have one focus located at the same place in space, and this is where the sun is. A planet in an elliptical orbit is depicted in Figure 2-12.

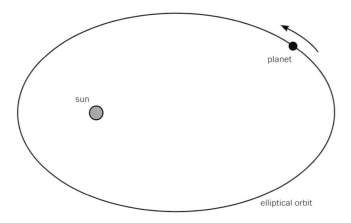

Figure 2-12. A planet in an elliptical orbit around the sun.

Second Law	A line drawn from the sun to any planet will trace out a region in space that has equal area for any equivalent length of time.

The Second Law is depicted in Figure 2-13. Since the sun is off-center, this law implies that the planets travel faster when they are closer to the sun.

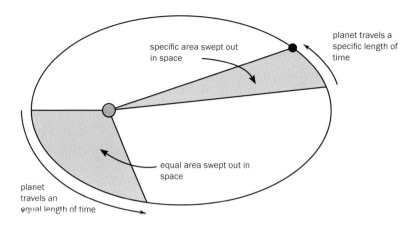

Figure 2-13. Equal areas are swept in space for equal periods of time.

Third Law   Any two planets will obey this law:

$$\left(\frac{T_1}{T_2}\right)^2 = \left(\frac{R_1}{R_2}\right)^3$$

where T_1 and T_2 are the planets' orbital periods, and R_1 and R_2 are their mean distances from the sun.

This Third Law of Kepler is a stunning example of mathematical modeling and is quite accurate. The equation can be expressed in a way that shows that the ratio of the square of the period to the cube of the mean distance is a constant. That is, this ratio has the same value for every planet. It can also be expressed such that the orbital period, T, for any planet is a function of the planet's mean distance from the sun, R. In equation form, this expression of the Third Law can be written as

$$T = k\sqrt{R^3}$$

or simply

$$T = kR^{3/2}$$

In these equations k is just a constant that depends on the units used for T and R. I wrote the equation in two ways to show that taking a square root of a quantity is equivalent to raising the quantity to the 1/2 power, which multiplies by the power of 3 already there to give the 3/2 power on R. I am not trying to go crazy with the math here. I just want

> Johannes Kepler viewed his discoveries of the mathematical order of nature as amazing revelations given to him by God. Some of the things Kepler worked on were very strange, such as his attempt to develop a theory of the spheres associated with the five regular Platonic solids and the mathematics of musical ratios developed by the Greeks. Although those ideas were abandoned, Kepler had the courage to look carefully at the astronomical data, and this led him to his discovery of the Laws of Planetary Motion.
>
> Read the prayer Kepler wrote at the end of his book *Harmonies of the World*:
>
> O Thou Who dost by the light of nature promote in us the desire for the light of grace, that by its means Thou mayest transport us into the light of glory, I give thanks to Thee, O Lord Creator, Who hast delighted me with Thy makings and in the works of Thy hands have I exulted. Behold! now, I have completed the work of my profession, having employed as much power of mind as Thou didst give to me; to the men who are going to read those demonstrations I have made manifest the glory of Thy works, as much of its infinity as the narrows of my intellect could apprehend. My mind has been given over to philosophizing most correctly: if there is anything unworthy of Thy designs brought forth by me – a worm born and nourished in a wallowing place of sins – breathe into me also that which Thou dost wish men to know, that I may make the correction: If I have been allured into rashness by the wonderful beauty of Thy works, or if I have loved my own glory among men, while I am advancing in the work destined for Thy glory, be gentle and merciful and pardon me; and finally design graciously to effect that these demonstrations give way to Thy glory and the salvation of souls and nowhere be an obstacle to that.
>
> – from Johannes Kepler, *Harmonies of the World* (1618)

to show how simple Kepler's Third Law really is. (If you don't know about fractional powers yet, don't worry. You will study them in depth when you get to Algebra 2, and we won't do any math with fractional powers in this course.)

Now I don't know about you, but when I see an equation that is as amazing and as simple as this a couple of things occur to me. First, Kepler's work as a scientist is first class. He figured this out from data collected in the era before telescopes, before calculators, before computers. This was only three years after Shakespeare died!

Second, this equation says something deep about the universe we live in. It has characteristics that can be modeled with simple mathematics that can be understood by high school kids! How do you think it got that way? Is it possible that a randomly evolving universe that occurred by chance with no plan could exhibit this kind of deep mathematical structure? I do not believe it is, and I am not alone. Many great scientists have called attention to the beautiful mathematical structure that appears everywhere in nature, and have called it either a mystery or evidence of God's handiwork. The fact that our solar system has the kind of beautiful, simple mathematical structure represented by Kepler's Third Law is strong evidence for an intelligent Creator. Kepler's discovery gives new meaning to David's poem in Psalm 19: "The heavens declare the glory of

God, and the sky above proclaims his handiwork." And remember our study of General Revelation in the previous chapter? Scripture tells us that nature itself reveals the divine Creator to us. Equations like this one are a testament to this truth.

Galileo Galilei (1564-1642), lived in Florence, Italy (Figure 2-14). He hypothesized that force was needed to *change* motion, not as impetus to *sustain* motion as Aristotle had taught. He discovered that all falling objects accelerate at the same rate (the acceleration of gravity, 9.80 m/s^2), which is mathematically very close to Isaac Newton's Second Law of Motion (our topic in the next chapter). He made significant improvements to the telescope and used the telescope to see the craters on the moon and sunspots, which provided powerful evidence that the heavens were not perfect and unchanging after all. In January, 1610 he used the telescope to discover four of the moons around Jupiter, which was clearly in conflict with the seven heavenly body idea.

Figure 2-14. Galileo Galilei.

He was fully on board with all the new science of the Copernican model, but he never did accept Kepler's discovery that the planets' orbits were elliptical rather than circular. Because of his Copernican views, Galileo was tried for heresy by the Church in 1633 and forced to recant his beliefs about the heliocentric model.

Figure 2-15. Isaac Newton.

The saga of the Copernican Revolution ends more or less with Galileo. Within a few decades of Galileo's death the heliocentric model of the planetary orbits was well established. But while we are studying the planets and gravity the whole story just isn't complete unless we mention two more key figures in the history of science.

Sir Isaac Newton (1643-1727) is perhaps the most celebrated mathematician and scientist of all time (Figure 2-15). He was English, as his title implies, and he was truly phenomenal. He held a famous professorship in mathematics at Cambridge University. He developed calculus. He developed the famous Laws of Motion, which we will examine in detail in the next chapter. He developed an entire theory of optics and light. He formulated the first quantitative law of gravity called the Law of

> Galileo's recantation speech is a fascinating example of what expectations were like in Galileo's day for such an occasion. Here it is:
>
> I, Galileo Galilei, son of the late Vincenzio Galilei of Florence, aged seventy years, being brought personally to judgment, and kneeling before you, Most Eminent and Most Reverend Lords Cardinals, General Inquisitors of the Universal Christian Commonwealth against heretical depravity, having before my eyes the Holy Gospels which I touch with my own hands, swear that I have always believed, and, with the help of God, will in future believe, every article which the Holy Catholic and Apostolic Church of Rome holds, teaches and preaches. But because I have been enjoined, by this Holy Office, altogether to abandon the false opinion which maintains that the Sun is the center and immovable, and forbidden to hold, defend, or teach, the said false doctrine in any manner...I am willing to remove from the minds of your Eminences, and of every Catholic Christian, this vehement suspicion rightly entertained towards me, therefore, with a sincere heart and unfeigned faith, I abjure, curse, and detest the said errors and heresies, and generally every other error and sect contrary to the said Holy Church; and I swear that I will never more in future say, or assert anything, verbally or in writing which may give rise to a similar suspicion of me; but that if I shall know any heretic, or anyone suspected of heresy, I will denounce him to this Holy Office, or to the Inquisitor and Ordinary of the place in which I may be. I swear, moreover, and promise that I will fulfill and observe fully all the penances which have been or shall be laid on me by this Holy Office. But if it shall happen that I violate any of my said promises, oaths, and protestations (which God avert!), I subject myself to all the pains and punishments which have been decreed and promulgated by the sacred canons and other general and particular constitutions against delinquents of this description. So, may God help me, and His Holy Gospels, which I touch with my own hands, I, the above named Galileo Galilei, have abjured, sworn, promised, and bound myself as above; and, in witness thereof, with my own hand have subscribed this present writing of my abjuration, which I have recited word for word.

Universal Gravitation. His massive work on motion, gravity and the planets, *Principia Mathematica*, was published in 1686. This work is one of the most important publications in the history of science.

Except for briefly in Chapter IV, we will not perform computations with Newton's Law of Universal Gravitation, and you do not need to memorize the equation for it. But let's look at it here briefly. The law is usually written as

$$F = G \frac{m_1 m_2}{d^2}$$

where G is a constant, m_1 and m_2 are the masses of any two objects (such as the sun and a planet) and d is the distance between the centers of the two objects.

Newton theorized that every object in the universe pulls on every other object in the universe, which is why his law is called the Law of *Universal* Gravitation. The equation above gives the force of gravitational attraction between any two objects in the universe. Amazingly, this equation is quite accurate, too! Notice that Newton's model depends on each object having mass, because the force of gravity has both masses in it and if either mass was zero the gravitational attraction would be zero.

While we are here looking at Isaac Newton we should pause and consider the relationship between his physical theories (including Law of Universal Gravitation and his Laws of Motion) and Kepler's mathematical theory of planetary motion. It turns out that Kepler's discovery about the elliptical orbits and the relationship between the period and mean radius of the orbit can be directly derived from Newton's theories, and Newton does derive them in *Principia Mathematica*. But Newton's equations apply much more generally than do Kepler's. As we will learn in the next chapter, Newton's Laws apply to all objects in motion – planets, baseballs, rockets – while Kepler's laws apply to the special case of the planets' orbits. If we consider this in light of my comments in Chapter I about the way theories work, we see that Newton's Laws explain everything Kepler's Laws explain, and more. This places Newton's theory about motion and gravity above Kepler's, so Newton's theories took over as the most widely-accepted theoretical model explaining motion. However, even though Newton's Laws ruled the scientific world for nearly 230 years, they do not tell the whole story.

This is where the German scientist **Albert Einstein** (1879-1955) comes in with his new general theory of relativity, published in 1915 (Figure 2-16). Einstein's theory explains gravity in terms of the curvature of space (or more precisely, "space-time") around a massive object, such as the sun or a planet. Fascinatingly, since Einstein's theory is about curving space, even phenomena without mass, such as rays of light, will be affected by it. Einstein noticed this and predicted that star light bends as it travels through space when it passes near a massive object such as the sun.

Figure 2-16. Albert Einstein.

Einstein became instantly world famous just a few years later when this prediction was stunningly confirmed. To test this hypothesis Einstein proposed photographing the stars we see near the sun during a solar eclipse. This has to be done during an eclipse, because otherwise it is broad daylight and we can't see the stars. Einstein predicted that the apparent position of the stars would shift a tiny amount relative to where they are when the sun is not near the path of the star light. British scientist Sir Arthur Eddington commissioned two teams of photographers to photograph the stars during the solar eclipse of 1919. After analyzing their photographic plates they found the star light to have shifted exactly the amount Einstein said it would. Talk about sudden fame – Einstein became the instant global rock star of physics when this happened! (And his puppy dog eyes contributed even more to his popularity!)

Just as Kepler's Laws were superseded by Newton's Laws and can be derived from Newton's Laws, Newton's Law of Universal Gravitation was superseded by Einstein's general theory of relativity and can be derived from general relativity. Einstein believed that his own theories would some day be superseded by an even more all-encompassing theory, but so far (after 97 years) that has not happened. The general theory of relativity remains today the reigning champion theory of gravity, our best understanding of how gravity works, and one of the most important theories in twentieth- and twenty-first-century physics.

CHAPTER II EXERCISES

Unit Conversions

Perform the following unit conversions. Express your results both in standard notation and in scientific notation using the correct number of significant digits. The unit conversion factors you need are all in Appendix A. On the first 20 problems you will use the standard method of multiplying conversion factors. The last four problems require somewhat different methods which you should try to figure out.

	Convert this Quantity	Into these Units
1	1,750 meters (m)	feet (ft)
2	3.54 grams (g)	kilograms (kg)
3	41.11 milliliters (mL)	liters (L)
4	7×10^8 m (radius of the sun)	miles (mi)
5	1.5499×10^{-12} millimeters (mm)	m
6	750 cubic centimeters (cm^3 or cc) (size of the engine in my old motorcycle)	m^3
7	2.9979×10^8 meters/second (m/s) (speed of light)	ft/s
8	168 hours (hr) (one week)	s
9	5,570 kilograms/cubic meter (kg/m^3) (average density of the earth)	g/cm^3
10	45 gallons per second (gps) (flow rate of Mississippi River at the source)	m^3/minute (m^3/min)
11	600,000 cubic feet/second (ft^3/s) (flow rate of Mississippi River at New Orleans)	liters/hour (L/hr)
12	5,200 mL (volume of blood in a typical man's body)	m^3
13	5.65×10^2 mm^2 (area of a postage stamp)	square inches (in^2)
14	32.16 ft/s^2 (acceleration of gravity, or one "g")	m/s^2
15	5,001 µg/s	kg/min
16	4.771 g/mL	kg/m^3
17	13.6 g/cm^3 (density of mercury)	mg/m^3

Chapter II
Motion and the Medieval Model of the Heavens

	Convert this Quantity	Into these Units
18	93,000,000 mi (distance from earth to the sun)	cm
19	65 miles per hour (mph)	m/s
20	633 nanometers (nm) (wavelength of light from a red laser)	in
21	5.015% of the speed of light	mph
22	98.6 degrees Fahrenheit (°F) (human body temperature)	degrees Celsius (°C)
23	50.0 °C (It gets this hot in northwest Australia.)	°F
24	4.3 light-years (lt-yr) (distance to the closest star, Proxima Centauri) (A light-year is the distance light travels in one year. You will have to work this out first.)	kilometers (km)

Answers
(A dash indicates that it would be either silly or incorrect to write the answer that way, so I didn't. Silly would be because there are simply too many zeros, or no zeros at all. Incorrect would be because we are unable to express the result this way and still show the correct number of significant digits.)

	Standard Notation	Scientific Notation
1	5,740 ft	5.74×10^3 ft
2	0.00354 kg	3.54×10^{-3} kg
3	0.04111 L	4.111×10^{-2} L
4	400,000 mi	4×10^5 mi
5	-	1.5499×10^{-15} m
6	0.00075 m³	7.5×10^{-4} m³
7	983,600,000 ft/s	9.836×10^8 ft/s
8	605,000 s	6.05×10^5 s
9	5.57 g/cm³	-
10	-	1.0×10^1 m³/min
11	60,000,000,000 L/hr	6×10^{10} L/hr
12	0.0052 m³	5.2×10^{-3} m³
13	0.876 in²	8.76×10^{-1} in²
14	9.802 m/s²	-
15	0.0003001 kg/min	3.001×10^{-4} kg/min
16	4,771 kg/m³	4.771×10^3 kg/m³

	Standard Notation	Scientific Notation
17	13,600,000,000 mg/m³	1.36×10^{10} mg/m³
18	–	1.5×10^{13} cm
19	29 m/s	2.9×10^{1} m/s
20	0.0000249 in	2.49×10^{-5} in
21	33,700,000 mph	3.37×10^{7} mph
22	37.0 °C	3.70×10^{1} °C
23	122 °F	1.22×10^{2} °F
24	–	4.1×10^{13} km

Motion Study Questions Set 1

1. A train travels 25.1 miles in 0.50 hr. Calculate the velocity of the train.

2. Convert your answer to the previous problem to km/hr.

3. How far can you walk in 4.25 hours if you keep up a steady pace of 5.0000 km/hr? State your answer in km.

4. For the previous problem, how far is this in miles?

5. On the German autobahn there is no speed limit and many cars travel at velocities exceeding 150.0 mi/hr. How fast is this in km/hr?

6. Referring again to the previous question, how long would it take a car at this velocity to travel 10.0 miles? State your answer in minutes.

7. An object travels 3.0 km at a constant velocity in 1 hr 20.0 min. Calculate the object's velocity, and state your answer in m/s.

8. A car starts from rest and accelerates to 45 mi/hr in 36 s. Calculate the car's acceleration, and state your answer in m/s².

9. A rocket traveling at 31 m/s fires its retro-rockets, generating a negative acceleration (it is slowing down). The rockets are fired for 17 s and afterwards the rocket is traveling at 22 m/s. What is the rocket's acceleration?

10. A person is sitting in a car watching a traffic light. The light is 14.5 m away. When the light changes color, how long does it take the new color of light to travel to the driver so that he can see it? State your answer in nanoseconds. (The speed of light in a vacuum or air, c, is one of the physical constants listed in Appendix A that you need to know.)

11. A proton is uniformly accelerated from rest to 80.0% of the speed of light in 18 hours, 6 minutes, 45 seconds. What is the acceleration of the proton?

12. A space ship travels 8.96 x 10^9 km at 3.45 x 10^5 m/s. How long does this trip take? Convert your answer from seconds to days.

13. An electron experiences an acceleration of 5.556 x 10^6 cm/s^2 for a period of 45 ms. If the electron is initially at rest, what will be its final velocity?

14. A space ship is traveling at a velocity of 4.005 x 10^3 m/s when it switches on its rockets. The rockets accelerate the ship at 23.1 m/s^2 for a period of 13.5 s. What is the final velocity of the rocket?

15. A more precise value for c (the speed of light) than the value given in Appendix A is 2.9979 x 10^8 m/s. Use this value for this problem. On a particular day the earth is 1.4965 x 10^8 km from the sun. If on this day a solar flare suddenly occurred on the sun, how long would it take an observer on the earth to see it? State your answer in minutes.

Answers
1. 22 m/s
2. 79 km/hr
3. 21.3 km
4. 13.2 mi
5. 241.4 km/hr
6. 4.00 min
7. 0.63 m/s
8. 0.56 m/s^2
9. −0.53 m/s^2
10. 48.3 ns
11. 3,680 m/s^2
12. 301 days
13. 2.5 x 10^3 m/s
14. 4.32 x 10^3 m/s
15. 8.3197 min

Motion Study Questions Set 2

1. Construct three simple graphs like those discussed in this chapter showing distance vs. time, velocity vs. time, and acceleration vs. time. Draw your graphs one beneath the other and line them up vertically so that the time intervals on each of them line up. Here is the situation: An SUV is at rest at a traffic signal. Then the car starts moving, accelerating uniformly to a certain velocity, and then continues at a constant velocity after that. Be sure to label the axes of your graphs.

2. Sketch the same three graphs for this situation: A pick-up truck is traveling

at a constant velocity. Then the driver slams on the brakes. Before the car stops, the driver releases the brakes so that the car continues to roll along at a constant (slower) velocity. Be sure to label the axes of your graphs.

3. Sketch the same three graphs for this situation: An MG convertible is coasting out of gas and gradually slowing down when it arrives at the entrance to a gas station. Next, the car begins rolling up the entrance ramp to the gas station, but the ramp is steep and the car slows down rapidly to a stop. Next, the driver fails to step on the brake, so after stopping at the top of the ramp the car immediately begins to roll backwards down the ramp. Note that there are only two time intervals here: before the ramp and on the ramp. (Ignore the up and down motion due to the ramp and any turning the car would do to turn into the gas station ramp. Just think about how far the car has traveled from the point where it ran out of gas.)

4. A space ship in space is cruising at a constant velocity. The captain switches on the retro-rockets, which begin slowing the vehicle. The captain never turns the rockets off, so after it comes to rest it immediately begins moving (still accelerating) in the opposite direction. Note that there are only two time intervals in this scenario.

5. The three graphs below represent three different scenarios involving an object in motion. For each one write a one-sentence description of the object's motion. In your descriptions use terms like "speeding up," "slowing down," "at rest," "backing up," "constant velocity," etc.

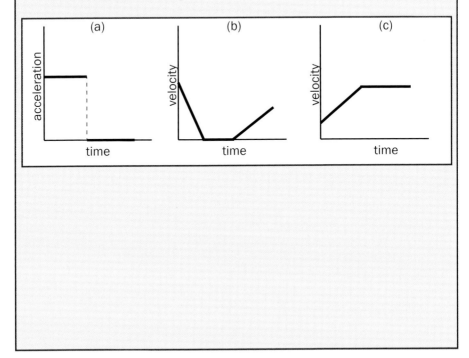

Answers

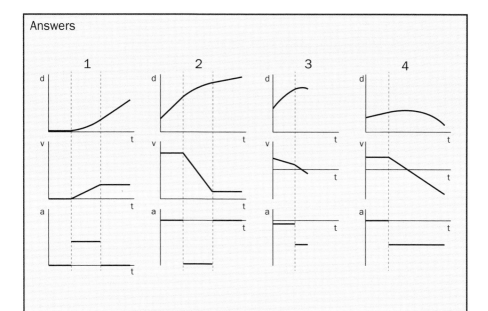

5. a) The object is speeding up (accelerating) uniformly, then it begins moving at a constant velocity. b) The object slows rapidly to a complete stop, stays at rest for a while, then begins accelerating again, moving in the same direction as before. c) The object is speeding up (accelerating) uniformly, then it begins moving at a constant velocity.

CHAPTER III
NEWTON'S LAWS OF MOTION

> **OBJECTIVES**
>
> Memorize and learn how to use these equations:
>
> $$a = \frac{F}{m} \qquad\qquad F_w = mg$$
>
> After studying this chapter and completing the exercises, students will be able to do each of the following tasks, using supporting terms and principles as necessary:
>
> 1. Define and distinguish between matter, inertia, mass, force and weight.
> 2. State Newton's Laws of Motion.
> 3. Calculate the weight of an object given its mass, and vice versa.
> 4. Perform calculations using Newton's Second Law of Motion.
> 5. Give several examples of applications of the Laws of Motion that illustrate their meaning.
> 6. Explain why the First Law is called the Law of Inertia.
> 7. Use Newton's Laws of Motion to explain how a rocket works.
> 8. Apply Newton's Laws of motion to application questions, explaining the motion of an object in terms of Newton's Laws.

MATTER, INERTIA, MASS AND FORCE

To study Sir Isaac Newton's three Laws of Motion one needs to have a clear understanding of three closely related and somewhat confusing terms, **matter**, **inertia**, and **mass**. **Matter** is simply the term we use for anything made of atoms or parts of atoms. Examples of matter are all around us, and include things like bananas, protons, carbon atoms, and planets. Examples of things that are *not* matter include light, radio waves, Beethoven's third symphony, justice, and the Holy Spirit. These are all real things, but they are not material objects.

Now, to set the stage for the next of these three terms, consider that all matter possesses certain properties. For example, one property you already know about is that matter takes up space. We quantify the amount of space that a material object occupies with the variable *volume*, and our measurements of the volume of an object are stated in certain units of measure. In the MKS system the units for volume are cubic meters, m^3.

Another property all matter possesses is **inertia**. The effect of this property is that a material object will always resist changes to its state of motion. Another way to say this is that to change the motion of an object a force is required. If an object is at rest and no net force is acting on it, it will remain at rest. The same applies to an object moving with a constant velocity; if there is no net force acting on it, it will continue moving with a constant velocity. In the previous chapter we saw that a changing velocity is called acceleration. Now you can see that to say an object's state of motion is changing is equivalent to saying its velocity is changing, which is also equivalent to saying that the object is accelerating. In summary, a net force on an object is required to make it accelerate. No net force, no acceleration. Without a net force on an object, the object's inertia will make it keep doing whatever it is presently doing.

Finally, **mass** is the quantitative variable we use to specify how much inertia an object has. In the MKS system the unit of measure for mass is the kilogram, kg. One way to emphasize the difference between inertia and mass is to say that inertia is a *quality* and mass is a *quantity*.

Now that we have a handle on matter, inertia and mass, there is one more term we need to discuss before getting to Newton's three famous Laws of Motion. This is the term **force**, or as you will see me write over and over, **net force**. A force is simply a push or a pull. The concept of force is fairly easy to understand, and Newton's Laws of Motion tell us very specific things about the way forces work in nature.

But to understand the laws we really need to have an understanding of *net* force. Consider the forces presently acting on me as I sit here at my computer writing this book. Because of the gravitational attraction of the earth, the earth is pulling me toward its center, which is *down* from my point of view. If this were the only force on me right now I would be accelerating downward, that is, falling. Luckily for me, there is another force pushing up on me from the chair I am sitting in. So although there are these two forces acting on me at present, they happen to be equal and directed in opposite directions, so they cancel each other out. This means there is no *net force* on me at the moment. A net force is a force that is not balanced out or cancelled out by another force.

NEWTON'S LAWS OF MOTION

Newton's Laws of Motion may be the three most famous scientific statements ever made. Accordingly, we must learn them.

First Law	An object at rest will remain at rest and an object in motion will continue moving in a straight line at a constant speed, unless it is compelled to change that state by forces acting on it.

The First Law applies when there is no net force acting on an object, and describes what objects will do in such circumstances. As I mentioned before, objects behave the way this law describes because of their inertia. For this reason the First Law is sometimes called the **Law of Inertia**.

In this book so far I have avoided using the term *speed*. Instead I have mostly used the term *velocity* for our discussion about motion. But the classic statement of Newton's First Law uses the term speed (Newton called it "uniform motion"), so I have kept it in there. The difference between the terms speed and velocity is a bit technical, but for those who are wondering I will address this issue briefly here. In the last chapter I defined velocity as the rate at which an object's distance is changing, but I also said we were only going to consider motion in a straight line. The term velocity actually involves not only *how fast* an object is moving, which is what we have discussed, but also the *direction* in which the object is moving. The term speed denotes only the "how fast" part of an object's motion, not the direction part. For our purposes, since we are concentrating on motion in a straight line, the terms speed and velocity are essentially synonyms.

Second Law	The acceleration of an object is proportional to the force acting on it, or $$a = \frac{F}{m}$$ where a is the acceleration of the object (m/s²), F is the net force on the object (N), and m is the object's mass (kg).

I must now inform you that by writing the Second Law as $a = F/m$ I have boldly written it differently than the way it is written in nearly every physics book in the world, which is $F = ma$. If you ask nearly anyone to state Newton's Second Law they will say $F = ma$. Some people reading this might wonder if I have I taken leave of my senses to restate one of the most famous equations in physics! No I haven't. I have a good reason for stating the law this way: This is the way Isaac Newton stated it! I don't know why everyone started saying $F = ma$. Mathematically, the equation $F = ma$ implies that force is dependent on acceleration, because F is in the position of the dependent variable in the equation. But we do not usually think of objects this way. When we think of objects accelerating, we usually think of the acceleration as being dependent on the force acting on the object. If a certain net force is applied to an object, the object will accelerate in direct proportion to the force. This is the way we think, and this is the way Newton said it. So this is the way I'm saying it.

The Second Law applies when there is a net force present, and it says what the object will do as a result. As we saw previously, when a net force is present on an object, the object will change its state of motion. In other words, the object will accelerate, and according to Newton's Second Law, the acceleration is in direct proportion to the force.

The MKS unit of measure for force is the newton (N), in honor of Sir Isaac Newton's contributions to science. This is the first derived unit we have encountered in this course. As you know, we are using the MKS system of units – meters, kilograms and seconds. Most of the other units of measure that we will learn are derived from combinations of these three main units. Since force is mass times acceleration (that is, $F = ma$), the unit for force is the product of the units for mass and acceleration, or

$$1 \text{ N} = 1 \text{ kg} \cdot \frac{\text{m}}{\text{s}^2}$$

Now for the Third Law, which is going to require some wordsmithing. The traditional way to state the Third Law is, "For every action there is an equal and opposite reaction." Newton himself wrote this law as, "To every action there is always opposed an equal reaction." The problem with these statements is that the words *action* and *reaction* do not mean anything like what we mean today when we use those words.

Newton meant these terms to refer to *forces*. They are not, in the way we use the terms today, actions, or reactions, or events or processes. So a better way to state the law now would be: For every force, there is an equal and opposite push-back force. That is, if object A pushes to the right on object B, object B pushes to the left with the same strength on object A. Of course, it is good for our students to know the traditional language. So I have developed a hybrid version of the Third Law that students should use in this course. Here it is:

| Third Law | For every action force, there is an equal and opposite reaction force. |

The Third Law describes the way objects act on *one another* when forces are present. If an object experiences a force – any force – a second object is pushing or pulling on the first object. The Third Law says that when this happens the second object also feels the same force on it from the first object pushing back. It is not possible for one object to push on another object without the second object pushing back in the opposite direction on the first one with the same amount of force. The Third Law applies all the time, whether the forces result in the acceleration of one or both of the objects involved, or the objects just remain motionless, like my chair and me in my example of sitting in my chair at my computer.

We will return soon to discussing the Laws of Motion and how to apply them. But before we do, let's get you started on using the two new equations that we are introducing in this chapter. We will begin with some examples using the Second Law.

A force of 8.61 nN is applied to a proton with a mass of 1.673×10^{-24} g. Determine the acceleration of the proton.

As always, begin by writing down the given information and converting all of the quantities into MKS units.

$$F = 8.61 \text{ nN} \cdot \frac{1 \text{ N}}{10^9 \text{ nN}} = 8.61 \times 10^{-9} \text{ N}$$

$$m = 1.673 \times 10^{-24} \text{ g} \cdot \frac{1 \text{ kg}}{1000 \text{ g}} = 1.673 \times 10^{-27} \text{ kg}$$

$$a = ?$$

Now write the equation, insert the values in MKS units and compute the result.

$$a = \frac{F}{m} = \frac{8.61 \times 10^{-9} \text{ N}}{1.673 \times 10^{-27} \text{ kg}} = 5.146 \times 10^{18} \frac{\text{m}}{\text{s}^2}$$

Since the given force has the least precision among the given quantities with 3 significant digits, we round this result to

ASPC

$$a = 5.15 \times 10^{18} \ \frac{m}{s^2}$$

A bullet from a rifle is traveling at 2,250 ft/s when it lodges in a tree, coming to rest 45 ms after impact. If the mass of the bullet is 65.5 g, determine the force of friction in the tree that stopped the bullet.

This is a two-part problem. We must first use the velocity and time information to determine the acceleration. Once we have that value we can use it in Newton's Second Law to determine the force that produced it.

Begin by writing down the givens for the acceleration calculation and performing the unit conversions.

$$v_i = 2{,}250 \ \frac{ft}{s} \cdot \frac{0.3048 \ m}{ft} = 685.8 \ \frac{m}{s}$$

$$v_f = 0$$

$$t = 45 \ ms \cdot \frac{1 \ s}{1000 \ ms} = 0.045 \ s$$

$$a = ?$$

Now write the equation, put in the MKS values, and compute the first result, keeping one extra digit of precision.

$$a = \frac{v_f - v_i}{t} = \frac{0 - 685.8 \ \frac{m}{s}}{0.045 \ s} = -15{,}240 \ \frac{m}{s^2}$$

In this intermediate result I kept an extra significant digit (two extra digits actually) over what I will need at the end. Notice that the acceleration is negative, meaning that the bullet slowed down. (This is pretty obvious from the problem, isn't it?)

Now we deal with the force calculation using the Second Law, beginning with the givens and the unit conversions to MKS units.

$$m = 65.5 \ g \cdot \frac{1 \ kg}{1000 \ g} = 0.0655 \ kg$$

$$F = ?$$

Now write down the Second Law, do the algebra and perform the computation.

$$a = \frac{F}{m}$$

$$F = ma = 0.0655 \text{ kg} \cdot \left(-15{,}240 \ \frac{\text{m}}{\text{s}^2}\right) = -998 \text{ N}$$

I have two comments to make here. First, the negative sign in front of the force simply means that the force is opposing the motion of the object, and thus is slowing it down. This is what friction always does. So write the negative sign if you wish, but since the problem is about a friction force, we know which direction friction forces point, so the negative sign is not really necessary. Suit yourself and don't worry about it.

Second, we need to round this result to 2 significant digits, since the time given in the problem only has 2 significant digits. However, rounding this value gives 1,000 N, which only indicates 1 significant digit. So we will have to express our result in scientific notation. In a value without a decimal, scientific notation is the only way to show that trailing zeros are significant. (In the old days there was another way to show that trailing zeros were significant – by putting a bar over the last significant zero. But this notation is not used much any more, so just stick to scientific notation when this situation arises.) Thus, our result is

$$F = 1.0 \times 10^3 \text{ N}$$

SHOWING UNITS OF MEASURE IN COMPUTATIONS

Now that I have worked a number of example problems in this text, I hope you have noticed that I always write the units with every quantity I write, every time I write it. To students your age this usually seems tedious and unnecessary, and many students resist doing it. But I want you to write them. Every time you write down a value of any kind, in any problem, I want you to write the units that go with it. Every time.

There are several reasons for this, actually, and we are not just being unnecessarily strict when we insist on it. First, just as with the overall problem-solving method that we are learning, successfully solving problems in physics depends a lot on having an orderly and reliable approach. You need to develop good habits, and using the problem-solving method presented in this book is one of those good habits you need to develop. Writing down the units is another one. I have seen literally hundreds of times when students lost points unnecessarily simply because they failed to develop these good habits. Without the good habits it is very easy to make silly mistakes.

A second reason is that physics problems get more complicated the farther you go in the subject. Keeping track of the units of measure you are working with can be a life saver in more complex problems, so now is the time to get in the habit of always writing them down. Third, by always writing down the units you will learn the units more thoroughly. Students sometimes forget what the units are for force or energy or power. They wouldn't forget so easily if they wrote them down every time they worked a problem. And finally, by writing down the units of measure you have a constant reminder in front of you to help you do the right thing. When you need to put the velocity of an object into an equation, writing in a value in m/s will be a little confirmation to you that you are doing the right thing, rather than, say, writing in the object's acceleration by mistake and not noticing what you did.

The bottom line is always write down the units. Always write them down. Always write them. Always.

WEIGHT

The term **weight** simply means the force with which gravity is pulling on an object. Weight is the force on an object due to gravity. The equation to calculate the weight of an object is

$$F_w = mg$$

where F_w is the weight of the object (N) and g is the acceleration due to gravity, which for earth is equal to 9.80 m/s² (at sea level). We can use this equation any time we have an object's mass and need its weight, or vice versa, which makes it quite important. The constant for g, the acceleration due to gravity on earth, is one of the constants listed in Appendix A that you must memorize.

As you recall, Galileo discovered that all falling objects accelerate at the same rate. This means that when released from rest, all objects experience the same acceleration in free fall. We designate this acceleration with the letter g. On earth at sea level, the average value of this acceleration due to gravity is $g = 9.80$ m/s². On the surface of earth, any object that falls freely will accelerate with an acceleration of 9.80 m/s².

Now, although it may take some getting used to, you should know that the weight equation above is just a special case of Newton's Second Law using g as the acceleration. The weight equation is a case in which it makes sense to write the Second Law as $F = ma$, with force as the dependent variable. An object's weight depends on the gravitational field it is in. When the astronauts were walking on the moon back in the late 1960s, they weighed less than they did on earth because the moon's gravitational attraction is not as great as the earth's. Thus, the astronauts' weights depended on the acceleration of gravity where they were. The same equation for weight would apply on the moon or other planets as well, but the value of g would be different on the moon or other planets.

A certain athlete weighs 225.00 lb. Determine the athlete's mass.

As always, begin by writing down the given information and converting the units to MKS units. This problem requires the use of a conversion factor we have not seen before. This conversion factor is listed in Appendix A, and is one you do not need to memorize.

$$F_w = 225.00 \text{ lb} \cdot \frac{4.45 \text{ N}}{1 \text{ lb}} = 1{,}001.3 \text{ N}$$

$m = ?$

THINKING ABOUT NEWTON'S LAWS OF MOTION
Applying Concepts

Learning to apply Newton's Laws of Motion to physical situations takes thought and practice. Students often get the laws confused, or choose one of the laws to describe a certain situation when another law would work better. Your study process for Newton's Laws of Motion should begin with memorizing the three laws verbatim. When answering an application question, you should always cite (i.e., quote) the law or part of a law which best describes a situation. Below are some tips and examples to aid you in selecting the best law for a particular situation and in crafting a good answer to a typical quiz question.

FIRST LAW

1. This law should be applied to situations where no acceleration is present, or, in other words, when objects are moving at a constant speed or sitting still.
2. This law is about what happens *without* a net force, so apply it when there is no net force present.
3. Another common way to apply this law is when something is moving at a constant speed, and then a collision or some other event occurs. If a piece of the moving object keeps moving, it does this because of its inertia, and because there is no outside force on the piece to slow it down and stop it.

Typical questions with appropriate answers:

Q: A space ship is traveling in space at 20,000 m/s when it runs out of fuel. As far as the captain can tell, there are no stars or other planets in front of him or any where else nearby. What will happen?

A: There are no forces from gravity or fuel to accelerate the ship. Thus, according to Newton's First Law, the ship will continue at traveling at 20,000 m/s in a straight line forever.

Q: A kid on a sled slides down a snow-covered hill but hits a rough spot of grass at the bottom. His sled stops suddenly, but the kid keeps right on going. How is this explained?

A: There was no force present on the kid to stop him so his inertia kept him going even when his sled stopped due to the friction force of the grass (Second Law). The Law of Inertia (Newton's First) states that unless a net force is present a moving object continues to move at a constant speed.

SECOND LAW

1. Apply this law whenever something is speeding up or slowing down. When faced with an application question, ask yourself if the object in question is speeding up or slowing down. If so, then $a = F/m$ applies.

2. Sometimes an object will stop for an instant when it changes direction while accelerating, such as when a ball is thrown straight up. The important point is that the ball was slowing before it stopped, and then speeding up after it stopped. Thus, the ball is accelerating the entire time and Newton's Second Law applies. Only if an object stays at rest for some period of time would the First Law apply.

Q: A woman accelerates her car to get on the freeway. How do Newton's laws describe this situation?

A: The car is accelerating according to Newton's Second Law, $a = F/m$. The force provided by the engine and the mass of the car determine how rapidly the car will accelerate.

Q: A boy is jumping up and down on a trampoline. He flies up in the air and is at the very top of his flight at the instant before he begins falling back down. How do Newton's laws apply to this moment?

A: Anything in the air close to earth is accelerated by gravity. The gravitational attraction of the earth produces a net force on the boy which accelerates him and causes him to fall.

THIRD LAW

1. This law is often misapplied. To apply it correctly requires a phrase such as "A pushes on B, and B pushes equally and oppositely on A." You must always refer to the *same two* things, each pushing equally and oppositely on the other.
2. Don't refer to this law just because there is a force mentioned in a question. Instead, look for language in the situation that sounds like two specific objects each pushing equally and oppositely on the other.
3. Remember that "action" and "reaction" do not refer to events or processes. They refer to forces, that is, *pushes* or *pulls*. Plain and simple.

Q: Carl got mad and slammed his fist into the car windshield and broke the windshield. Unfortunately, upon striking the windshield Carl's knuckles were broken, too. Use Newton's laws to describe what happened.

A: Newton's Third Law states that for every action force there is an equal and opposite reaction force. When Carl's fist hit the windshield hard enough to break it, the windshield hit Carl's fist just as hard. Both the windshield and Carl's fist were broken by these equal and opposite forces.

Q: Jimmy jumped to the left off of the stern of his canoe into the lake. Jimmy's leap caused the canoe to shoot off to the right. Describe this situation using Newton's laws.

A: Newton's Third Law states that for every action force there is an equal and opposite reaction force. Jimmy pushed the canoe to the right and the canoe pushed Jimmy to the left.

Continue with writing down the equation, doing the algebra and computing the result.

$$F_w = mg$$

$$m = \frac{F_w}{g} = \frac{1{,}001.3 \text{ N}}{9.80 \frac{\text{m}}{\text{s}^2}} = 102.2 \text{ kg}$$

Since the value of g we are using has 3 significant digits, this determines the precision of our result, which must also have 3 significant digits. Rounding accordingly, we have

$m = 102$ kg

APPLYING NEWTON'S LAWS OF MOTION

If you really *understand* Newton's Laws of Motion, you ought to be able to apply them to different circumstances, explaining what happened in terms of one of the laws. I will illustrate this with a few examples, but before I do, here are some guidelines to follow. First, ask yourself if the situation is about an object that has a net force on it and is accelerating, or if it is about an object that does not have a net force on it and is maintaining its state of motion. If there is no net force and no acceleration, apply the First Law. In any scenario that depicts objects at rest and remaining at rest, or moving at a constant velocity and continuing to do so, the First Law applies. Do not apply the First Law to any scenario which depicts the state of motion as changing. In short, if the velocity is constant, the First Law applies.

If the velocity is not constant, the Second Law applies. If any object is depicted as accelerating, apply the Second Law, which says that the acceleration of the object will be directly proportional to the net force applied to it.

If the scenario refers to two objects pushing or pulling on each other, whether they are moving or not, apply the Third Law. Remember that the action-reaction pair is always a pair of forces, not anything else like actions, processes or events. So you must always apply this law by saying something like, "A pushes B, and B pushes back equally on A." Do not make the mistake of using events, movements or actions as the action force or the reaction force. Remember, when Newton used the term *action* in his Third Law the term did not have the same meaning that it has today. Newton was referring to *forces*, that is, *pushes* or *pulls*.

Please take the time to read the box on the previous pages called, "Thinking About Newton's Laws of Motion." Then you will be ready to study the following examples of applying Newton's Laws of Motion to specific scenarios involving motion and forces.

An asteroid speeding through space comes close to a star. When it does, it begins to speed up and turn in its course. Use one of Newton's Laws of Motion to comment on what is happening. (Note that since a force is required to change the direction of a moving object, turning is an example of acceleration.)

The asteroid is accelerating because of the attraction of the star's gravitational field. Since there is a force (gravity) on the asteroid, its velocity is changing. That is, it is speeding up and changing direction. The acceleration of the asteroid will be according to Newton's Second Law, $a = F/m$. The acceleration of the asteroid will be in proportion to the strength of the gravitational attraction of the star.

While driving, a man wearing glasses fell asleep at the wheel and ran into a tree. The man was wearing his seat belt, and was unharmed, but his glasses flew right off of his face. Use one of Newton's Laws of Motion to comment on what happened.

The glasses were moving at the same velocity the man and car were moving. While the man and the car experienced forces that made them stop (according to Newton's Second Law) there was no force on the glasses. In the absence of a force the inertia of the glasses made them continue moving in a straight line at a constant speed until they hit the windshield, just as Newton's First Law says.

Susie wants to go hunting with her dad. Her father decides to let her try to shoot his .308 calibre rifle at a target out in the country. She places the rifle against her shoulder and fires at the target. The rifle fires the bullet, and Susie feels the painful kickback of the gun on her shoulder.

The firing of the bullet and the kickback of the gun are explained by Newton's Third Law of Motion. The gun pushed the bullet (action force), and the bullet pushed the gun equally and oppositely (reaction force). The force of the gun on the bullet is very high, and caused the bullet to accelerate out of the gun barrel. The force of the bullet on the gun is equal, and in the opposite direction. This high kickback force causes the gun to hit Susie's shoulder hard.

HOW A ROCKET WORKS

Rockets are a great way to see Newton's Third Law in operation. In outer space there is no friction or air, so normal methods of accelerating or stopping a vehicle on earth cannot be used. Cars accelerate by using the friction between the tires and the road to speed the car up or to slow it down. Propeller-driven airplanes and boats push against the air or water. These vehicles all use Newton's Third Law. The tires push against the road, the road pushes against the tires in the opposite direction. The boat propeller pushes against the water, and the water pushes against the propeller, equally and in the opposite direction.

But in space there is nothing there to push against. No road, no water, no air. Rockets speed up or slow down by using Newton's Third Law of Motion just like every other vehicle does, but a rocket has to do it without anything to push against.

A rocket engine works by throwing the mass of its own burnt fuel (the products of combustion) out of the rocket engine with a massive amount of force. In the parlance of rocketry, this force is called **thrust**. The combustion of the fuel is really only needed to get the fuel to fly out of the rocket with a great deal of force, producing a large thrust. In Third Law terms, we can say, "The rocket engine pushes the fuel." And the rocket pushes the fuel very *hard*! This is the action force in the Third Law. The reaction force is, "The fuel pushes the rocket engine." This push is *just as hard*. Since the rocket engine is, of course, connected to the rest of the space ship, the force of the fuel pushing on the rocket engine accelerates the space ship according to the Second Law. The action-reaction pair is illustrated in Figure 3-1.

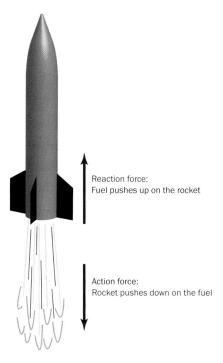

Figure 3-1. Action-reaction pairs associated with a rocket.

To slow down, the space ship again has nothing to push against, and so must again use a rocket engine. However, in this case the ship will use **retro-rockets**, which are small rocket engines that point toward the front of the space ship. When the "engine pushes the fuel" from the front of the space ship, the "fuel pushes the engine" from the front, against the motion of the ship, thus slowing the ship down. As we saw in the example computations earlier, the retro-rockets would cause a negative acceleration, and a corresponding negative force. That is, the force is on the front of the ship, against its motion, slowing it down.

CHAPTER III EXERCISES

Newton's Second Law Practice Problems

Convert everything to MKS units before you start.

1. Consider a car with a mass of 1,880 kg. How much force does it take to give this car an acceleration of 1.50 m/s^2?

2. A softball has a mass of 188.4 g. Determine its weight.

3. A certain baseball player can hit the ball with a force of 250.0 N. If the ball has a mass of 144,000 mg, what will its acceleration be?

4. A rocket has an acceleration of 2.3 m/s^2 at launch. If the engines are putting out 230,000 N of thrust (force), what is the rocket's mass?

5. A car accelerates at a rate of 0.0022 mi/hr^2. If it has a mass of 2.2 Mg, what is the force produced by the engine?

6. A woman weighs 125.1 lb. Determine her mass.

7. The same woman (same mass) weighs 17.9 lb on the moon. Determine the acceleration of gravity on the moon, g_m.

8. A gun accelerates a bullet from rest to 125.0 m/s in 22.00 ms. The exploding gunpowder provides a force of 142.0 N. What is the mass of the bullet? State your answer in grams.

9. How much force is required to accelerate a 4.5 kg shot-put from rest to 8.00 mi/hr in 500 ms?

10. A space ship is cruising at 2,500.0 km/hr with the engines off. The engines fire for 8.000 s with a thrust of 45,450 N bringing the ship up to its new speed of 2,750 km/hr. What is the mass of the ship?

11. A child's toy car has a mass of 166 g and uses a force of 0.0450 N to accelerate the car. If the car begins from rest, how fast will it be going after 2.1 s?

12. A proton has a mass of 1.673 x 10^{-18} µg. It is accelerated from rest to 0.05000% of the speed of light in 455 ns. Determine the force required to make the proton do this. Express your answer in GN.

13. A large ship has a mass of 6.548 Gg and is traveling at 8.35 mi/hr. The ship reverses its propellers and comes to a stop in 0.288 min. Determine the force the propellers had to produce to do this.

14. A mouse is creeping along at 3.5 cm/s when a cat attacks it. The mouse accelerates from its initial speed to a new speed of 18.5 cm/s in 220 ms. What was the acceleration of the mouse, in m/s²?

15. A missile with a mass of 45,500 kg is accelerated from rest to 55 m/s in 6.4 seconds.
    a. What was the acceleration of the missile?
    b. What was the thrust of the missile's rockets while it was being launched?

16. How large of a horizontal force must be exerted on an 8.5 g bullet to give it an acceleration of 18,500 m/s²?

Answers
1. 2,820 N
2. 1.85 N
3. 1,740 m/s²
4. 1.0×10^5 kg
5. 6.0×10^{-4} N
6. 56.8 kg
7. 1.40 m/s²
8. 24.99 g
9. 30 N
10. 5,230 kg
11. 0.57 m/s
12. 5.52×10^{-25} GN
13. 1.41×10^6 N
14. 0.68 m/s²
15. a) 8.6 m/s² b) 3.9×10^5 N
16. 160 N

Newton's Laws of Motion Study Questions

1. A space ship is in outer space firing its rockets and accelerating. What will the space ship do immediately after the rockets cease firing?

2. A space ship is near a planet, heading toward it, when it switches off its rockets. What will happen? According to Isaac Newton why does this happen? According to Albert Einstein why does it happen?

3. According to Newton's Second Law of Motion, what is necessary to change the velocity of a moving object?

4. Distinguish between matter and mass.

5. Distinguish between mass and inertia.

6. Give four new examples applying the Law of Inertia (Newton's First Law) to

actual situations.

7. Why is the First Law called the Law of Inertia? What is inertia and how does it relate to this law?

8. Answer each of the following questions by arguing from Newton's Laws of Motion.
   a. If you start a pendulum swinging it will eventually come to a stop. Why? (Be careful with your answer. Sure, gravity is acting on the pendulum to make it swing. But is gravity what makes the pendulum slow down and stop?)
   b. Reconsidering the previous question, suggest a way to set up a pendulum so that it would almost never slow down, and would swing for a very long time. (Sorry, perfect perpetual motion is not possible, but think of a design that would come as close as possible.)
   c. You are an astronaut and are traveling in a space ship from earth to Mars. Once you get up to speed and leave earth's gravitational field you have a long journey ahead. During this part of the journey, will your rocket engines need to be on full, medium, low, or off? Why?
   d. As an astronaut, when do you ever need to have your engines on? Do you even need engines?
   e. If Newton's First Law is correct, why do we have to keep the engine on while we are driving down the highway at a constant speed?

9. Give two examples applying Newton's Third Law to actual situations in which objects remain at rest. In each case identify the action force, the reaction force, and the results of each of these forces.

10. Give two examples applying Newton's Third Law to actual situations in which the two objects do not remain at rest, but one or both of them accelerate instead. In each case identify the action force, the reaction force, and the results of each of these forces.

11. You and another space traveler are stranded in space at a dead stand still because your space ship had a fuel leak while at rest and you lost all your fuel. Your companion notices that your ship is pointed exactly toward the space station where you need to go, and suggests that you quickly heave your massive copy of *Les Miserables* (which you haven't finished reading) out the back window of the space ship. Outraged, you suggest that your companion take a flying leap out the back window himself. Will either one of these actions really help? If so, which would help more? Answer these questions by arguing from Newton's Laws of Motion.

12. For the Soul of Motion Lab, identify the explanatory variable and the response variable. Also identify two realistic possibilities for lurking variables.

CHAPTER IV
VARIATION AND PROPORTION

> **OBJECTIVES**
>
> After studying this chapter and completing the exercises, students will be able to do each of the following tasks, using supporting terms and principles as necessary:
>
> 1. Identify and graph linear and nonlinear variation.
> 2. Identify the constant of proportionality in a physical equation.
> 3. Identify direct, inverse, square, inverse square and cubic proportions when seen in equations.
> 4. Identify direct, square and inverse proportions when seen graphically.
> 5. Identify dependent and independent variables in physical equations.
> 6. Graph functional relationships, making proper use of dependent and independent variables.
> 7. Normalize non-essential constants and variables in a given physical equation, and describe how the remaining two variables vary with respect to each other.
> 8. Describe the way variations occur between variables in physical equations, including:
>   a. area of a triangle
>   b. area of a circle
>   c. volume of a sphere
>   d. gravitational potential energy
>   e. kinetic energy
>   f. pressure under water
>   g. force of gravitational attraction
>   h. Charles' Law
>   i. Boyle's Law
>
> Nota Bene:
>
> The purpose of this study is to learn about how physical quantities vary in equations, not to understand completely all of the physical equations used as examples. The student should gain some understanding of the physical relationships involved, but detailed study of the physics involved will occur later in the course or in future courses.

THE LANGUAGE OF NATURE

The fact that nature can be modeled with mathematics is utterly fascinating. As Galileo famously said, "The laws of nature are written in the language of mathematics." One of the key ways scientists describe the mathematical properties in the laws of nature is in terms of the different proportionalities that are present. In other words, they describe how one variable in an equation will vary with respect to another variable. This chapter is about learning how to see these relationships in equations and appreciate what they indicate about what one variable will do if another variable is manipulated.

The presentation in this chapter is short. This is because there is not that much content to learn. Most of the time spent in this chapter will come not from learning new laws and equations, but from working through a set of exercises that will immerse you in the principles and language of variation and proportionality, with both analytic and graphical exercises. These exercises are in the Variation and Proportion Study Packet presented at the end of the chapter.

INDEPENDENT AND DEPENDENT VARIABLES

As every algebra student knows, the generic equation for a line is

$y = mx + b$

This equation is called the slope-intercept form of an equation for a line, because the constant m is the slope of the line and the constant b is the y-intercept of the line. In this equation, the variable x is called the **independent variable**, and the variable y is called the **dependent variable**, because the value of y depends on the value we choose for x. In general, we write our equations so that the dependent variable is alone on the left side, and all other variables and constants are on the right side.

When graphing a line or curve relating two variables, say p and r, we often speak of "a graph of p vs. r." The phrase "p vs. r" means that p is the dependent variable, represented by the scale on the vertical axis of the graph, and r is the independent variable, represented by the scale on the horizontal axis of the graph. Referring to "r vs. p" would call to mind a different graph, with the variables on the axes switched. The order matters in this expression. The graph of "p vs. r" has p on the vertical axis and r on the horizontal axis.

It is important that you think this through and get this down. In this course you will repeatedly be asked to generate a graph representing the relationship between two variables, both in the exercises and in your lab reports. It is quite common for students to see a statement like "develop a graph of p vs. r" and fail to consider which of these variables should be placed on the vertical axis and which on the horizontal axis.

Now, if a line passes through the origin, the y-intercept b is zero. The equation for the line then becomes

$y = mx$

Many of the equations in physics are linear functions like $y = mx$. We have already learned several of them, such as $d = vt$ and $F_w = mg$. In many physical linear equations like these the y-intercept is zero; a graph would have a line that goes through the origin. These are examples of direct variation, as described in more detail below.

COMMON TYPES OF VARIATION

Quantities in nature often exhibit common types of variation with respect to other quantities. This is part of what Galileo was talking about, and if you think about it, it is quite wonderful that nature behaves this way. Just as an aside, I wrote about this a bit back in Chapter II when we were considering the work of Johannes Kepler. The regular, mathematical way nature behaves, from planets and stars down to quarks and molecules, is stunning evidence that nature was created and is maintained each moment by the wise hand of an intelligent and rational Creator. Notice that I am not claiming that the mathematical order in nature is *proof* for God. (I doubt there are such proofs.) But it is strong evidence, because it is unreasonable to think that mathematical order like this can come about by chance from random bunches of matter and energy, just like it would be unreasonable to suppose that if you drop all your loose ASPC papers off the roof of a building they might organize themselves neatly into a binder when they hit the sidewalk down below. And it is nice to know that there is strong evidence in creation itself for the faith we have in the God who created it.

Now going back to the common types of variation found in physics, we need to know how to describe these common types of variation using appropriate language. The graphs in Figure 4-1 show five common types of variation. Each of these is a graph of y vs. x, with k as a generic constant. The constant is usually called the **constant of proportionality**. There are other very common types of variation in addition to those shown in Figure 4-1. You will learn about those in future math or science classes.

One of our objectives is to use correct scientific language to describe the variation going on between the variables in an equation. Here are the correct phrases one should

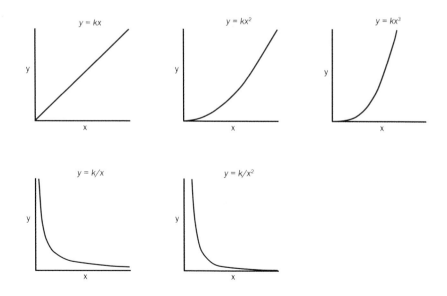

Figure 4-1. Five common types of variation.

use to describe each of the types of variation represented in the figure. The prepositions are important.

$y = kx$	y varies directly with x, or y varies in direct proportion to x
$y = kx^2$	y varies as the square of x, or y varies as x squared
$y = kx^3$	y varies as the cube of x, or y varies as x cubed
$y = k/x$	y varies inversely with x
$y = k/x^2$	y varies inversely with the square of x

This last type of variation is often called an "inverse square law." There are many quantities in nature that follow inverse square laws.

Here are a couple of examples of how to apply this language using equations we have already learned.

In the weight equation, how does the weight of an object vary with respect to the acceleration of gravity?

The weight equation is

$$F_w = mg$$

In this equation, the acceleration of gravity, g, is the independent variable, weight, F_w, is the dependent variable, and the mass, m, is the constant of proportionality. Thus, this equation represents the same kind of variation as in the first graph of Figure 4-1, and F_w varies directly with g.

In the standard equation for calculating uniform acceleration, how does acceleration vary with time?

The equation is

$$a = \frac{v_f - v_i}{t}$$

Time, t, is the independent variable and acceleration, a, is the dependent variable. Since the time is in the denominator without an exponent, this variation is like the fourth graph in Figure 4-1 and a varies inversely with t.

NORMALIZING EQUATIONS

Often equations involve many variables and constants, such as Newton's Law of Universal Gravitation that we saw in the last chapter. We frequently want to consider only two of the variables to see how they vary with respect to one another, while holding everything else constant. To do this we "**normalize**" the equation and rewrite it as a proportion. To normalize an equation, take all of the constants and all of the extra variables and set them equal to "1." Write what is left as a proportion.

For example, let's say we want to consider the variables E and V in the equation

$$E = \tfrac{1}{2}CV^2$$

Don't worry what this equation is about; we will not be studying it in this course, although if you take physics again as an upperclassman you probably will. All we want to do here is use the equation to learn about variation.

We want to know how E, the dependent variable, varies with respect to V, the independent variable, if everything else were to be held constant. Normalizing the equation means we set the 1/2 and the variable C each equal to 1, and write what remains as a proportion. Doing so gives

$$E \propto V^2$$

This expression is read "E is proportional to V squared." You can see that when we write a normalized equation we have to replace the equals (=) sign with the mathematical symbol meaning "is proportional to" ($\propto$). This is because although the proportionality is the same, the two sides are not strictly equal any more. With the normalized expression it is easier to see that this expression has the same form as $y = x^2$, which is graphed in the second graph of Figure 4-1. So we would say that "E varies as the square of V."

Be sure to notice that any exponent that is present on the independent variable is a crucial part of the proportionality and will appear in the normalized expression. Exponents are not variables, nor are they constants, so they must be retained. Only the variables and the constants are replaced with "1," and only these vanish from the normalized expression.

The graph of a normalized expression will have the same basic shape as the original equation; it will only be stretched horizontally or vertically. This means normalizing a complicated equation helps us to see how the variation between the two variables of interest will work, because the nature of the relationship between the two variables is preserved while making the equation (which becomes a proportion) simpler and easier to analyze.

Consider the following equation from the field of fluid dynamics:

$$Q = \frac{-kA}{\mu}\frac{(P_b - P_a)}{L}$$

Assuming that we are interested in L as the independent variable and Q as the dependent variable, normalize this equation and describe how Q varies with respect to L, assuming all else is held constant.

To normalize, we set all constants and variables other than the ones under consideration equal to 1. Treating the parenthetical sum as a single variable and normalizing gives

$$Q \propto \frac{1}{L}$$

From this expression we can see that Q varies inversely with L.

Chapter IV — Variation and Proportion

CHAPTER IV EXERCISES

The Variation and Proportion Study Packet

Completion of this packet counts as a quiz grade. For full credit do all of the activities and graphs with great care, thoroughness and neatness.

Packet Requirements:

- Use a separate sheet of graph paper for each activity.
- Answer each question on a separate line on your graph paper.
- Each graph must include a minimum of five plotted points, and must be accompanied by the table of values used to make the plot.
- Make it neat.
- Use a straightedge for all axes.
- Choose scales to fit the data well so that the graph is well proportioned and large enough to read easily.
- Label all axes and scales with the variable name and units of measure. Make sure your scales are linear.
- Title each graph.
- Label each curve as the normalized curve or the regular curve.
- All graphs should be approximately the size of an index card (3" x 5").

The packet consists of the following activities:

- Activity 1: How does the area of a triangle vary with its height?
- Activity 2: How does the area of a circle vary with its radius?
- Activity 3: How does the volume of a sphere vary with its radius?
- Activity 4: How does the gravitational potential energy of an object vary with its height?
- Activity 5: How does the kinetic energy of an object vary with its velocity?
- Activity 6: How does pressure vary with depth under water?
- Activity 7: How does the force of gravitational attraction between two objects vary with the distance between their centers?
- Activity 8: At constant pressure, how does the volume of a gas vary with its temperature?
- Activity 9: At constant temperature, how does the volume of a gas vary with its pressure?

Activity 1. How does the *area* of a triangle vary with its *height*, if all else is held constant?

1. What is the relation for the area of a triangle?
2. What are the two key variables that need to be compared in this activity?
3. Which variable is the independent variable, and which is the dependent variable?

4. After combining and normalizing all non-essential variables and constants, what is the expression relating area to height?
5. Select any value other than 2 to use for the base of a triangle. Using this value for the base, choose several values for the height of the triangle and calculate the area of the triangle for each height. Enter all of these in a table of values.
6. Prepare a graph of area vs. height using the values you computed in the previous step.
7. Compute another table of values for the normalized expression from step 4. Treat the proportional sign as an equals sign for this.
8. Graph the normalized equation on the same set of coordinate axes you used for the graph in step 6.
9. Describe the similarities and the differences between the two "curves" on your graph.
10. Answer the main question (the title) for this activity.

Activity 2. How does the *area* of a circle vary with its *radius*, if all else is held constant?

1. What is the relation for area of a circle?
2. What are the two key variables that need to be compared in this activity?
3. Which variable is the independent variable, and which is the dependent variable?
4. After combining and normalizing all non-essential variables and constants, what is the expression relating area to radius?
5. Choose several values for the radius of the circle and calculate the area of the circle for each radius. Enter all of these in a table of values.
6. Prepare a graph of area vs. radius using the values you computed in the previous step.
7. Compute another table of values for the normalized expression from step 4. Treat the proportional sign as an equals sign for this.
8. Graph the normalized equation on the same set of coordinate axes you used for the graph in step 6.
9. Describe the similarities and the differences between the two curves on your graph.
10. Answer the main question for this activity.

Activity 3. How does the *volume* of a sphere vary with its *radius*, if all else is held constant?

1. What is the relation for volume of a sphere? Look it up if you don't know it.
2. What are the two key variables that need to be compared in this activity?
3. Which variable is the independent variable, and which is the dependent variable?
4. After combining and normalizing all non-essential variables and constants,

Chapter IV — Variation and Proportion

what is the expression relating volume to radius?
5. Choose several values for the radius of the sphere and calculate the volume of the sphere for each radius. Enter all of these in a table of values.
6. Prepare a graph of volume vs. radius using the values you computed in the previous step.
7. Compute another table of values for the normalized expression from step 4. Treat the proportional sign as an equals sign for this.
8. Graph the normalized equation on the same set of coordinate axes you used for the graph in step 6.
9. Describe the similarities and the differences between the two curves on your graph.
10. Answer the main question for this activity.

Activity 4. How does the *gravitational potential energy* (E_G) of an object vary with its *height*, if all else is held constant?

The equation for E_G is

$$E_G = mgh$$

where
E_G is the gravitational potential energy in joules (J),
m is the mass of an object in kilograms (kg),
g is the acceleration of gravity equal to 9.80 m/s², and
h is the height of the object in meters (m).

1. What are the two key variables that need to be compared in this activity?
2. Which variable is the independent variable, and which is the dependent variable?
3. After combining and normalizing all non-essential variables and constants, what is the expression relating E_G to height?
4. Select a value to use for the mass of the object for this activity. Using this value, choose several values for the height of the object and calculate the E_G of the object for each height. Enter all of these in a table of values.
5. Prepare a graph of E_G vs. height using the values you computed in the previous step.
6. Compute another table of values for the normalized expression from step 3. Treat the proportional sign as an equals sign for this.
7. Graph the normalized equation on the same set of coordinate axes you used for the graph in step 5.
8. Describe the similarities and the differences between the two "curves" on your graph.
9. Answer the main question for this activity.

ASPC

Activity 5. How does the *kinetic energy* (E_K) of an object vary with its *velocity*, if all else is held constant?

The equation for E_K is

$$E_K = \tfrac{1}{2} m v^2$$

where
E_K is the kinetic energy in joules (J),
m is the mass of an object in kilograms (kg), and
v is the velocity of the object in meters per second (m/s).

1. What are the two key variables that need to be compared in this activity?
2. Which variable is the independent variable, and which is the dependent variable?
3. After combining and normalizing all non-essential variables and constants, what is the expression relating E_K to velocity?
4. Select a value other than 2 kg to use for the mass of the object for this activity. Using this value, choose several values for the velocity of the object and calculate the E_K of the object for each velocity. Enter all of these in a table of values.
5. Prepare a graph of E_K vs. velocity using the values you computed in the previous step.
6. Compute another table of values for the normalized expression from step 3. Treat the proportional sign as an equals sign for this.
7. Graph the normalized equation on the same set of coordinate axes you used for the graph in step 5.
8. Describe the similarities and the differences between the two curves on your graph.
9. Answer the main question for this activity.

Activity 6. How does *pressure* under water vary with *depth*, if all else is held constant?

The equation for pressure under any liquid is

$$P = \rho g h$$

> The Pascal is the derived unit we use for pressure in the MKS system. If you multiply all the units together for the variables in this equation you get kg/ms². Thus, 1 Pa = 1 kg/ms².

where
P is the pressure in pascals (Pa),
ρ (that is, the Greek letter rho, the r in the Greek alphabet) is the density of the liquid in kg/m³,
g is the acceleration of gravity equal to 9.80 m/s², and
h is the depth under the liquid in meters (m).

For water, ρ = 998 kg/m³ (at room temperature).

Chapter IV   Variation and Proportion

1. What are the two key variables that need to be compared in this activity?
2. Which variable is the independent variable, and which is the dependent variable?
3. After combining and normalizing all non-essential variables and constants, what is the expression relating pressure to depth?
4. Using the value for the density of water given above, choose several values for the depth under water and calculate the pressure for each depth. Enter all of these in a table of values.
5. Prepare a graph of pressure vs. depth using the values you computed in the previous step.
6. Compute another table of values for the normalized expression from step 3. Treat the proportional sign as an equals sign for this.
7. Graph the normalized equation on a *separate* set of coordinate axes.
8. Describe the similarities and the differences between the curves in the two graphs.
9. Answer the main question for this activity.

Activity 7. How does the *force* of gravitational attraction vary with the *distance* between the centers of two objects?

The equation for the force of gravitational attraction between any two objects is given by Newton's Law of Universal Gravitation:

$$F = G\frac{m_1 m_2}{d^2}$$

The units of measure for this constant may look strange, but when G is placed in the equation all of the units cancel except for newtons, which are the appropriate units for force.

where
F is the force in newtons (N),
G is a constant equal to 6.67 x 10^{-11} Nm^2/kg^2,
m_1 and m_2 are the masses of the two objects in kilograms (kg), and
d is the distance between the centers of the two objects in meters (m).

1. What are the two key variables that need to be compared in this activity?
2. Which variable is the independent variable, and which is the dependent variable?
3. After combining and normalizing all non-essential variables and constants, what is the expression relating the gravitational force between two objects to the distance between them (that is, between their centers)?
4. Select values to use for the two masses of the objects for this activity. Using these values, choose several values for the distance between the objects and calculate the force of attraction for each distance. Enter all of these in a table of values.
5. Prepare a graph of force vs. distance using the values you computed in the previous step. (Hint: Convert all the force values in your table of values so that they have the same power of 10. Then scale your vertical axis using

this power of 10.)
6. Compute another table of values for the normalized expression from step 3. Treat the proportional sign as an equals sign for this.
7. Graph the normalized equation on a *separate* set of coordinate axes.
8. Describe the similarities and the differences between the curves in the two graphs.
9. Answer the main question for this activity.

Activity 8. How does the *volume* of a gas vary with its *temperature*?

This activity is different from those you have completed so far. Instead of graphing an equation and the normalized version of it and comparing them, you will prepare a graph of predicted and experimental values, similar to the one you made for the experiment on Newton's Second Law. In this activity we will do a class demonstration in which we take data measuring the volume and temperature of a gas (air) as it changes temperature. Using the values from this demonstration and the equation for Charles' Law, you will prepare a predicted curve of volume (V) vs. temperature (T). Then, on the same set of axes, you will graph the data from the class demonstration.

Charles' Law for gases can be written as:

$$V = \left(\frac{V_i}{T_i}\right) T$$

> The Charles' Law equation does not require us to use MKS units for the volume. Cubic centimeters will be more convenient for our demonstration, so that is what we will use.

where
V is the volume of the gas in cubic centimeters (cm^3),
T is the temperature of the gas in kelvins (K),
V_i is the initial gas volume in cubic centimeters (cm^3), and
T_i is the initial gas temperature in kelvins (K).

If a certain quantity of a gas (a specific number of gas molecules) is placed in a container which holds the gas pressure constant, Charles' Law shows how volume will vary with temperature. The values for V_i and T_i will be the conditions of the gas at the start of the demonstration. Charles' Law applies to so-called ideal gases, which you will learn more about when you get to chemistry. Air behaves like an ideal gas in our demonstration as long as we keep the temperature below about 40°C.

1. If one wishes to compare how volume changes with temperature, what are the two key variables which need to be compared? Which of these is the independent variable, and which is the dependent variable?
2. After combining and normalizing all non-essential variables, what is the expression relating volume to temperature?
3. How are these variables related to one another (i.e., what kind of propor-

tion, etc.)?
4. As the instructor demonstrates the change of volume of a gas with temperature, take volume and temperature data in your lab journal. Document the materials, experimental set-up and the procedure just as you would for an experiment. The Charles' Law Demonstration notes below will guide you in what data to take and what calculations need to be performed. You will need to follow the notes to get your data and computations set up in a table of experimental values.
5. Use the V_i and T_i values from the demonstration (in units of cm³ and K) and the Charles' Law equation to calculate predicted volume values for each of the temperature values recorded during the demonstration. As shown in the example table below, these values can be entered in the same table with the experimental data.
6. Prepare a graph of V vs. T. On the same set of axes graph your predicted values and your experimental values. Be sure to label which curve is which. Reproduce your data table next to your graph. Use temperatures in degrees Celsius for the graph (even though for the calculations you must use temperatures in kelvins, as explained below). Also, I encourage you to use software such as Microsoft Excel to prepare the graph for this activity. Your graph will look much nicer and will be much more accurate if you do.

Charles' Law Demonstration
Notes for Activity 8

Use these notes to compute the values you need to prepare the graph for this activity. However, the volume calculations are a bit complicated to understand. It is not necessary that you understand everything about where the equations came from. For those who are interested and want to learn where our equations came from, these notes explain the whole thing. If you find it confusing or would rather not get into it, that's fine. Just use the equations from the notes to fill in the table and prepare your graphs. At a minimum, each student must collect data for the activity, use equations to calculate the values needed for the graphs of predicted and experimental volume vs. temperature, and prepare the graphs.

A few words about the experimental set-up are necessary here. In the demonstration a small volume of air will be trapped in an upside down buret. A buret is a glass tube marked in increments of 0.1 mL (which equals 0.1 cm³), and normally used for measuring volumes of liquid. We will use a buret that has been cut down to a much smaller size. It will be placed upside down in a beaker of automotive antifreeze (ethylene glycol). The antifreeze will be cold, because it will spend the night before the demo in a freezer. With the air trapped in the buret which is immersed in the cold liquid, we will place a thermocouple up in the buret in the air space at the top. A thermocouple is an electrical probe that can measure temperature. We will then slowly heat up the antifreeze in the beaker, reading the temperature and volume of air on the buret as the air warms up.

Because the buret is upside down we will not be able to measure the initial volume of the air (V_i) directly. This will require some fancy mathematics to get an approximation for what the initial volume is. Additionally, since the buret is upside down the readings of air volume we take as the air expands will be read with numbers on the buret that are getting smaller. We will have to take all of this into account to figure out the volumes of air we have in the buret at different temperatures.

I will start with the table we are going to use. Make a table like the one below in your lab journal. You will need about 10 or 15 rows in the table for all the data. The green columns are where you enter your data from the demonstration. The values that go in the other four columns must be calculated with the equations I will present next.

T_C (°C) Demo Data	T_K (K)	V_{buret} (mL=cm³) Demo Data	ΔV (cm³)	$V_{experimental}$ (cm³)	$V_{predicted}$ (cm³)
	$T_i =$		0	$V_i =$	$V_i =$

First, enter the data from the demo in your table. Next, convert of all of the temperature values from degrees Celsius (T_C) to kelvins (T_K). Enter all of the temperatures in kelvins in the table. To do this use this conversion equation:

$$T_K = T_C + 273.2$$

Use this equation to convert all of your temperature values into kelvins for the second column in the table.

Next, we will obtain the values for ΔV. The Greek letter delta (Δ) means *change* when used in math and science. I use the symbol ΔV to mean the total change in gas volume from when the demo started. So for example, assume the buret reading was 49.8 mL when the demo began. Assume you have a data value in your table with a buret reading V_{buret} of 46.1 mL. The value of ΔV for this data point will be 49.8 cm³ - 46.1 cm³ = 3.7 cm³. So, the ΔV values are all calculated using this equation:

$$\Delta V = \left(\text{first reading of } V_{buret}\right) - V_{buret}$$

Use this equation to calculate the values of ΔV for the fourth column in the table. The first ΔV is zero as I have shown in the table above.

Next we need to determine the value of the initial air volume in the buret, V_i. Now, note that the actual air volume in the buret, V, always equals the initial volume V_i, plus the total change in volume from V_i, which is ΔV. In other words, after we determine the value of V_i, the equation to use to compute all of the experimental values of V is

Chapter IV — Variation and Proportion

$$V_{experimental} = V_i + \Delta V$$

Use this equation to compute all of your experimental values for V after you determine V_i. These experimental values go in the fifth column in the table.

To get V_i we will substitute the expression from the previous equation for V into the Charles' Law equation to get

$$V_i + \Delta V = \left(\frac{V_i}{T_i}\right)T$$

Now we can solve this equation for V_i, which gives us

$$V_i = \frac{\Delta V}{\left(\dfrac{T}{T_i} - 1\right)}$$

Use this equation to compute V_i using the data from a temperature around 35 °C.

I suggest that if you really want to follow what's going on here (which you should if you intend to study science or engineering in college) you should work out the algebra for yourself to verify this new equation for V_i. (It won't take long and it will be good for you.) This equation allows us to get a good approximation for V_i. Simply pick one of the data points and insert the values of ΔV, T, and T_i into the equation and you've got a good estimate for V_i. Voila! Make sure you use temperature values in kelvins for this. I usually use one of the last data points for this calculation. Somewhere around 35°C (that is, 308 K) would be good.

Finally, now that we have V_i, we will use it with T_i and the values of T to calculate the predicted value of the volume for each temperature. For this calculation we will, of course, use Charles' Law, which is what this investigation is all about.

$$V_{predicted} = \left(\frac{V_i}{T_i}\right)T$$

Use this equation to compute the predicted values for V. These go in the sixth column of the table.

The result of our work will be a graph of volume vs. temperature, showing both predicted and experimental values of volume for each value of temperature (two curves on the same pair of axes).

Activity 9. How does the volume of a gas vary with its pressure?

In Activity 8 we looked at Charles' Law for gases, which relates the volume and temperature of an ideal gas. Another law for ideal gases is Boyle's Law, which relates together the volume and the pressure of a gas. Boyle's Law for gases can be written as:

$$V = \frac{V_i P_i}{P}$$

where
V is the volume of the gas in cubic meters (m³),
P is the pressure of the gas in pascals (Pa),
V_i is the initial gas volume in cubic meters (m³), and
P_i is the initial gas pressure in pascals (Pa).

If a certain quantity of a gas (a specific number of gas molecules) is placed in a container which holds the gas temperature constant, Boyle's Law shows how volume will vary with pressure. The values for V_i and P_i will be the conditions of the gas at the start of the event or experiment.

1. If one wishes to compare how volume changes with pressure, what are the two key variables which need to be compared?
2. Which of these is the independent variable, and which is the dependent variable?
3. After combining and normalizing all non-essential variables, what is the expression relating volume to pressure?
4. How are these variables related to one another (i.e., what kind of proportion, etc.)?
5. To graph the variation of volume with pressure, let's imagine that we are out in the middle of a deep lake with a large balloon and a kit of SCUBA diving gear. Let's take our balloon and fill it with air so that it is the size of a basketball. This balloon is the size of a basketball without stretching. All we have to do is fill it; no extra pressure is required. Now let's take our balloon down below the water. Your job is to use Boyle's Law to calculate the volume of the balloon at various pressures from atmospheric pressure at the surface of the water down to the pressure at a depth of 50.0 m. So your lowest pressure value is at the surface, the highest pressure value is the pressure at 50.0 m deep. You can pick several pressures in between and calculate the volume for them as well.

   Here are the values you need. A sphere the size of a basketball has a volume of 0.0071 m³ (a diameter of 9.39 inches for a regulation basketball, FYI.) This is V_i. Atmospheric pressure, P_i, is 101.3 kilopascals, or 101.3 kPa. At a depth of 50.0 m, the pressure under the water is 590.3 kPa. Prepare a table of values for pressures ranging from 101.3 kPa to 590.3 kPa, and the corresponding volumes. Graph the changes in the volume as the pressure is increased from atmospheric pressure at the surface to pressure at final depth.

6. Just for fun, let's now find out how big this balloon is at this depth. To do this we will make the slightly silly assumption that the balloon can shrink right along with the air inside it and remain spherical. Take the final volume you calculated for a depth of 50.0 m and use the equation below to calculate the radius of the balloon at the 50.0 m depth. Then double it to get the diameter. Give your final result in inches.

$$r = \left(\frac{3V}{4\pi}\right)^{\frac{1}{3}}$$

CHAPTER V
ENERGY

> **OBJECTIVES**
>
> Memorize and learn how to use these equations:
>
> $$E_G = mgh \qquad E_K = \tfrac{1}{2}mv^2 \qquad v = \sqrt{\frac{2E_K}{m}} \qquad W = Fd$$
>
> After studying this chapter and completing the exercises, students will be able to do each of the following tasks, using supporting terms and principles as necessary:
>
> 1. State the Law of Conservation of Energy.
> 2. Describe how energy can be changed from one form to another, including:
>   a. different forms of mechanical energy (kinetic, gravitational potential, elastic potential)
>   b. chemical potential energy
>   c. electrical energy
>   d. elastic potential energy
>   e. thermal energy
>   f. electromagnetic radiation
>   g. nuclear energy
>   h. acoustic energy
> 3. Briefly define each of the types of energy listed above.
> 4. Describe two processes by which energy can be transferred from one object to another (work and heat), and the conditions that must be present for the energy transfer to occur.
> 5. Describe in detail how energy from the sun is converted through various forms to end up as energy in our bodies, or as energy used to run appliances in your home or machines in industry.
> 6. Calculate kinetic energy, gravitational potential energy, work, heights, velocities, and masses from given information using correct dimensional analysis.
> 7. Define friction.
> 8. Using the Law of Conservation of Energy and equations for W, E_K and E_G, calculate and graph the energy an object has at various stages in a mechanical system.
> 9. Describe how gravitational potential energy and kinetic energy vary with respect to height, mass and velocity.
> 10. Using the pendulum as a case in point, explain the behavior of ideal and actual systems in terms of mechanical energy.
> 11. Explain how friction affects the total energy present in a mechanical system.

WHAT IS ENERGY?

Defining energy is tricky. Dictionaries usually say, "the capacity to do mechanical work," which is not particularly helpful. So we are not going to try to define it accurately, we are just going to accept that energy exists in the universe, it was put there by God, and it exists in many different forms. It is fairly obvious that a bullet traveling at 2,000 ft/s has more energy than a bullet at rest. This is why the high speed bullet can kill but the bullet at rest cannot. This study is mainly about tracking energy as it changes from one form to another, and calculating the quantities of three particular forms of energy.

THE LAW OF CONSERVATION OF ENERGY

The **Law of Conservation of Energy** is,

> Energy can be neither created nor destroyed, only changed in form.

Energy can be in many different forms in different types of substances, such as in the molecules of gasoline, in the waves of a beam of light, in heat radiating through space, in moving objects, in compressed springs, or in objects raised vertically on earth. As different physical processes occur such as digesting food, throwing a ball, operating a machine, heating due to friction, or accelerating a race car, energy in one form is being converted into some other form. Energy might be in one form in one place, such as in the chemical potential energy in the muscles of your arm, and be converted through a process like throwing a ball to become energy in another form in another place, like kinetic energy in the ball.

Strictly speaking, the Law of Conservation of Energy is violated if nuclear processes are considered. This is because of the so-called mass-energy equivalence discovered by Albert Einstein and forever enshrined in his famous equation, $E = mc^2$. In this equation E stands for energy and m stands for mass. As you can see, the constant of proportionality is the speed of light squared, a *very* large number. This explains why nuclear reactions, such as take place in stars and in nuclear weapons, give off such a huge amount of energy. A tiny quantity of mass multiplied by c^2 results in a lot of energy. During nuclear reactions mass is actually converted into energy, creating energy when none was there before. For this reason we now say that the Law of Conservation of Energy is only considered to hold across the board if the energy equivalence of mass is included. However, since most of the problems we encounter in physics and chemistry don't involve nuclear reactions, the standard form of the Law of Conservation of Energy stated above works just fine.

FORMS OF ENERGY

Here are some common forms energy can take:

GRAVITATIONAL POTENTIAL ENERGY

This is the energy an object possesses because it has been lifted up in a gravitational field. If such an object is released and allowed to fall, the gravitational potential energy will convert into kinetic energy. (The term *potential* in the name of this form of energy indicates that the energy is stored and has the capability of converting into another form of energy when released. There are other forms of energy listed below that use this term for the same reason.)

KINETIC ENERGY

This is the energy an object possesses because it is in motion. The faster an object is moving, the more kinetic energy the object has.

ELECTROMAGNETIC RADIATION

This is the energy in electromagnetic waves traveling through space, or through media such as air or glass. This type of energy includes all forms of light, as well as radio waves, microwaves, and a number of other kinds of radiation. We will study electromagnetic waves in detail in a later chapter.

CHEMICAL POTENTIAL ENERGY

This energy is in the chemical bonds of molecules. We will study bonding and molecules next semester. In the case of substances that burn, the chemical potential energy in the molecules can be released in large quantities as heat and light when the substance is burned, making these substances useful as fuel.

ELECTRICAL ENERGY

This is energy flowing in wires, such as from a power station to your house to power your appliances.

THERMAL ENERGY

This is a vague term we will define better in the next chapter. It basically means the energy in hot things that got there by heating them.

ELASTIC POTENTIAL ENERGY

This is the energy contained in any object that has been stretched (like a rubber band or a hunter's bow) or compressed (like a compressed spring).

NUCLEAR ENERGY

This is energy released from the nuclear processes of fission (when the nuclei of atoms are split apart) or fusion (when atomic nuclei are fused together). As I mentioned previously, these processes convert mass into energy.

ACOUSTIC (SOUND) ENERGY

This is the energy carried in sound waves, such as from a person's voice, the speakers in a sound system or the noise of an explosion. Since sound waves are moving air molecules, this is really a special form of kinetic energy.

WORK

Work is a mechanical process by which energy is transferred from one object to another. Objects do not possess work energy, like they do other forms of energy. Instead one object does work on another by applying a force to it and moving it a certain distance. When one object does work on another, energy is transferred from the first object to the second. We will study work in more detail later in the chapter.

HEAT ENERGY

Heat is energy in transit, flowing by various means from a hot substance to a cooler substance when a difference in temperature is present. We will study heat in more detail in the next chapter. As with work, objects do not possess heat.

Let's look at a common example of energy changing from one form into others. We all know what happens when a person lights a firecracker. (It explodes!) What forms of energy are present during the explosion, and where did all this energy come from? As shown in Figure 5-1, the energy released in the explosion was the chemical potential energy in the molecular bonds of the chemicals inside the firecracker. When these chemicals burn they release a lot of energy. (Burning is a type of chemical reaction that we will examine when we get to chemistry later in the course.) And as you already know, an exploding firecracker gives off a flash of light and heat (both are forms of electromagnetic radiation), a loud bang, and the fragments of the firecracker are blown all over the place. Thus, the chemical potential energy in the powder inside the firecracker was converted into several different kinds of energy during the explosion.

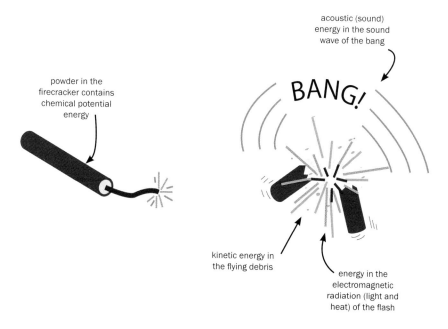

Figure 5-1. Chemical potential energy in a firecracker is converted into other forms of energy when the firecracker explodes.

Now let's consider how the Law of Conservation of Energy would apply to this explosion. All of the energy present in the chemicals before the explosion is still present in various forms after the firecracker explodes. We could represent the conservation of energy in a sort of equation like this:

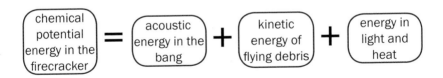

We will not do any calculations this complex in this course. But later in this chapter we will learn how to use the principle of conservation of energy to solve problems involving just two of the forms of energy we have seen so far.

THE ENERGY TRAIL

Much of the energy we depend on here on earth comes to us from the sun. As we track the forms this energy takes in its journey from the sun to, say, the energy in our bodies, we might call this the "energy trail." (This way we can have fun yelling Yee-Haw! while we are studying this.) We will follow the energy trail beginning with the sun, through different processes of conversion, and arriving at different places where energy is commonly found.

The sun's energy is produced by fusion reactions as the nuclei of hydrogen atoms fuse together to form helium. This is the same nuclear reaction as the reaction in a nuclear bomb. When referring to this energy being produced on the sun we will simply call it nuclear energy. The energy leaves the sun as electromagnetic radiation and travels through space to us. Some of this electromagnetic radiation is captured by plants and converted into chemical potential energy in the molecules in the plants, which eventually become the foods we eat. In ancient eras in the earth's history many vast forests were buried and the plant matter was converted into what we now call "fossil fuels" (crude oil, coal, and natural gas). The energy in the molecules of these fuels is chemical potential energy which can be converted into heat energy when the fuels are burned.

Your task is to be able to describe the energy conversions each step of the way from the sun all the way to your breakfast cereal or your computer. Tables 5-1 through 5-4 below illustrate a few examples of following the energy from the sun to different places it can end up here on earth. When asked to outline one of these pathways in the "energy trail" you should always list two things for each step of the way: (1) Where the energy is, and (2) what form the energy is in.

Chapter V

Energy

Where is the energy?	The Sun	Electro-magnetic waves in space	Plants on earth	Breakfast cereal	Muscles in the human body	Stretched bow	Flying arrow
What form is the energy in?	Nuclear energy	Electro-magnetic radiation	Chemical potential energy	Chemical potential energy	Chemical potential energy	Elastic potential energy	Kinetic energy

Table 5-1. Energy transformations from the sun to a flying arrow, assuming the archer was on a vegetarian diet.

Where is the energy?	The Sun	Electro-magnetic waves in space	Plants on earth	Chicken feed	Muscles in the bodies of chickens	Muscles in the human body	Moving kid on skate-board
What form is the energy in?	Nuclear energy	Electro-magnetic radiation	Chemical potential energy	Chemical potential energy	Chemical potential energy	Chemical potential energy	Kinetic energy

Table 5-2. Energy transformations from the sun to a kid on a skateboard, assuming the kid was eating chicken.

Where is the energy?	The Sun	Electro-magnetic waves in space	Plants on earth	Fossil fuel (crude oil, coal, natural gas)	Spinning gas turbine generator at power station	Wires from the power station to your house	Heat from the coils in the toaster
What form is the energy in?	Nuclear energy	Electro-magnetic radiation	Chemical potential energy	Chemical potential energy	Kinetic energy	Electrical energy	Electro-magnetic radiation

Table 5-3. Energy transformations from the sun to the heat from a toaster in your house.

Where is the energy?	The Sun	Electro-magnetic waves in space	Plants on earth	Fossil fuel (coal)	Heat from burning coal	Steam in the boiler	Moving train
What form is the energy in?	Nuclear energy	Electro-magnetic radiation	Chemical potential energy	Chemical potential energy	Heat	Thermal energy	Kinetic energy

Table 5-4. Energy transformations from the sun to a moving steam locomotive.

CALCULATIONS WITH ENERGY

Two important forms energy can take in mechanical systems are gravitational potential energy, E_G, and kinetic energy, E_K. The gravitational potential energy an object possesses depends on how high up it is, and the kinetic energy of an object depends on how fast it is moving. Gravitational potential energy is calculated as

$$E_G = mgh$$

where m is the mass (kg), $g = 9.80$ m/s², h is the height (m), and E_G is energy in joules (J).

Notice that if you know how much gravitational potential energy an object has, and its mass, you can solve this equation for h to find out how high the object is above the ground. Notice also that the gravitational potential energy of an object is directly proportional to its height. When calculating gravitational potential energy, the energy you calculate will always depend on the location you choose to use as your zero reference for the height. This zero reference might be sea level, or the ground, or the floor of your classroom, or a table top. It doesn't matter, because the E_G an object has is always relative to where $h = 0$ is. Usually the most logical location for $h = 0$ is clear from the context.

The equation for gravitational potential energy gives us another example of a derived MKS unit, joules. Multiplying the units together for the terms on the right side of the E_G equation, we can see that a joule is made up of fundamental units as follows:

$$1 \text{ J} = 1 \text{ kg} \cdot \frac{m}{s^2} \cdot m = 1 \frac{\text{kg} \cdot m^2}{s^2}$$

A golf ball has a mass of 45.9 g. While climbing a tree near a driving range, little Janie finds a golf ball stuck in a branch 9.5 ft above the ground. What is the gravitational potential energy of the golf ball at that height?

Start by writing the givens and doing the unit conversions.

$$m = 45.9 \text{ g} \cdot \frac{1 \text{ kg}}{1000 \text{ g}} = 0.0459 \text{ kg}$$

$$h = 9.5 \text{ ft} \cdot \frac{0.3048 \text{ m}}{\text{ft}} = 2.90 \text{ m}$$

$$E_G = ?$$

Now write the equation and complete the problem.

$$E_G = mgh = 0.0459 \text{ kg} \cdot 9.80 \frac{m}{s^2} \cdot 2.90 \text{ m} = 1.30 \text{ J}$$

These calculations were all done with one extra significant digit. The given height only has 2 significant digits, so now we round our final result to 2 digits.

$$E_G = 1.3 \text{ J}$$

Chapter V

Energy

An ant carries a grain of sugar up the side of a building to its nest on the roof. The mass of the grain of sugar is 0.0356 μg. After it has been carried to the roof, the E_G in the grain of sugar is 1.91 nJ. How high is the ant nest?

Write the givens and do the unit conversions.

$$m = 0.0356 \text{ μg} \cdot \frac{1 \text{ g}}{10^6 \text{ μg}} \cdot \frac{1 \text{ kg}}{1000 \text{ g}} = 3.56 \times 10^{-11} \text{ kg}$$

$$E_G = 1.91 \text{ nJ} \cdot \frac{1 \text{ J}}{10^9 \text{ nJ}} = 1.91 \times 10^{-9} \text{ J}$$

$$h = ?$$

Now write the equation, solve for h, and compute the result.

$$E_G = mgh$$

$$h = \frac{E_G}{mg} = \frac{1.91 \times 10^{-9} \text{ J}}{3.56 \times 10^{-11} \text{ kg} \cdot 9.80 \frac{\text{m}}{\text{s}^2}} = 5.47 \text{ m}$$

Every value in this computation has 3 significant digits, as does this result, so the problem is complete.

Now let's look at another important form of energy we will use in calculations. Kinetic energy is calculated as

$$E_K = \tfrac{1}{2} mv^2$$

where m is the mass (kg), v is the velocity (m/s), and E_K is the kinetic energy in joules (J).

The units for kinetic energy are joules, just as with all other forms of energy. Having completed the study of variation and proportion in the previous chapter, you can see that kinetic energy varies in direct proportion to the mass of an object, and it varies as the square of the object's velocity.

Notice here that if you know how much kinetic energy an object has, and its mass, you can solve this equation for v to find out how fast the object is moving. Since the algebra to do this may be less familiar to students in this course, you may want to just go ahead and memorize the equation for velocity as a function of kinetic energy. This equation is

$$v = \sqrt{\frac{2E_K}{m}}$$

An electron with a mass of 9.11 x 10⁻²⁸ g is traveling at 1.066% of the speed of light. Determine the amount of kinetic energy the electron has, and state your result in nJ.

Start by writing the givens and doing the unit conversions.

$$m = 9.11 \times 10^{-28} \text{ g} \cdot \frac{1 \text{ kg}}{1000 \text{ g}} = 9.11 \times 10^{-31} \text{ kg}$$

$$v = 0.01066 \cdot 3.00 \times 10^8 \frac{\text{m}}{\text{s}} = 3.198 \times 10^6 \frac{\text{m}}{\text{s}}$$

$$E_K = ?$$

Now compute the kinetic energy.

$$E_K = \tfrac{1}{2} mv^2 = 0.5 \cdot 9.11 \times 10^{-31} \text{ kg} \cdot \left(3.198 \times 10^6 \frac{\text{m}}{\text{s}}\right)^2 = 4.658 \times 10^{-18} \text{ J}$$

The problem statement requires the result to be in units of nanojoules (nJ), so perform this conversion.

$$4.658 \times 10^{-18} \text{ J} \cdot \frac{10^9 \text{ nJ}}{\text{J}} = 4.658 \times 10^{-9} \text{ nJ}$$

Both the mass and the speed of light values have 3 significant digits, so this result needs to be rounded to 3 significant digits, giving

$$E_K = 4.66 \times 10^{-9} \text{ nJ}$$

A kid fires a plastic dart from a dart gun. The mass of the dart is 21.15 g and its kinetic energy is 0.3688 J when it comes out of the dart gun. Determine the velocity of the dart.

Write the givens and do the unit conversions.

$$m = 21.15 \text{ g} \cdot \frac{1 \text{ kg}}{1000 \text{ g}} = 0.02115 \text{ kg}$$

$$E_K = 0.3688 \text{ J}$$

$$v = ?$$

Now complete the problem.

$$v = \sqrt{\frac{2 E_K}{m}} = \sqrt{\frac{2 \cdot 0.3688 \text{ J}}{0.02115 \text{ kg}}} = 5.905 \frac{\text{m}}{\text{s}}$$

Both the mass and the kinetic energy values have 4 significant digits, so the result has been rounded to 4 significant digits.

WORK

The way an object acquires kinetic energy or gravitational potential energy is that another object or person or machine does **work** on it. Work is the way mechanical energy is transferred from one machine or object to another. Work is a form of energy, but objects don't possess work. Work is the process by which energy is transferred from one mechanical system to another. Work is defined as the energy it takes to push an object with a certain (constant) force over a certain distance, and it is calculated as

$$W = Fd$$

where F is the force on an object (N), d is the distance it moves (m), and W is the work done on the object in joules (J).

Notice from this equation that since work is energy, the units here come out to be

$$1\,\text{J} = 1\,\text{N} \cdot \text{m}$$

You should take a moment to convince yourself that the units here are the same as the units described above right after the E_G equation.

If I push a person on a bicycle over a certain distance, I delivered energy to the person on the bicycle equal to the pushing force times the distance pushed. That work energy that came from the pusher (me) is now in the kinetic energy of the person on the bicycle (assuming there was no friction).

There is another important thing to note about work. The force applied to an object and the distance the object travels have to point in the same direction. If a person lifts a bucket of water, then work was done on the water. The force is applied vertically and the bucket moved vertically, so the work done to lift a bucket of water is the force required to lift it, which is the weight of the bucket, times the distance it was lifted. But a person carrying a bucket of water down the road is not doing any work on the bucket. This is because the force on the bucket to hold it up is vertical, but the distance the bucket is moving is horizontal. These do not point in the same direction. In fact they are at right angles to one another, and no work is done on the bucket of water.

A handy problem solving tip to keep in mind is that the force it takes to lift an object is equal to its weight. Recall that you can always calculate the weight of an object from its mass as

$$F_w = mg$$

An elevator in a skyscraper has a mass of 904.9 kg. Inside the elevator are three people whose masses are 67.8 kg, 55.9 kg, and 75.1 kg. Determine how much work the elevator

87

motor has to do to lift this elevator and the people inside it from the ground floor up to the 47th floor, which is 564 ft above the ground floor. State your result in kJ.

Write the givens and do the unit conversions.

$m = 904.9 \text{ kg} + 67.8 \text{ kg} + 59.9 \text{ kg} + 75.1 \text{ kg} = 1{,}103.7 \text{ kg}$

$d = 564 \text{ ft} \cdot \dfrac{0.3048 \text{ m}}{\text{ft}} = 171.9 \text{ m}$

$W = ?$

As I wrote just above, the force required to lift an object is equal to its weight. So next we need to compute the weight of the elevator and the people.

$F_w = mg = 1{,}103.7 \text{ kg} \cdot 9.80 \dfrac{\text{m}}{\text{s}^2} = 10{,}820 \text{ N}$

My calculator had a lot more digits in it than this, but I see that several of the pieces of given information have 3 significant digits, and I only need one extra digit for intermediate calculations, so I rounded to 4 digits.

Now complete the problem.

$W = Fd = F_w d = 10{,}820 \text{ N} \cdot 171.9 \text{ m} = 1{,}860{,}000 \text{ J}$

This result has been rounded to 3 significant digits. As a last step we need to convert this value to kilojoules, as required by the problem statement.

$W = 1{,}860{,}000 \text{ J} \cdot \dfrac{1 \text{ kJ}}{1000 \text{ J}} = 1{,}860 \text{ kJ}$

APPLYING CONSERVATION OF ENERGY

When an object is thrown or fired straight up from the ground it leaves the ground with a certain amount of E_K. As it goes up this E_K is converted to E_G. At the top of its flight all of the energy it had at the ground in E_K will have been converted into E_G. We can use the Law of Conservation of Energy, along with the equations for E_G and E_K, to determine how high the object will go. The same thing works in reverse. An object at a certain height has E_G. If the object is then released, then as it falls the E_G is gradually and continuously converted into E_K. Just before it hits the ground all of the E_G it had at the top has been converted into E_K. We can use the Law of Conservation of Energy, along with the equations for E_G and E_K to find out how fast the object will be going just before it strikes the ground.

In all of the problems we do in this course involving conservation of energy we will ignore friction. In reality, friction is always present in any so-called *mechanical system*, such as moving objects or machines. I will say a few words at the end of this chapter about the effect friction has on mechanical systems, but in our computations we will ignore it. Many physical systems can be approximated pretty well even if friction is ignored. In the conservation of energy experiment you will do (the Hot Wheels experiment) friction is low enough that the experimental velocity you measure should agree fairly well with the prediction.

Let us now look at a simple example of the conservation of energy in action. Figure 5-2 illustrates the application of the Law of Conservation of Energy to a person lifting a bucket and letting it drop. When a person lifts a bucket vertically the person does work on the bucket. To compute this work, the force to lift the bucket is the weight of the bucket, and the distance involved is the height it is lifted, so the work done on the bucket by the person is

$$W = F_w h$$

Since the weight, F_w, is equal to mg, this equation can be written

$$W = F_w h = mgh$$

Energy was transferred from the person (the chemical potential energy in the person's muscles) to the bucket, and the bucket now has gravitational potential energy equal to

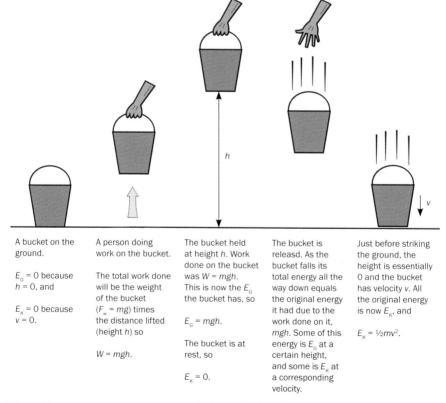

Figure 5-2. Conservation of energy applied to a lifted and falling bucket.

$E_G = mgh$

Right here we can see the conservation of energy at work. The work done by the person to lift the bucket is *mgh*. Where did that energy go? It went into the E_G the bucket has at the top, which is *mgh*, the same amount of energy. If the person releases the bucket, then as the bucket falls the gravitational potential energy begins to convert to kinetic energy. At any point as the bucket is falling, energy is conserved, which means that the total energy the bucket has is still the same as the energy it started with, but some of the energy is in kinetic energy and some of it is in gravitational potential energy. At the instant before the bucket hits the ground there will be no more gravitational potential energy because the height then is zero, so all of the energy originally given to the bucket by the work done on it is in the kinetic energy of the bucket.

CONSERVATION OF ENERGY PROBLEMS

In the bucket example above, we considered a case in which the bucket had only gravitational potential energy (which it obtained by work being done on it) and then had fallen all the way to the ground where it had only kinetic energy. But more generally, we can consider problems where an object is rising or falling and has both gravitational potential energy and kinetic energy at all times. So we are now going to consider this more general scenario. In this course we will limit our conservation of energy problems to include only kinetic energy and gravitational potential energy.

Energy considerations such as those in the bucket example allow us to use the conservation of energy as a tool in solving problems. Before we get to the examples, here is the basic rule to keep in mind when you are setting up a problem using the Law of Conservation of Energy. After writing down the givens and converting to MKS units, the next thing to write is a general conservation of energy equation. This equation adds together all of the forms of energy present in the system at the beginning of the problem on the left, and all of the forms of energy present at the end of the problem on the right, and sets them equal. This type of equation is called an *energy balance*, and can be thought of as an energy accounting equation. Since energy is conserved, all of the energy present in different forms at the beginning of the problem has shifted around into other forms at the end of the problem. But the energy we had to start with is all still around somewhere, because of the Law of Conservation of Energy, and we are simply accounting for where it all went.

The general conservation of energy equation for a rising or falling object in the absence of friction is

$E_{Gi} + E_{Ki} = E_{Gf} + E_{Kf}$

in which the subscripts *i* and *f* stand for *initial* and *final*. This equation says that the sum of the E_G and E_K at the beginning of the problem (the *initial* forms of energy) must equal the sum of these two forms of energy at the end of the problem (the *final* forms).

After setting up the general energy equation, cross out any terms equal to zero and solve the equation that remains for the energy term that relates directly to the unknown you are solving for. Heights relate to E_G because E_G is a function of the height. Velocities

Chapter V   Energy

relate to E_K because E_K is a function of velocity. After determining the amount of energy related directly to the unknown height or velocity, use the appropriate equation for E_G or E_K to determine this unknown quantity.

A stone with a mass of 136 g falls straight down from a bridge toward the river water below. When the stone is 15.5 ft above the water its velocity is 22.115 m/s. How fast was the stone falling when it was 27.25 ft above the water?

Think of this problem chronologically. The stone is *falling*, so the time when the stone is 27.25 ft above the water comes before the time when it is 15.5 ft above the water. This means that we should regard the height of 27.25 ft as the initial height and 15.5 ft as the final height. We are given the stone's velocity at the 15.5 ft height, so this velocity is the final velocity. Thus, the variable we must solve for, the velocity at 27.25 ft, is the initial velocity.

Now, begin by writing the givens and doing the unit conversions.

$$m = 136 \text{ g} \cdot \frac{1 \text{ kg}}{1000 \text{ g}} = 0.136 \text{ kg}$$

$$h_i = 27.25 \text{ ft} \cdot \frac{0.3048 \text{ m}}{\text{ft}} = 8.306 \text{ m}$$

$$h_f = 15.5 \text{ ft} \cdot \frac{0.3048 \text{ m}}{\text{ft}} = 4.724 \text{ m}$$

$$v_f = 22.115 \, \frac{\text{m}}{\text{s}}$$

$$v_i = ?$$

Now write the general conservation of energy equation.

$$E_{Gi} + E_{Ki} = E_{Gf} + E_{Kf}$$

The unknown velocity, v_i, relates to the initial kinetic energy, E_{Ki}. So the next thing to do is to solve for E_{Ki} in the energy equation.

$$E_{Ki} = E_{Gf} + E_{Kf} - E_{Gi}$$

Now we have to determine each of the three energy quantities on the right side of this equation using the given information.

$$E_{Gf} = mgh_f = 0.136 \text{ kg} \cdot 9.80 \, \frac{\text{m}}{\text{s}^2} \cdot 4.724 \text{ m} = 6.296 \text{ J}$$

$$E_{Kf} = \tfrac{1}{2} mv_f^2 = 0.5 \cdot 0.136 \text{ kg} \cdot \left(22.115 \, \frac{\text{m}}{\text{s}}\right)^2 = 33.26 \text{ J}$$

$$E_{Gi} = mgh_i = 0.136 \text{ kg} \cdot 9.80 \, \frac{\text{m}}{\text{s}^2} \cdot 8.306 \text{ m} = 11.07 \text{ J}$$

Now compute the initial kinetic energy.

$$E_{Ki} = E_{Gf} + E_{Kf} - E_{Gi} = 6.296 \text{ J} + 33.26 \text{ J} - 11.07 \text{ J} = 28.49 \text{ J}$$

Finally, use this kinetic energy to compute the initial velocity.

$$v_i = \sqrt{\frac{2 E_{Ki}}{m}} = \sqrt{\frac{2 \cdot 28.49 \text{ J}}{0.136 \text{ kg}}} = 20.5 \, \frac{\text{m}}{\text{s}}$$

Several of the pieces of given information have 3 significant digits. Notice that I used 4 significant digits for each of the intermediate calculations and rounded the result to 3 digits at the end.

A roller coaster car with a mass of 2,188.5 kg is racing along on a flat, horizontal stretch of track. The car arrives at the base of a hill moving at 19.00 ft/s and heads up the hill. When the car's velocity has dropped to 7.500 ft/s, how high will the car be above the flat track where it started?

Write the givens and do the unit conversions. We will use the flat track at the bottom of the hill as our reference for height. This means that h_i and E_{Gi} are both equal to zero.

$m = 2{,}188.5 \text{ kg}$

$h_i = 0$

$v_i = 19.00 \, \dfrac{\text{ft}}{\text{s}} \cdot \dfrac{0.3048 \text{ m}}{\text{ft}} = 5.791 \, \dfrac{\text{m}}{\text{s}}$

$v_f = 7.500 \, \dfrac{\text{ft}}{\text{s}} \cdot \dfrac{0.3048 \text{ m}}{\text{ft}} = 2.286 \, \dfrac{\text{m}}{\text{s}}$

$h_f = ?$

Now write the general conservation of energy equation.

$$E_{Gi} + E_{Ki} = E_{Gf} + E_{Kf}$$

As I said above, $E_{Gi} = 0$. The unknown height, h_f, relates to the final gravitational potential energy, E_{Gf}. So the next thing to do is to solve for E_{Gf} in the energy equation, with $E_{Gi} = 0$.

$$E_{Gf} = E_{Ki} - E_{Kf}$$

Now we have to solve for each of the two energy quantities on the right side of this equation using the given information.

$$E_{Ki} = \tfrac{1}{2} m v_i^2 = 0.5 \cdot 2{,}188.5 \text{ kg} \cdot \left(5.791 \, \frac{\text{m}}{\text{s}}\right)^2 = 36{,}696 \text{ J}$$

$$E_{Kf} = \tfrac{1}{2} m v_f^2 = 0.5 \cdot 2{,}188.5 \text{ kg} \cdot \left(2.286 \, \frac{\text{m}}{\text{s}}\right)^2 = 5{,}718 \text{ J}$$

Now compute the final gravitational potential energy.

$$E_{Gf} = E_{Ki} - E_{Kf} = 36{,}696 \text{ J} - 5{,}718 \text{ J} = 30{,}978 \text{ J}$$

Finally, use this gravitational potential energy to compute the height.

$$E_{Gf} = mgh_f$$

$$h_f = \frac{E_{Gf}}{mg} = \frac{30{,}978 \text{ J}}{2{,}188.5 \text{ kg} \cdot 9.80 \, \frac{\text{m}}{\text{s}^2}} = 1.444 \text{ m}$$

The given values each have 4 or 5 significant digits, but the value of g we used only has 3 significant digits. This means that g is the limiting factor on the precision we can state in our result. So we will round our result to 3 digits. When computing the intermediate energy values I needed to use at least 4 significant digits. I wrote some of those results with 5 digits simply because it was easier to do. Carrying around extra digits never hurts and they are just going to be rounded off at the end anyway. Rounding to 3 digits we have our final result.

$$h_f = 1.44 \text{ m}$$

NUCLEAR ENERGY CALCULATIONS (JUST FOR FUN!)

Back at the beginning of the chapter I mentioned Einstein's equation for mass-energy equivalence, $E = mc^2$. Some of you real science geeks out there were probably itching to see what a real calculation with this equation would look like. Well, why not do one? This type of calculation is not on the Objectives List for this chapter. This exploration is just for fun. You will be amazed at the result!

A common nuclear process you might hear people talking about is "nuclear decay." Nuclear decay is a natural process in which the nucleus of an atom basically falls apart, one little bit at a time. The elements that exhibit nuclear decay are said to be *radioactive*. During each step in the decay process of an atom, some kind of particle comes flying out of the nucleus, carrying away mass and kinetic energy. (This stream of particles flying out of a radioactive material is what nuclear radiation is.) As decay happens, the atom repeatedly mutates from its original identity, changing from one element to another as it goes through the decay process. The decay continues until the atom has turned into lead (usually), at which point the decay process stops.

There are three different kinds of nuclear decay, named with the first three letters of the Greek alphabet, α (alpha), β (beta), and γ (gamma). In α-decay the nucleus of an atom fires out an α-particle, which contains two protons and two neutrons. (We are going to run into α-particles again in Chapter XI.) This particular cluster of particles is identical to the nucleus of a helium atom, and we can symbolize an alpha particle as $^{4}_{2}He$. This type of notation is used in nuclear physics to represent any particle or atom that contains protons, neutrons or both. The upper number indicates the total number of protons and neutrons in the particle. (Together these are called *nucleons*, because they are the particles located in the nucleus of an atom.) The lower number indicates the number of protons in the particle.

There are two other kinds of nuclear decay. In β-decay the nucleus fires out a β-particle, which is an electron. This is a nice trick, since there *are no electrons* in the nucleus of an atom! Nevertheless, it happens. The way this works is that the mass of a neutron is a teeny bit higher than the mass of a proton. When β-decay occurs, a neutron in the atom's nucleus mutates into a proton (I know – weird!), and the difference in mass comes flying out of the atom as an electron with kinetic energy. Finally, in γ-decay the nucleus fires out a photon, which is a massless particle of light energy. (All these types of decay happen around you all the time and you have never even noticed. The world *just might* be stranger than you thought!)

Now that you know these basics, let's look at an example of nuclear decay and see how mass gets turned into energy. Radium is a radioactive element that can undergo α-decay. When an atom of radium goes through α-decay it turns into a radon atom. An atom of radium has 88 protons and 138 neutrons, so we symbolize it as $^{226}_{88}Ra$. Radon has 86 protons and 136 neutrons and is symbolized as $^{222}_{86}Rn$. We can write the decay process in the form of an equation as follows:

$$^{226}_{88}Ra \rightarrow {}^{222}_{86}Rn + {}^{4}_{2}He$$

Let's look at the masses and energies involved in this nuclear process. Looking up the masses for these three particles in a standard reference source, we find the masses to be:

Radium $m = 3.753242 \times 10^{-25}$ kg

> Radon $m = 3.686691 \times 10^{-25}$ kg
>
> alpha $m = 0.06646481 \times 10^{-25}$ kg
>
> The sum of the masses of the radon nucleus and the alpha particle is 3.753156×10^{-25} kg. This is less than the original radium nucleus mass by 8.6×10^{-30} kg. (We don't get to keep all our significant digits when we subtract like this.) The Law of Conservation of Energy tells us the missing mass has been converted into the kinetic energy the α-particle has when it comes flying out of the atom. Let's now figure out how fast this α-particle will be going. Using Einstein's equation $E=mc^2$ and $c = 3.00 \times 10^8$ m/s, we find the missing mass is equal to 7.7×10^{-13} J of energy. Finally, loading this energy and the α-particle mass into the velocity equation we get a velocity of 15,000,000 m/s, or *9,300 miles per second*, which is 5.0% of the speed of light! Yikes! Those ordinary, everyday α-particles are moving with incredible speed!

THE EFFECT OF FRICTION ON A MECHANICAL SYSTEM

In this course we are not considering friction in the calculations we do. However, in all real mechanical systems friction plays a significant role. Friction is caused when parts of the system rub against each other, or when parts of the system move through a fluid such as air or water. Just as when you rub your hands together on a cold day, friction always results in heating. When the parts of a mechanical system such as a machine get warm, heat flows from the warm parts into the cooler surrounding environment. (We will look more at how this happens in the next chapter.) This heat energy flowing out of the system is energy that used to be in the system.

When heat energy flows out of a system due to friction the Law of Conservation of Energy still applies. No energy is created or destroyed. However, the energy remaining in the system is reduced by the amount of energy that has flowed out of the system due to heating from friction. A scientist or engineer may refer to energy "lost" due to friction. This does not mean the energy was destroyed or ceased to exist, just that it flowed out of the system as heat and is no longer available as energy in the system. The net effect, of course, is that things slow down as energy gradually leaves the system as heat due to friction.

ENERGY IN THE PENDULUM

A swinging pendulum provides us with one final example of the conservation of energy at work. To begin, note that because of friction between the swinging pendulum and the air and at the pivot at the top of the pendulum any actual pendulum will lose energy to the environment as heat. This is why any actual free-swinging pendulum will always come to a stop.

But let's imagine a perfect pendulum, one that loses no energy due to friction. We will call this an *ideal pendulum*. In an ideal pendulum no energy would leave the "system" (the swinging pendulum) as heat and the pendulum would just keep on swinging without slowing down. (Actually, it's a bit more complicated because of the rotation of the earth, so even in a vacuum with a magnetic pivot bearing the pendulum

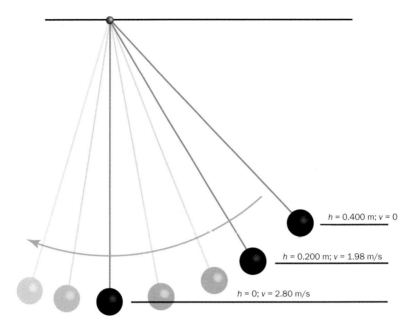

Figure 5-3. Conservation of energy in a swinging pendulum.

would still slow down, so don't start getting visions of a perpetual motion machine! But our imaginary ideal pendulum is also free from such influences.)

From what you know about the forms of energy and energy conservation, you can probably already see how energy transformation will work in this ideal pendulum. As shown in Figure 5-3, we will let the height of the pendulum when it is at rest (that is, not swinging) be our reference for height measurements. This means when the pendulum is straight down its height is zero and its gravitational potential energy is also zero. When the pendulum swings up to its highest point it momentarily comes to rest. At this moment its velocity is zero and so its kinetic energy is zero. Put these facts together and let the pendulum start swinging.

Because of the conservation of energy, the pendulum always has the same amount of energy no matter where it is. No energy is leaving the system, and no one is adding any energy to the system (that is, not after we get it started). Thus, the total energy in the system always equals the sum of the E_G and the E_K, and this sum must always add up to the same value no matter where the swinging pendulum is.

Just to run through a quick calculation, let's say the mass at the end of the pendulum is 2.00 kg and we lift it up 0.400 m above its lowest point to release it. The total E_G at this starting point will be *mgh*, which gives $E_G = 7.84$ J. The kinetic energy here is zero, so now we know that the pendulum has a total of 7.84 J of energy no matter where it is.

How fast will it be going when it is halfway down, 0.200 m high? Well, the E_G at that position will be 3.92 J, so the E_K will be 7.84 J – 3.92 J = 3.92 J. Using this kinetic energy value in the velocity equation gives a velocity of 1.98 m/s. At the bottom all of the energy (7.84 J) is kinetic energy, so the velocity will be 2.80 m/s. As you can see, if you know how high the pendulum is at any point, you can determine how fast it is moving, and vice versa.

CHAPTER V EXERCISES

Classroom Examples

1. Water is pumped into a high water tower. If the total mass of the water is 1.00×10^5 kg and the tower is 240 feet high, what will the gravitational potential energy (E_G) be of the water in the tower?

2. A bullet of mass 25 g is fired at a velocity of 556 ft/s. How much kinetic energy (E_K) does the bullet have?

3. A man lifts a bucket of sand 75 cm above the ground. If the bucket of sand has a mass of 12,500 g, how much work did the man do?

4. Referring again to the previous problem, after the bucket of sand has been lifted, how much E_G will the sand have?

5. If the man releases the bucket of sand and lets it drop, what will its velocity be the instant before it strikes the ground?

6. A boy carries a water balloon up to the top of a ladder to let it drop. The mass of the water balloon is 255.8 g and the ladder is 10.4 ft tall.
   a. How much E_G does the balloon have at the top of the ladder?
   b. How fast will the balloon be going just before it splats on the ground?

Answers
1. 72,000,000 J
2. 360 J
3. 92 J
4. 92 J
5. 3.8 m/s
6. a) 7.95 J b) 7.88 m/s

Energy Calculations Set 1

1. A load of building materials is hoisted to the top of a building under construction. If the total mass of the material is 1.31×10^3 kg and the building is 177.44 feet high, what will the gravitational potential energy (E_G) be of the material at the top of the building?

2. A car of mass 2,345 kg is traveling at a speed of 31 mph. How much kinetic energy (E_K) does the car have?

3. A woman lifts a bucket of water 61.7 cm above the ground. If the bucket of water has a mass of 17.5 kg, how much work did the woman do?

4. How much E_G does the bucket have after being lifted?

5. If the woman releases the bucket of water and lets it drop, what will its velocity be the instant before it strikes the ground?

6. A kid shoots an arrow straight up. The arrow has a mass of 122 g and leaves the ground with a velocity of 13.75 m/s. How high will it go?

7. A girl drops a stone into the water from a bridge. The stone has a mass of 325 g and the bridge is 36.1 m above the water. How fast will the stone be moving just before it hits the water?

8. A worker slides a carton across the floor. The force of friction between the carton and the floor is 735 N. If the worker pushes the carton 26 m, how much work did he do?

Answers
1. 694,000 J
2. 230,000 J
3. 106 J
4. (This answer is top secret!)
5. 3.48 m/s
6. 9.65 m
7. 26.6 m/s
8. 19,000 J

Energy Questions Set 2

1. A carpenter hauls twenty 80.0-pound bundles of shingles up onto a roof which is 8.5 m above the ground.
   a. Compute the total mass of the shingles using the conversion factor for pounds to newtons found in Appendix A, and then the weight equation to get the mass.
   b. Compute the work the carpenter had to do to get the shingles up onto the roof. State your answer in joules (J).
   c. Using the equation for gravitational potential energy (E_G) compute the E_G of the shingles while they are on the roof.
   d. If the entire stack of shingles slides off of the roof, how much kinetic energy will the stack have at the following times:
      i. At the moment it first begins to slide
      ii. Just before it hits the ground.
   e. Compute how fast the stack will be falling just before it hits the ground.

2. A car weighing 3,193 lb rests at the top of a hill, then begins to roll down the hill (the engine is off). Assume it rolls to the bottom of the hill with negligible friction.
   a. If the hill is 16 m high compared to the flat road at the bottom, compute the E_G of the car while it is at the top of the hill.

b. After the car rolls to the bottom of the hill, where will the energy be which was in the E_G of the car while it was at the top of the hill?
c. Compute the E_K of the car when it reaches the bottom of the hill.
d. Compute the velocity of the car when it reaches the bottom of the hill.
e. Explain how conservation of energy relates to this problem.
f. Explain specifically how friction would change the results of the problem in a more realistic example.

3. Use the same car and the same hill from the preceding problem.
   a. Compute the E_G the car will have when it has rolled half way down the hill.
   b. Compute the E_K the car will have when it is half way down the hill.
   c. Compute the velocity of the car when it is half way down the hill.
   d. Explain how conservation of energy relates to these calculations.

4. Use the same car and the same hill from the preceding two problems.
   a. Beginning with the car's starting point at the top of the 16.0 m hill, make a table and calculate the following values at increments of two meters at every height from 16.0 m down to 0.000 m: (a) the E_G of the car, (b) the E_K of the car, and (c) the velocity of the car.
   b. Draw three graphs by hand showing the relationships between the variables in the previous part, E_G vs. height, E_K vs. height, and velocity vs. height. You have 9 values to plot on each of the three graphs. Choose the scales on your axes appropriately so that the values you need to graph fit comfortably on the coordinate plane. Use identical scales for the two energy graphs.

Answers (Some extra significant digits are included to help you check your work.)
1. a) 724 kg (This value has one extra significant digit.) b) 6.0 x 10^4 J c) 6.0 x 10^4 J d) 0 J, 6.0 x 10^4 J e) 13 m/s
2. a) 230,000 J c) 230,000 J d) 18 m/s
3. a) 114,000 J (This value has one extra digit.) b) 114,000 J (one extra digit) c) 13 m/s

Energy Questions Set 3

1. Consider a large, ideal (that is, frictionless) pendulum with a steel ball weighing 27.05 lb on the end. This ball is lifted 185 cm and released. Since the pendulum is frictionless, the ball will now swing back and forth forever.
   a. How much work was done to lift the ball?
   b. What was the E_G of the ball after it was lifted, but before it had started falling?
   c. At the bottom of the ball's pathway, what will its E_G be?
   d. At the bottom of the ball's pathway, what will its E_K be?
   e. How fast will the ball be going at the bottom?
   f. How much E_G and E_K will the ball have when it is half way down?

g. How fast will the ball be going when it is half way down?
h. How fast will the ball be going when it is 90.0 % of the way down?
i. Use friction and energy considerations to explain the difference between this pendulum and an actual pendulum.

2. A group of water pumps operated by the city pump water up into a water tower 197 feet high. The water tower holds 6.016 x 10^6 kg of water. Determine the amount of mechanical work the pumps must do to fill the water tower.

3. Imagine a new frictionless roller coaster which uses magnetic levitation so that the cars float above the rails without actually touching them. Imagine also that the aerodynamic design of the cars is so good that there is essentially no air friction. The car has a mass of 5,122 kg. From the top of a 25.0 m-hill the cars roll down a valley where the lowest point is 2.50 m above the ground, and then back up to the top of a lower hill, 18.0 m above the ground. Assuming the roller coaster begins at rest at the top of the first hill, determine how fast it will be traveling
   a. When it reaches the bottom of the valley.
   b. When it reaches the top of the second hill.

4. A boy weighing 104.6 lb runs up the steps two at a time. There are 13 steps, each one 16.5 cm high.
   a. How much work did he do to get to the top of the steps?
   b. If he steps over the hand rail and drops back down to the ground, how fast will he be moving just before he lands?

5. An object with mass m = 351 g is sliding on a frictionless surface at 500.00 cm/s when it begins going up a ramp. What will the height, h, be when it stops?

6. A roller coaster car weighing 4,294 lb is moving at 27.89 ft/s when it goes up a small hill. Ignoring friction, if the hill is 7.710 ft high, how fast will the roller coaster be going as it tops the hill?

7. A man uses a mallet to try to ring the bell at a county fair. The sliding puck goes straight up 6.500 m from the ground to the bell. The mass of the puck is 950.0 g. If there is no friction and the puck is going 1.000 m/s when it strikes the bell, what was its speed as it left the ground? For this problem, use the more precise value for g, g = 9.803 m/s^2.

Answers
1. a) 223 J b) 223 J c) 0 J d) 223 J e) 6.02 m/s f) 111 J (112 J is also acceptable) g) 4.26 m/s h) 5.71 m/s
2. 3.54 GJ
3. a) 21.0 m/s b) 11.7 m/s
4. a) 998 J b) 6.48 m/s
5. 1.28 m

6. 5.12 m/s
7. 11.33 m/s

Energy Questions Set 4
General Conservation of Energy Problems

In all of the following problems assume there is no friction. After writing down the givens and converting units, write an energy balance equation for each problem. Then proceed to work out the solution.

1. A car is parked at the top of a hill when its parking brake fails. The car weighs 3,420.1 lb and the top of the hill is 31.0 ft above the level road below. How fast will the car be going when it is still 6.50 ft above the road at the bottom?

2. An elevator with mass m = 2,200 kg is in a mine shaft when the elevator cable snaps. When the elevator is 33.87 m above the bottom it is falling at 17.88 m/s. How fast was the elevator going when it was 37.00 m above the bottom?

3. A tiny sample of carbon with a mass of 1.873×10^{-2} mg is placed in a vacuum chamber and thrust upward with a spring mechanism. When the sample has risen 2.177 cm from its launch its velocity is 202.75 cm/s. What will its velocity be when it has risen 7.50 cm above the launch location?

4. A microscopic oil drop with a mass of 9.0022 µg is held in a container in a vacuum and released from rest. When released, the oil drop is 9.0125 cm above the bottom of the container. How high above the bottom of the bottle will the oil drop be when its velocity is 85.160 cm/s?

5. A kid is on the roof of his clubhouse. He throws a brick straight down with a velocity of 1.562 ft/s. The brick weighs 4.1843 lb. When the brick hits the ground its velocity is 24.75 ft/s. How high is the roof of the clubhouse?

Answers
1. v_f = 12.1 m/s
2. v_i = 16 m/s
3. v_f = 1.75 m/s
4. h_f = 5.31 cm
5. h_i = 2.89 m

CHAPTER VI
HEAT AND TEMPERATURE

> **OBJECTIVES**
>
> Memorize and learn how to use these equations:
>
> $$T_C = \frac{5}{9}(T_F - 32°) \qquad T_K = T_C + 273.2$$
>
> After studying this chapter and completing the exercises, students will be able to do each of the following tasks, using supporting terms and principles as necessary:
>
> 1. Define and distinguish between heat, internal energy, thermal energy, thermal equilibrium, specific heat capacity, and thermal conductivity.
> 2. State the freezing/melting and the boiling/condensing temperatures for water in °C, °F, and K.
> 3. Convert temperature values between °C, °F, and K.
> 4. Describe and explain the three processes of heat transfer: radiation, conduction and convection.
> 5. Describe how temperature relates to the internal energy of a substance and to the kinetic energy of its molecules.
> 6. Explain the kinetic theory of gases, and use the kinetic theory of gases to explain why the pressure of a gas inside a container is higher when the gas is hotter and lower when the gas is cooler.
> 7. Apply the concepts of specific heat capacity and thermal conductivity to explain how common materials such as metals, water, and thermal insulators behave.

TEMPERATURE SCALES

There are a lot of temperature scales around because scientists in different countries have been investigating materials at different temperatures for a long time. We will use only three different scales in this course, the Fahrenheit scale you have been using all of your life (if you grew up in America), and the two main SI scales, the Celsius scale and the Kelvin scale.

For historical reasons, the Celsius and Fahrenheit scales use the strange term "degrees." This is odd when you think about it. We don't use a term like this with any other units of measure. The Kelvin scale does not use this term. A temperature of 300 K is read as, "three hundred kelvins."

A degree on the Celsius scale is the same as a "degree" on the Kelvin scale; the degrees are the same size. But the Kelvin scale is an absolute scale, which means there are no negative temperatures on the Kelvin scale. The temperature of 0.0 K is referred to as "absolute zero." According to the laws of physics as we understand them, this temperature cannot be achieved. A region might be refrigerated down to a small fraction of 1.0 K, but nothing can ever reach 0 K. (And nothing can ever be less than 0 K.)

The temperatures for freezing/melting or boiling/condensing of water are used a lot in scientific work and you should know them. They are summarized in Table 6-1.

Phase Change	Degrees Fahrenheit (°F)	Degrees Celsius (°C)	Kelvins (K)
Boiling or Condensing	212	100	373.2
Freezing or Melting	32	0	273.2

Table 6-1. Freezing and boiling points for water in three temperature scales.

TEMPERATURE UNIT CONVERSIONS

The only computations we are going to do in this chapter are converting temperature values from one system of units to another. This is easy if you get the equations down and pay attention to your algebra. To convert a temperature in degrees Fahrenheit (T_F) into degrees Celsius (T_C) use this equation:

$$T_C = \frac{5}{9}(T_F - 32°)$$

To convert a temperature in degrees Celsius into kelvins (T_K) use this equation:

$$T_K = T_C + 273.2$$

Memorize these two equations. The first one enables you to convert a temperature value in degrees Fahrenheit into degrees Celsius. The second one enables you to convert from degrees Celsius to kelvins. If you want to convert temperature values the other way around, simply use a little algebra and solve the first equation for T_F. Ditto for converting from kelvins to degrees Celsius. I advise against memorizing the other forms of these conversion equations because they looks so similar to the ones above that they are easily confused. Just remember these two and use algebra when you need to.

In fact, as examples, I will do just this, beginning with a temperature in kelvins and computing its value in the other two scales.

One note is in order about significant digits for these calculations. Figuring out the significant digits with these two equations can be confusing, so when we do temperature conversions we will use a special rule. Our special rule for this kind of calculation is simply this: When performing temperature conversions state every answer with one decimal place.

The melting point of aluminum is 933.5 K. Express this temperature in degrees Celsius and degrees Fahrenheit.

103

First, solve the T_K equation for T_C to allow converting the given value into degrees Celsius.

$$T_K = T_C + 273.2$$

$$T_C = T_K - 273.2$$

From this we calculate the Celsius value as

$$T_C = T_K - 273.2 = 933.5 - 273.2 = 660.3 \ °C$$

Next, do the algebra to solve the T_C equation for T_F. Do this very carefully. The algebra is not difficult, but students often do it incorrectly.

$T_C = \frac{5}{9}(T_F - 32°)$. Now multiply both sides by $\frac{9}{5}$.

$\frac{9}{5}T_C = T_F - 32°$. Now add 32° to both sides to get

$$T_F = \frac{9}{5}T_C + 32°$$

With this equation calculate the Fahrenheit value.

$$T_F = \frac{9}{5}T_C + 32° = \frac{9}{5} \cdot 660.3 \ °C + 32° = 1,220.5 \ °F$$

DEFINITIONS OF COMMON TERMS

Before we get to definitions, a word is in order about the difference between atoms and molecules. We will study these in detail next semester, but for now you should know that in some substances, like water and oxygen gas, the atoms are bonded together in little groups called molecules. In substances like this the smallest particle of the substance is a molecule. So it makes no sense to refer to an atom of water. The smallest particle of water is a water molecule, represented by the formula H_2O. There are three atoms in this molecule (two hydrogen atoms and one oxygen atom). In other substances like gold and helium gas, the atoms don't do this, but remain alone. So here it is appropriate to speak of an atom of gold or an atom of helium gas. The bottom line for now is that in the discussion that follows I will use the terms atom and molecule interchangeably, meaning the smallest particle of the substance.

The terms used in this study are subtle, so study these carefully. **Heat** is energy in transit, flowing from one substance to a cooler substance. No substance ever possesses heat. A substance can possess other forms of energy, but heat is the term for energy in the process of flowing from one substance to another because of a difference in temperature. **Thermal energy** is a very informal and relative term, so it is only useful when speaking generally. It does not have any real technical use. Basically, thermal energy is energy an object possesses because it has been heated. The term is only really used to compare substances to room temperature. So we might speak of boiling water as having thermal energy because it was heated on a stove. But if an object was taken from a freezer where its temperature was $-20°C$ and heated up to $-15°C$, we probably would not say that it has thermal energy, because even though it was heated it is still very cold relative to room temperature.

The **internal energy** of a substance is the total of all of the kinetic energies possessed by the atoms or molecules of a substance. Atoms or molecules are constantly in motion, vibrating or translating, or both. Atoms in solids cannot fly around, so they vibrate in place, but atoms in gases are free to translate, or zoom around (see the box for the details). Either way, since atoms are always moving they always possess kinetic energy. If you added up the kinetic energy of every particle in a certain substance, that would be its internal energy. This term is much more precise than "thermal energy," and can actually be computed. Although it is somewhat complicated to explain, the "temperature" of a substance is very closely related to its internal energy. The higher the temperature, the higher the internal energy.

Now that you know about internal energy you can understand what *absolute zero* is, the zero temperature on the Kelvin temperature scale. The temperature of a substance

HOW FAST ARE AIR MOLECULES MOVING?

When we say that air molecules are zipping around all the time, how fast are they actually going? The answer is rather shocking. At room temperature an average air molecule is moving at about 1,100 mph, or 500 m/s! This speed is 1.5 times the speed of sound, also called Mach 1.5. Of course, this is just the average speed; half the molecules are moving faster than that! This sounds bizarre.

But it gets weirder, so hold on to your hat. We know there are *lots* of molecules in the air, so if they are moving this fast, don't they bump into each other a lot? YES! In every cubic centimeter of air there are about 2.7×10^{19} molecules. Do you have *any idea* how many molecules this is? If there were this many people on earth there would be over 4,800 people *per square foot* everywhere on the planet, including the oceans!

With this many molecules all moving so fast, the average distance a molecule travels before colliding with another molecule is much smaller than the wavelength of light, and an average molecule will collide 5 times every nanosecond (billionth of a second). Sprightly little fellers, huh? In sum, have a look at a little chunk of air in front of you about the size of a sugar cube. In this little space there are 27,000,000,000,000,000,000 molecules and they each bump into another molecule 5,000,000,000 times per second moving at Mach 1.5 between every hit.

God's world is jaw-dropping amazing. But heck, this is just boring old air. How cool do you think the DNA molecules are inside your body? But you'll have to wait for biology to learn about that.

varies directly with its internal energy. A temperature of 0 kelvins (absolute zero) means no internal energy at all, or in other words, the atoms would be standing still! Atoms can't move any slower than standing still, and thus, there is no temperature lower than absolute zero. As far as we know, there is no place in the universe where the temperature is absolute zero, although physicists in low-temperature research labs have succeeded in achieving temperatures of only a few millionths of a degree above absolute zero. (As you might imagine, very weird things happen at temperatures that low.)

Finally, we have the concept of **thermal equilibrium**. The laws of thermodynamics say that if a substance is at a different temperature than its environment heat will flow from the warmer of the two into the cooler of the two. We are all familiar with this principle, even if we don't know the actual law. (FYI, it is the Second Law of Thermodynamics that says heat must flow this way.) Thermal equilibrium is the state in which an object is at the same temperature as its environment, and when this happens heat flow ceases.

HEAT TRANSFER PROCESSES

There are three processes by which heat can transfer from one place to another. As you read, refer to the diagrams in Figure 6-1 which illustrate these processes. The first heat transfer process is **conduction**. This process occurs in solids. In a solid the atoms or molecules of the substance vibrate in place but they are not free to flow around. Imagine a great number of atoms in a substance attached together in a flexible grid, as if they were connected to one another by springs. If one side of the substance was heated, the internal energy of the atoms there goes up, which means they begin vibrating more vigorously because their kinetic energy is increasing. This vibration will be transferred to the other atoms nearby, because the atoms are all linked together, as atoms always are in a solid. The kinetic energy in these vibrations will continue to spread to atoms farther and farther away from the location where the material is being heated. As atoms gain more kinetic energy, their temperature goes up. This is heat flow by conduction.

A well-known example of conduction is the way heat transfers in the metal of a frying pan. Initially the atoms in the pan are all at room temperature (and are thus at thermal equilibrium with the room). When the center of the pan is heated the atoms there gain internal energy and begin vibrating more vigorously. These vibrations are passed atom to atom and the temperature increases more and more, farther and farther away from the location where the source of heat is. Eventually the vibrations get all the way out to the handle of the pan and the handle itself gets hot, even though it is nowhere near the heat source.

The second heat transfer process is **convection**. This process occurs in fluids, that is, liquids and gases. The particles of a fluid are free to flow around and mix and mingle and collide with other particles. Particles with high internal energy are hot and will be moving very rapidly, because the internal energy is the kinetic energy of the particles. These particles will mix with cooler particles and when they do they collide with them and transfer some of their high energy to the lower energy particles. This is just like balls colliding on a pool table. When a fast ball hits a slow one the fast one slows down and the slow one speeds up as kinetic energy is transferred from the ball with more energy to the one with less. Unlike pool balls, particles in liquids and gases never stop

Chapter VI — Heat and Temperature

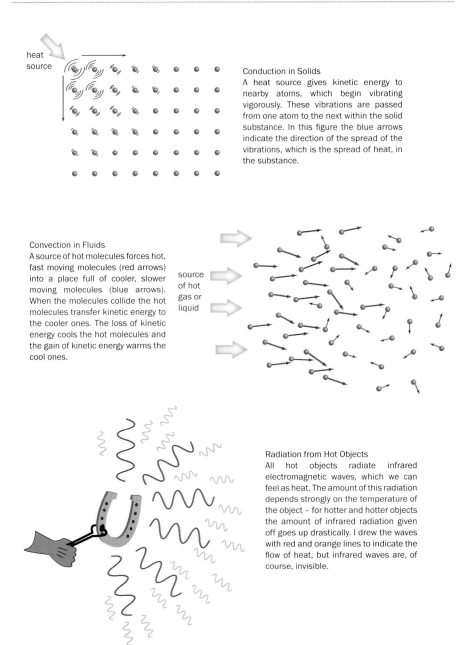

Conduction in Solids
A heat source gives kinetic energy to nearby atoms, which begin vibrating vigorously. These vibrations are passed from one atom to the next within the solid substance. In this figure the blue arrows indicate the direction of the spread of the vibrations, which is the spread of heat, in the substance.

Convection in Fluids
A source of hot molecules forces hot, fast moving molecules (red arrows) into a place full of cooler, slower moving molecules (blue arrows). When the molecules collide the hot molecules transfer kinetic energy to the cooler ones. The loss of kinetic energy cools the hot molecules and the gain of kinetic energy warms the cool ones.

Radiation from Hot Objects
All hot objects radiate infrared electromagnetic waves, which we can feel as heat. The amount of this radiation depends strongly on the temperature of the object – for hotter and hotter objects the amount of infrared radiation given off goes up drastically. I drew the waves with red and orange lines to indicate the flow of heat, but infrared waves are, of course, invisible.

Figure 6-1. The three heat transfer processes.

moving, so over time all the particles share the energy evenly. When this happens, they are all at the same temperature (thermal equilibrium).

An example of convection is when hot air from a central heating system is blown into a room by a blower fan, as is the case in most contemporary homes. The hot air molecules move around mixing with the cooler molecules, colliding with them and exchanging energy so the cooler molecules begin moving faster, which means they are hotter. The same process is going on between water molecules when the heating system in a swimming pool circulates warm water into the pool, mixing it in with the cool water.

The third heat transfer process is **radiation**. Radiation is the term for heat energy transferred by electromagnetic waves. As we saw in the previous chapter, electromagnetic waves travel through space and through the atmosphere to carry energy from the sun to the earth. Light, infrared radiation, ultraviolet radiation, microwaves, X-rays – these are all the same thing, electromagnetic radiation (or waves). The only thing that distinguishes these different terms for radiation is the *wavelength* of the waves. When we consider heat transfer, we are mainly talking about *infrared* (IR) radiation, which consists of invisible electromagnetic waves with wavelengths in the range of roughly 750 to 2500 nanometers.

In the next chapter we will look at why this electromagnetic heat energy is called "infrared radiation," and we will discuss how infrared radiation relates to other electromagnetic radiation such as radio waves and light. But here it is important to point out that the hotter an object is, the more infrared energy it radiates. When you feel the heat radiating from a hot object such as a branding iron it is the IR radiation you are feeling. When you hold your hands in front of a fire, it is the IR radiation from the coals you are feeling. Your eyes can't see it the way they can visible light, but your skin can feel it. When you feel warm in the sun even on a cool day, it is the IR radiation from the sun that is warming you. When the inside of your car gets hot it is because of IR radiation from the sun striking the car and then transferring into the car by the radiation passing through the glass and by conduction through the metal. Figure 6-2 depicts all three heat transfer processes as each of them contribute to the warming of the interior of a car in the sun.

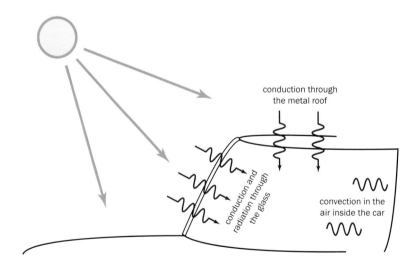

Figure 6-2. Heat transfer processes at work heating the interior of a car.

KINETIC THEORY OF GASES

Now that you know that the internal energy in a gas relates to the temperature of the gas, and that this energy is stored in the gas in the form of kinetic energy in each of the molecules of gas, you are ready to understand what causes gas pressure. This explanation is called the **Kinetic Theory of Gases**.

According to the Kinetic Theory of Gases, the atoms or molecules of a gas in a container are zooming around inside the container all the time at high speed. The hotter the gas is, the more kinetic energy the particles have and the faster they are zooming around. There are an incredible number of particles in even a small container, and as they zoom around, the individual particles of this huge population of particles are constantly striking the walls of the container. These collisions between the particles and the container walls cause the effect we call *pressure*. If the temperature goes up, the particles have higher internal energy, meaning higher kinetic energy, so they are moving faster and striking the container walls harder and more frequently, creating higher pressure. We can summarize this theory of gas behavior this way:

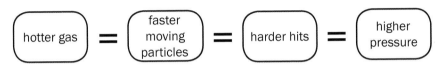

A nice example that many of you have seen occurs when you buy a Mylar birthday balloon at the grocery store on a hot day. The balloon vendor fills the balloon with an appropriate amount of helium inside the cool grocery store. When you carry the balloon outside into the blazing heat the balloon swells up so tightly it appears it is ready to burst from the high pressure. Then you put it inside a car with the air conditioner blasting away. The very cold air from the car A/C is so cold the pressure in the balloon decreases significantly, even way below what it was inside the store. This causes the balloon to shrink and collapse and fall down on the floor of the car. When you arrive home and take the balloon out of the car, the hot outside air warms up the helium, those atoms pick up speed again and begin colliding once again at high speed on the inside of the balloon causing the pressure to spike back up. All of this is caused by the atoms in the helium gas gaining kinetic energy when they warm up and losing kinetic energy when they cool.

SPECIFIC HEAT CAPACITY

The **specific heat capacity** of a substance is a property of the substance that relates to how the material behaves when it is absorbing or releasing heat energy. The specific heat capacity (or simply, heat capacity) of a substance is defined as the amount of heat energy that must transfer into or out of 1 gram of the substance to change its temperature by 1 degree Celsius.

A substance with a low specific heat capacity will change temperature very easily and quickly because very little energy must be added or removed to change its temperature. Metals are like this. A very small amount of heat will heat up a metal a lot, and removing only a small amount of heat will cause the temperature of a metal to drop a lot. A substance with a high specific heat capacity will change temperature only very slowly because it must absorb or release a lot of energy to change its temperature. Water is like this. (In fact, water has a great number of very unique properties, and this is one of them.) It takes a lot of heat energy to warm up a pan of water even a little bit. And to cool water down requires removing a lot of heat from it just to cool it down a little bit. For comparison, the specific heat capacity of water is about ten times greater than that of copper, and about five times greater than that of aluminum.

THERMAL CONDUCTIVITY

Like specific heat capacity, **thermal conductivity** is another property of materials that relates to how materials behave when they are absorbing or releasing heat energy. The thermal conductivity of a substance is a measure of how well a material transfers heat within its own atomic structure by conduction. Metals have a very high thermal conductivity, so heat conducts readily through a metal. (This is because the atoms in a metal are arranged in a crystal structure, a subject we will take up when we get to chemistry.) On the other hand, building insulation used inside the walls of buildings and houses has a very low thermal conductivity, so heat transfers through it very, very slowly. These materials are called **thermal insulators**. Other materials with a low thermal conductivity are glass and plastic. In fact, high thermal conductivity in materials seems typically to accompany high electrical conductivity. We all know that metals conduct electricity the best. Well, they conduct heat the best, too.

The effect of different thermal conductivities in materials is demonstrated dramatically by watching what happens when ice cubes are placed on blocks of plastic and aluminum. As shown in Figure 6-3, the block on the left is made of some kind of composite wood and plastic material. The block on the right is made of aluminum. The ice cubes were placed on the blocks at the same time and the photo was taken after about 30 seconds. Which material has a higher thermal conductivity? What do you think the two blocks would feel like if you touch them? Which direction is the heat flowing?

DISTINGUISHING BETWEEN SPECIFIC HEAT CAPACITY AND THERMAL CONDUCTIVITY

It is important for you to spend enough time on this topic so you can confidently distinguish between the two properties we have been discussing, specific heat capacity and thermal conductivity. Students often get these properties confused. It will help for us to consider the actual values of these quantities for some common materials.

Specific heat capacity has to do with the *quantity* of heat it takes to warm a substance, or the quantity of heat that must be removed from a substance to cool it down. As you can see in Figure 6-4, the heat capacity of water is huge compared to just about everything else. Also notice that the specific heat capacity of metals is quite low compared to water.

The huge heat capacity of water is a major factor in regulating the temperatures on earth. (The atmosphere is another major factor.) About 70% of the earth is covered by water, primarily in the oceans. During the day the radiation from the sun warms the oceans, but a lot of energy must be added to all this water to warm it. At night the

Figure 6-3. High thermal conductivity, illustrated by the block on the right, which is made of aluminum.

thermal energy in the water can escape into the atmosphere, allowing the water to cool. But again, a lot of heat energy has to be removed from water for it to cool down. The result is that the ocean water covering the earth does not change temperature very much. Since there is so much water on the earth, the steady temperature of the water helps regulate the temperature of the atmosphere, and thus of the entire planet. Without the oceans, the temperature swings from day to night on earth would be huge, and complex life on earth could not survive.

By contrast, thermal conductivity is about how well heat can flow *through* a substance by conduction. As shown in Figure 6-5a, the thermal conductivity of metals completely dominates everything else. Since the thermal conductivities of all other materials are so small compared to metals, I made a separate chart in Figure 6-5b for the nonmetal substances so that you can compare them. Notice that the thermal

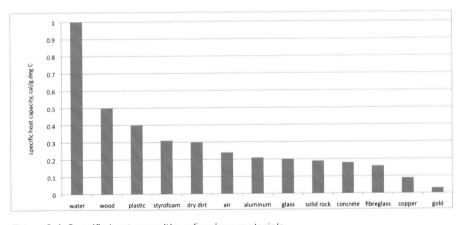

Figure 6-4. Specific heat capacities of various materials.

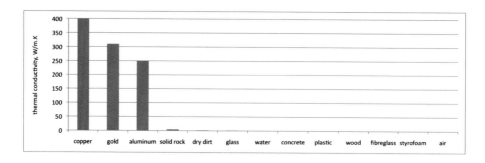

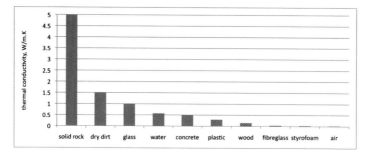

Figures 6-5a (top) and 6-5b (bottom). Thermal conductivities of various materials.

conductivities of water and air are miniscule, because heat transfers through fluids mainly by convection, not by conduction, as we discussed previously.

The data in these charts suggest some things you will want to think about when addressing questions about specific heat capacity and thermal conductivity. Since the specific heat capacity of water is so much greater than the heat capacity of anything else, the presence of water will be the major factor in questions about how fast things heat up or cool down, or about how much energy must be added to a substance to warm it up or removed from it to cool it down. And since the thermal conductivity of metals is so much greater than the thermal conductivity of anything else, the presence of metal will be the major factor in how rapidly heat will flow through a substance by conduction.

Chapter VI Exercises

Temperature Unit Conversions

Fill in the missing values in the table. Warning: It is very common for students to think these conversions are easy (they are), complete this exercise, and commence to get the conversions wrong on quizzes. So take care that you memorize the equations and constants correctly and that you are careful with your algebra.

	°F	°C	K
1		32.0	
2	56.5		
3			455.0
4	−17.9		
5		−41.6	
6			79.0

Answers

	°F	°C	K
1	89.6	32.0	305.2
2	56.5	13.6	286.8
3	359.2	181.8	455.0
4	−17.9	−27.7	245.5
5	−42.9	−41.6	231.6
6	−317.6	−194.2	79.0

Heat Transfer and Kinetic Theory Study Questions

1. Imagine a closed glass jar (like a Mason jar) full of ice water sitting on a table in the middle of a metal building (say, a tool shed). The metal building is in a clearing in the sun. Starting with the sun, identify every heat transfer process that happens for energy to transfer from the sun where it starts to the ice where it ends up.

2. People who live in places where summer temperatures are much higher than winter temperatures have to adjust the amount of air in their car tires a few times per year. Explain why this is, and identify when air must be added (raising the pressure) and when air must be released (lowering the pressure).

3. When driving on a long trip I like to take along a vacuum bottle full of hot coffee to drink on the road. A vacuum bottle consists of two metal bottles,

one inside the other. They are connected together only at the top, and the air is removed from between the bottles, creating a vacuum between them. A thick plastic lid opens to the inside of the inner bottle where the hot coffee is kept.

    a. Explain why the vacuum bottle will keep the coffee hot for a lot longer than just about any other type of container.

    b. Even with a vacuum bottle, the hot coffee inside will eventually reach thermal equilibrium with the outside. Explain the heat transfer mechanisms involved in the heat getting out.

    c. Explain why the bottle would not work nearly as well if the vacuum seal were to fail, allowing air to fill the space between the inner and outer bottles.

    d. Explain why the vacuum bottle would work just as well if the contents were iced tea instead of hot coffee.

4. Cans of spray paint include this warning on the label: "CONTENTS UNDER PRESSURE. Avoid prolonged exposure to sunlight or heat sources that may cause bursting. Do not store above 120°F." Explain why heat sources or sunlight could cause the can to burst.

5. Consider a Styrofoam ice chest full of ice and canned drinks sitting in the sun at the beach on a hot summer day. Explain the heat transfer processes involved in bringing the contents of the ice chest into thermal equilibrium with the environment. In your answer, be sure to consider the fact that initially there is no liquid in the ice chest, but as the ice begins to melt there will be lots of water in there.

Specific Heat Capacity and Thermal Conductivity Study Questions

In responding to the following questions you will need to consider whether the situation described has to do primarily with specific heat capacity or with thermal conductivity. You may also need to refer to the graphs in Figures 6-4 and 6-5.

1. If you take a large glass of water and an iron horseshoe that have the same mass, both at room temperature, and place them in the sun for five minutes, what will happen to the temperatures of the water and the horseshoe? Which one will warm up faster and why? After sunset, which one will cool off faster and why?

2. If you were going to try to stay warm in an unheated deer blind on a frosty morning, which would you rather have on the floor in front of you, a big bowl of steaming hot water or a chunk of iron with the same mass and temperature as the water? Why?

3. When a person bakes a casserole in a glass baking dish to take to a party she will often cover the dish with aluminum foil while transporting it in the car to the party. If her purpose is to keep the casserole hot, which would be

better, to cover the dish with foil or to cover it with a tight fitting plastic lid? Explain your reasoning.

4. Think about the previous question some more. Despite what you wrote for an answer, people *do* use aluminum foil to cover hot food. Also, thermal blankets for keeping people warm in cold weather emergencies are also lined with metal foil. Given the high thermal conductivity of metals, why is this? (The answer must have to do with something other than conduction!)

5. In northern states houses are commonly covered in wood siding, but in southern states brick homes are more common. Assuming that the thermal properties of brick are similar to those of solid rock, why is this?

6. Suppose you were shopping and saw a bag of aluminum "ice" cubes for sale. The bag says they save water, are reusable, and great for ice chests because they are so light weight. Would such a product be a good idea or a lousy one? Explain your reasoning.

7. A computer chip can generate a fair bit of heat when it is busy working. Try to explain why electronics manufacturers often mount components like computer chips on aluminum blocks called "heat sinks."

8. If you placed a brass belt buckle and a cup of coffee in a freezer, which one would reach 0°C the quickest and why? (Assume they weigh about the same.)

9. Inside of a hot car on a summer day the temperature can get so high that some objects inside the car can burn your skin. Let's say you slide into the hot car wearing shorts. The surface of a cloth seat will not burn your skin, but the buckle of the seat belt will. This seems odd, since they should be at thermal equilibrium inside the car, and thus at the same temperature. Try to explain why the metal buckle will burn your skin while other materials at the same temperature will not.

10. Explain why a koozie helps to keep a cold beverage in a can to stay cold longer. (The term "koozie" may not be in your dictionary, but your internet search engine will show you what koozies are if you don't know.)

11. Explain why the same koozie could be used to keep a hot beverage in a can to stay hot longer (if anyone ever wanted to drink a hot beverage from a can).

12. Explain why a cold beverage will stay cold longer in a glass bottle than in an aluminum can.

13. The Gulf Stream is a massive ocean current of warm water that moves from the Gulf of Mexico up the Atlantic seaboard, northeast across the Atlantic

Ocean, and right up and past the United Kingdom. By the time this water reaches the UK its temperature is still about 8°C warmer than surrounding waters. Use this information to explain the fact that even though London is 8 degrees of latitude farther north than Toronto, the average temperature in London in January is 7°C and the average temperature in Toronto in January is -2°C.

14. Referring back to the questions about the vacuum bottle in the previous section, years ago the inner container in vacuum bottles used to be made of glass. (These fell out of favor for practical reasons. Whenever a kid dropped his lunch box the glass vacuum bottle inside would get smashed.) Explain why the glass vacuum bottle would keep beverages hot even longer than a metal vacuum bottle does.

15. If you were going to build a house in a location with an extremely cold climate, would it make more sense (from an energy point of view) to build the walls of concrete or of wood (assuming an equal thickness of either one)? In your response, explain what the building materials are supposed to accomplish as far as heat energy is concerned.

16. If you were going to build a house in a location with an extremely hot climate, would it make more sense (from an energy point of view) to build the walls of stones or of mud bricks, which when dried would have similar thermal properties to those of dirt (assuming an equal thickness of either one)? In your response, explain what the building materials are supposed to accomplish as far as heat energy is concerned.

17. Consider the previous question again, and imagine that you live in the nineteenth century and will have no air conditioning in your house. Explain why the house would be more comfortable with a dirt floor than with a concrete floor.

18. Refer again to Figure 6-3. Explain why the ice cube on the right is melting so fast.

19. Referring yet again to Figure 6-3, the water forming from the cube on the right is coming from the bottom of the ice cube, where it is in contact with the aluminum block. For the ice to melt it must first be warmed up to 0°C. What do you think is happening to the temperature in the center or at the top of the ice cube?

20. Although it is a bit difficult to see, Figure 6-5b indicates that the thermal conductivity of fibreglass insulation is actually slightly *higher* than the thermal conductivity of air. So why is this material used in walls for insulation? Why not just leave the walls empty except for the air?

CHAPTER VII
WAVES, SOUND AND LIGHT

OBJECTIVES

Memorize and learn how to use these equations:

$$v = \lambda f \qquad \tau = \frac{1}{f}$$

After studying this chapter and completing the exercises, students will be able to do each of the following tasks, using supporting terms and principles as necessary:

1. Define what a wave is.
2. On a graphical representation of a wave, identify the wave parameters and parts: crest, trough, amplitude, wavelength and period.
3. Define the frequency and period of a wave.
4. Describe the following five wave phenomena, giving examples of each:
   a. Reflection
   b. Refraction
   c. Diffraction
   d. Resonance
   e. Interference
      i. Constructive interference
      ii. Destructive interference
5. Give examples of longitudinal, transverse, and circular waves and the media in which they propagate.
6. Define infrasonic and ultrasonic, and give examples of these types of sounds.
7. Define infrared and ultraviolet, and give examples of these types of radiation.
8. Calculate the velocity, frequency, period and wavelength of waves from given information.
9. Given the frequency of a wave, determine the period, and vice versa.
10. List at least five separate regions in the electromagnetic spectrum, in order from low frequency to high frequency.
11. State the frequency range of human hearing in hertz (Hz), and the wavelength range of visible light in nanometers (nm).
12. State the six main colors in the visible light spectrum in order from lowest frequency to highest.
13. Identify the relations: frequency and pitch; amplitude and volume.
14. Explain how waves of different frequencies (harmonics) contribute to the timbre of musical instruments.

THE ANATOMY OF WAVES

A **wave** is a disturbance in space and time that carries energy from one place to another. A wave can be a single pulse, such as a tsunami caused by an earthquake, or a continuous "train" of waves, such as a beam of light or sound waves from a horn. As shown in Figure 7-1, the tops of a wave are called **crests** (or **peaks**), the bottoms are called **troughs**. The height of the wave from the center line to a crest is called the **amplitude**. The amplitude relates to how much energy the wave is carrying, and the units will depend on what kind of wave it is. (We will not deal with calculations involving the amplitude in this course.) Louder sounds, brighter lights, and more destructive earthquakes are all examples of waves with higher amplitudes.

The waves shown in Figure 7-1 are often called *sine waves*, and waves possessing this very typical shape are said to be *sinusoidal*. We will encounter this term again in a future chapter, and you will learn the mathematical details when you take trigonometry.

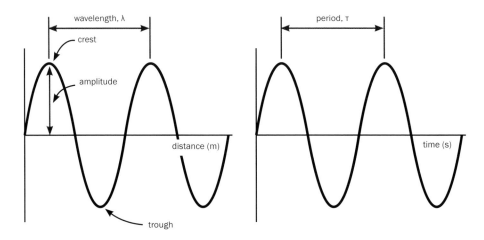

A wave depicted in space at one instant in time. The distance between crests is the wavelength, measured in meters.

A wave depicted in time as it passes one point in space. The distance between crests is the period, measured in seconds.

Figure 7-1. The parts of a wave.

We can think of the length of a wave in distance (spatial) units or time units. If we use distance, the length of one cycle of a wave is called the **wavelength**, measured in meters. In time units the length of one cycle of the wave is called the **period**, which is measured in seconds and is the amount of time it takes for the wave to complete one cycle. We came across this same definition before when we did the Pendulum Lab. When you are looking at a graph of a wave be sure to notice whether the horizontal axis is labeled distance or time, because the units on the horizontal axis will either be meters or seconds, respectively. Whether we are speaking of the period or the wavelength, these variables are represented by the length from one crest of the wave to the next crest.

CATEGORIZING WAVES

We could divide all people into two categories, male or female. A different way to categorize all people would be in the three categories of child, adolescent or adult. Similarly, there are different ways to categorize all waves.

One way to categorize waves is by whether the wave needs a medium or not in which to propagate. We use the word *propagate* instead of travel or move when we are talking about waves, because waves don't simply travel from place to place. They also spread out as they go and the term propagate connotes this better. **Mechanical waves** need a **medium** (matter of some kind) in which to propagate. Most waves are like this. Sound does not travel in a vacuum, but it does travel in media such as air and water. By contrast, **electromagnetic waves** can travel in the vacuum of empty space without a medium. This is how energy gets here from the sun. Electromagnetic waves can also travel in some media – light travels in air and through glass, radio waves travel in air and through wooden walls.

Another way to categorize all waves is according to the relationship between the direction of oscillation that is causing the wave and the direction the wave is propagating. There are three basic possibilities. In **transverse waves** the oscillation is perpendicular to the direction of propagation. Light and waves on strings are good examples of transverse waves. A transverse wave on a string is depicted in Figure 7-2, and transverse electromagnetic waves are depicted later in the chapter in Figure 7-13. The red arrows in Figure 7-2 are there to illustrate that the direction of wave propagation is perpendicular to the up and down motion of the device vibrating the string.

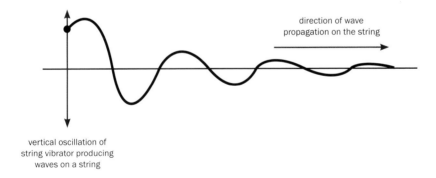

Figure 7-2. A depiction of a transverse wave on a string, in which the oscillation and propagation are perpendicular.

With **longitudinal waves** the oscillating motion causing the wave is parallel to the direction of propagation. Sound waves are an example, and considering how sound is produced by a loudspeaker pushing back and forth is a good way to think about this and remember it. Longitudinal sound waves are depicted in Figure 7-3. The wave train is actually a succession of pressure fluctuations in the air above and below atmospheric

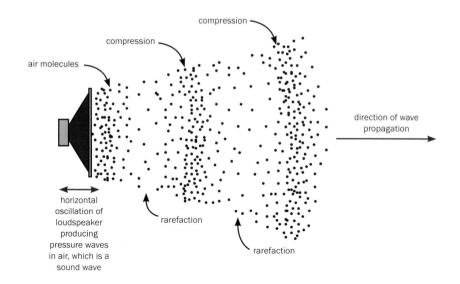

Figure 7-3. An example of a longitudinal sound wave produced by a loudspeaker, in which the oscillation and propagation are parallel.

pressure. We will discuss sound waves in more detail, along with the other terms I placed in this diagram, later in this chapter.

The third type of wave is **circular waves**, mainly represented by waves on water. In circular waves the oscillating action is moving in a circular fashion. Because of this, a floating object on water will move in a circular pattern as the waves pass by underneath it, as illustrated in Figure 7-4.

WAVE CALCULATIONS

There are two important equations for you to learn for our work with waves. The first is the equation relating the velocity at which the wave propagates to the frequency and wavelength of the wave:

$$v = \lambda f$$

where v is the velocity of the wave propagation (m/s), λ ("lambda," the Greek letter l) is the wavelength of the wave (m), and f is the frequency of the wave (Hz).

The **frequency** of a wave is the number of cycles the wave completes in one second. The MKS units for frequency are cycles per second. We have a name for this unit, which is "hertz" (Hz). So if a wave completes 5,000 cycles per second, its frequency is 5,000 Hz or 5 kHz.

As you can see from this equation, the velocity of a wave varies directly with its wavelength and directly with its frequency. But physically, it is better to think of the wavelength as depending on the other two variables in this equation. This is because the wave velocity actually depends on the medium in which the wave is propagating.

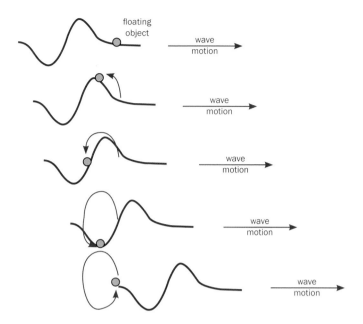

Figure 7-4. A water wave causing a floating object to loop around in a circular fashion as the wave goes by.

Also, the frequency depends on the source of the wave; whatever is causing the wave is oscillating at a certain rate, and this rate determines the frequency of the wave being produced. So since the wave velocity is determined by the medium, and the frequency is determined by the source of the wave, this means the wavelength is the variable that really depends on the other two. So solving the wave equation for the wavelength, we have

$$\lambda = \frac{v}{f}$$

Normalizing this we can see that wavelength varies inversely with frequency. In other words, when one goes up, the other goes down. Knowing this will be particularly useful when we study light (below).

Determine the wavelength of a 1.500 kHz sound wave. (This frequency is right around the pitch people make when they whistle.) Assume that sound travels at 342 m/s in air.

Write the givens and do the unit conversions.

$$f = 1.500 \text{ kHz} \cdot \frac{1000 \text{ Hz}}{\text{kHz}} = 1{,}500 \text{ Hz}$$

$$v = 342 \; \frac{\text{m}}{\text{s}}$$

$$\lambda = ?$$

Now complete the problem.

$$v = \lambda f$$

$$\lambda = \frac{v}{f} = \frac{342 \; \frac{\text{m}}{\text{s}}}{1500 \text{ Hz}} = 0.228 \text{ m}$$

Although the frequency is given with 4 significant digits, the wave velocity is given with only 3 significant digits, so my result is stated with 3 significant digits.

Determine the frequency of the light produced by a laser operating at a wavelength of 488 nm in air.

Write the givens and do the unit conversions.

$$\lambda = 488 \text{ nm} \cdot \frac{1 \text{ m}}{10^9 \text{ nm}} = 4.88 \times 10^{-7} \text{ m}$$

$$v = 3.00 \times 10^8 \; \frac{\text{m}}{\text{s}}$$

$$f = ?$$

Now complete the problem.

$$v = \lambda f$$

$$f = \frac{v}{\lambda} = \frac{3.00 \times 10^8 \; \frac{\text{m}}{\text{s}}}{4.88 \times 10^{-7} \text{ m}} = 6.148 \times 10^{14} \text{ Hz}$$

Both the wavelength and the speed of light values have 3 significant digits, so we need to round this result to 3 significant digits.

$$f = 6.15 \times 10^{14} \text{ Hz}$$

As mentioned above, the length of time it takes for any oscillating system, including a wave of some kind, to complete one full cycle is called the period, measured in seconds. The period of a wave relates directly to the frequency. Since the units for

Chapter VII
Waves, Sound and Light

frequency are cycles per second, the reciprocal of the frequency would have units of seconds per cycle. This time value, the number of seconds in one cycle, is the period of the wave, although when we speak of the period or use it in equations we just call the units "seconds." We use the Greek letter τ ("tau," the t in the Greek alphabet) to represent the period of a wave or of any other oscillating system. Since period and frequency are reciprocals, this gives us our second equation,

$$\tau = \frac{1}{f}$$

You have already been using the speed of light in air or a vacuum, $c = 3.00 \times 10^8$ m/s, in calculations from previous chapters. Since all light is electromagnetic waves, and since our main wave equation has the velocity in it, you will now be using the speed of light even more frequently in calculations. You should take special note at this point that radio waves propagate at the speed of light, not at the speed of sound. They are not sound waves (you can't hear them); they are electromagnetic waves. Microwaves and X-rays are also electromagnetic waves, so any problem involving any of these will use the speed of light for the velocity.

Radio station KUT FM in Austin broadcasts a carrier signal at 90.5 MHz. Determine the wavelength and period of this wave. State the period in nanoseconds.

As I mentioned above and will discuss more later in the chapter, all radio waves (FM, AM, short wave, etc.) are part of the electromagnetic spectrum and propagate at the speed of light. Now that you know this, write the givens and do the unit conversions.

$$f = 90.5 \text{ MHz} \cdot \frac{10^6 \text{ Hz}}{\text{MHz}} = 9.05 \times 10^7 \text{ Hz}$$

$$v = 3.00 \times 10^8 \, \frac{\text{m}}{\text{s}}$$

$$\lambda = ?$$

$$\tau = ?$$

The problem requires us to perform two separate calculations. Let's calculate the wavelength first.

$$v = \lambda f$$

$$\lambda = \frac{v}{f} = \frac{3.00 \times 18^8 \, \frac{\text{m}}{\text{s}}}{9.05 \times 10^7 \text{ Hz}} = 3.31 \text{ m}$$

All of the values in the problem have 3 significant digits, as does this result. Now compute the period.

$$\tau = \frac{1}{f} = \frac{1}{9.05 \times 10^7 \text{ Hz}} = 1.105 \times 10^{-8} \text{ s} \cdot \frac{10^9 \text{ ns}}{\text{s}} = 11.05 \text{ ns}$$

This value has been converted to nanoseconds, as required, but it has 1 extra significant digit. Rounding to the required 3 significant digits gives the final result.

$\tau = 11.1$ ns

WAVE PHENOMENA

There are several phenomena common to all types of waves. I usually call these the "wave behaviors," and we are going to look at five of them. For some of these descriptions it is helpful to think of the wave as a ray, like the thin line of light we are familiar with in a laser beam. In some of the illustrations below I have shown rays as arrows pointing in the direction of propagation.

REFLECTION

Waves **reflect** off surfaces or objects. As the wave or ray approaches the surface it is called the **incident ray**. The ray that has reflected off the surface is called the **reflected ray**. The **law of reflection** states that the angle of incidence equals the angle of reflection. These two angles are defined with respect to a line perpendicular to the reflecting surface. This line is called the *normal line*, because in mathematics and physics the word *normal* means perpendicular. Figure 7-5 depicts a reflective surface with the normal line shown (the dashed line), and a ray of light striking the surface at the angle of incidence. Using standard notation we designate the angle of incidence as θ_i and the angle of reflection as θ_r. (The Greek letter θ, "theta," which is the "th" sound, is used throughout mathematics and physics to represent measures of angles.) So we can write the law of reflection as $\theta_i = \theta_r$.

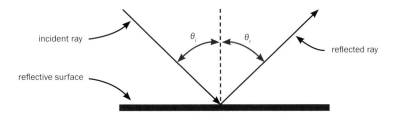

Figure 7-5. Wave reflection.

REFRACTION

All waves can **refract**, but we see it often with light. The speed of light in a plastic prism or glass or water is slower than the speed of light in air. When the light enters the new medium it changes speed instantly. This change of speed causes the light to change direction, or refract. Figure 7-6 shows a ray of red laser light coming straight down in the air and refracting as it crosses the boundary from air into a

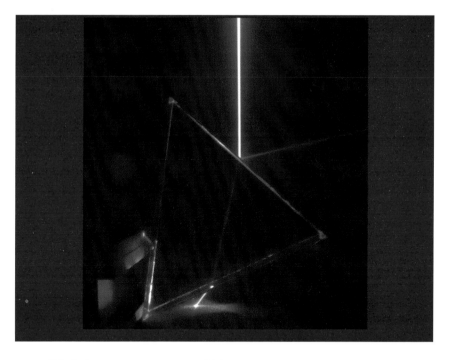

Figure 7-6. Refraction, illustrated by a 633 nm laser passing from air into a plastic prism (the triangle), and from the plastic prism into the air at the bottom. Some light can be seen reflecting off the upper side of the prism.

plastic prism. (Some of the light also reflects off the prism, obeying the law of reflection, and heads off to the right in the photograph.) When the laser light reaches the other side of the prism, at the bottom, and comes back out into the air you see it refract again, bending sharply to the left at the bottom of the photo. The amount the light bends depends on both of the media in which the wave is propagating. In the case of the photo in Figure 7-6 these media are air and plastic. The amount of refraction also depends slightly on the color of the light, which depends on the wavelength. White light is a combination of all different colors, and when it goes through a prism the different colors refract different amounts, so the white light splits into the different colors, as in a rainbow.

The familiar image of a straw in a glass of water, and the way it appears to be broken, is because of the refraction of light as it passes from air to water, reflects off the straw under the water, and refracts again as it passes from water to air. The light we see reflected from the underwater part of the straw has refracted twice and is propagating in a different direction that the light reflecting from the upper part of the straw. The light we see reflecting from the upper part never went under water and never refracted.

DIFFRACTION

Diffraction occurs when a wave bends around the corner of some obstruction. A familiar example of diffraction is that a person's voice can be heard around the edge of a barrier like the corner of a building because of the sound waves bending around the barrier. Without diffraction it would be very difficult to hear what people were saying, because sound waves would travel straight out of our mouths without spreading out, like the beam of a flashlight. And just as the flashlight creates a spot of light, the sound wave would create only a spot of sound. (The flashlight beam diffracts too, but the wavelengths of light are so short that diffraction of light is difficult to perceive.) As a result of diffraction a laser beam can be arranged to create what we call an *interference pattern.* This effect is illustrated in the section on wave interference below.

A really cool example of the bending of waves from diffraction occurs when laser light is beamed through a diffraction grating, as shown in Figure 7-7. In this figure an orange laser beam composed of the light from 633 nm red and 543 nm green lasers hits a diffraction grating. A diffraction grating is a glass plate with very fine grooves etched into its surface (70 grooves per mm for the grating used to make the image shown, which is actually not very many for a diffraction grating). The two laser beams can be seen at the bottom of the image, and are combined in the plastic cube. The orange beam that emerges strikes the grating in the center of the image and the red and green beams fan out separately beyond the grating. The grating splits the light into many beams, each of them bending more and more away from the center. You can see also that the two colors are bent different amounts, just as with refraction. Many of the separated beams are hitting the white screen I placed in the beam path, but you can also see distant red and green dots where other beams are hitting the wall in the back of the room.

Figure 7-7. Combined red and green laser beams are split apart and bent by a diffraction grating.

RESONANCE

When an original wave and its reflection in a medium consistently line up with each other a **standing wave** is formed. To understand what a standing wave is, consider a wave train continuously propagating in a medium. As the waves reach a boundary such as a wall, they reflect back into the region where the original waves are, so we have the original waves and the reflected waves mingling together in the same space. If the dimensions of the medium are just right, this mixing of original and reflected waves can produce an effect where the peaks and troughs of the original and reflected waves line up and stay lined up and the wave appears to stand still as a result. Since the wave appears to stand still, it is called a standing wave.

When a standing wave occurs the original and reflected waves essentially add together, creating a much stronger wave, that is, one with a much higher amplitude. This is the phenomenon called **resonance**. Most of the time, such as with sound or light waves, we cannot see the waves. But on a vibrating string we can, as shown in Figure 7-8. The wave on the string is not really standing still, it just appears to be.

Figure 7-8. A standing wave on a vibrating string strung across my classroom.

For such a standing wave to occur the dimension of the medium in which the wave is propagating (the length of the string in this case) must be equal to an integer number of half-wavelengths. For a standing wave to occur with a sound wave in a room, at least one of the dimensions of the room (height, width or depth) must equal an integer number of half-wavelengths of the sound wave. Resonance is what is happening when someone plays a wind instrument, when you blow across the top of a bottle, or when you find the sweet notes that resound while you are humming in a tile and glass shower stall.

Another fascinating example of resonance producing standing waves is the pattern formed on the surface of a liquid when the container is vibrating, as in the vibrating coffee in a coffee cup in Figure 7-9. To make this pattern for the photograph, I had to make the table vibrate at just the right frequency so that the coffee cup sitting on the table would vibrate the coffee, producing waves on the surface of the coffee that would turn into standing waves.

INTERFERENCE

One of the most historically significant wave phenomena is **interference**. It was interference patterns produced by light in 1801 that first gave scientists strong evidence that light was a wave phenomenon. (Before that everyone accepted Isaac

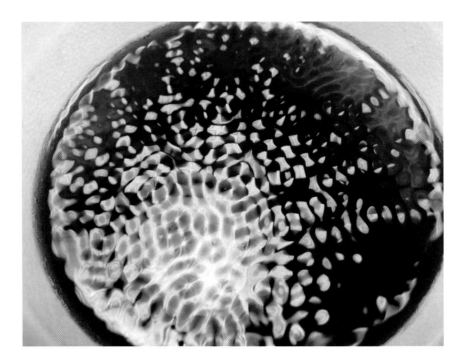

Figure 7-9. Standing waves on the surface of coffee in a cup.

Newton's view that light was made of particles.) Interference occurs when two different waves arrive at the same place at the same time. The two waves will always add together, but the result depends on how the two waves are aligned with each other. As illustrated in Figure 7-10, if two waves arrive at the same place **in phase**, with their crests and troughs lined up, the energies of the waves will add together, producing an effect stronger than either of the individual waves. This effect is called **constructive interference**. Constructive interference with sound waves produces a loud spot. With light waves constructive interference causes a bright spot of

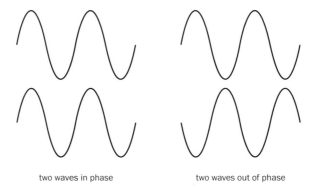

Figure 7-10. Waves in phase and out of phase.

light. By contrast, if one wave has to travel farther than the other one it will arrive later and can be **out of phase** with the first one so that the crests of one wave line up with the troughs of the other. This is called **destructive interference**, because when the waves add together they will partially or completely cancel each other out. With sound waves this creates a dead spot where the sound cannot be heard at all. With light waves a dark spot is created where there is no light.

An example of interference is the pattern of alternating light and dark spots called an **interference pattern** that results when laser light diffracts around a very thin obstruction placed in the beam, as shown in Figure 7-11. The obstruction basically splits the beam, creating two beams from one. Then because of diffraction the two beams begin spreading out rapidly. When the two beams strike the screen shown in the photograph in Figure 7-11 they create a long horizontal region of alternating light and dark spots. This interference pattern also illustrates something you should notice about diffraction. If you didn't know about this diffraction pattern what would you expect to see if a thin obstruction was in the laser beam? You would expect to see the beam spot with a thin dark line through it, the shadow of the obstruction. Not only do we have a bright spot in the center instead of a shadow, the light from the diffracted beam is spread out into a wide strip of light.

Figure 7-11. Interference pattern caused by placing a thin obstruction in the beam of a 543-nm laser.

As illustrated in Figure 7-12, this interference pattern is caused by waves diffracting around the edges of the obstruction. When the two beams hit the target there are places where the two light beams arrive in phase (the bright spots) and other places where they arrive out of phase (the dark spots).

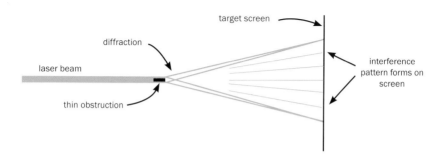

Figure 7-12. Placing a thin obstruction in the path of a laser beam causes diffraction and interference.

SOUND WAVES

As I mentioned previously, sound waves are mechanical waves. That is, they propagate in a medium (matter) of some kind. We most often think of sound waves as propagating in air, but we have all heard sound waves under water, too. Sound waves can also travel in other media, which is why you can hear sounds through closed doors or walls. Sound waves cannot travel in a vacuum, which means all the explosions and screeching space fighter planes we see in science fiction movies are absurd, because there is no sound in space. (Sorry.)

In air, sound waves are created by causing the pressure in the air to fluctuate above and below atmospheric pressure, as illustrated back in Figure 7-3. The regions of higher pressure are called *compressions*. Regions where the pressure is a bit lower than atmospheric pressure are called *rarefactions* (pronounced "RARE-uh-faction"). A sound wave in air is a continuous train of successive compressions and rarefactions.

In air, sound waves travel at a velocity of around 342 m/s, depending on temperature, humidity, and pressure. (You don't have to memorize this value.)

FREQUENCIES OF SOUND WAVES

Humans can hear sound waves in the frequency range of about **20 Hz to 20 kHz**. Frequencies lower that this are called **infrasonic**. (The Latin prefix *infra* means below.) Frequencies higher than this range are called **ultrasonic**. (The Latin prefix *ultra* means above or beyond.) You should be careful not to confuse these terms with the terms subsonic and supersonic, which refer to the velocity of a moving object like a jet airplane, and not to the frequencies of sound waves. (Subsonic refers to a velocity slower than the speed of sound in air; supersonic refers to a velocity greater than the speed of sound.)

Table 7-1 lists a few reference points for frequencies of common musical notes and sounds.

41 Hz	The lowest note on a bass guitar.
82 Hz	The lowest note on a guitar.
440 Hz	The A above middle C on a keyboard, often called "A 440" and used as a reference for tuning pianos and other concert instruments.
100 - 3000 Hz	The frequencies in the human voice lie primarily in this range.
1000 Hz	A low whistle.
6 kHz - 8 kHz	This is the range for the "s" sounds we make between our teeth when we say a word like "Susan." This sound is called *sibilance*.

Table 7-1. Reference frequencies of some well-known sounds.

Perfect human hearing extends up to around 20 kHz, but by middle age few people can hear sounds above 15 kHz. This is natural as our ear drums stiffen up over time and do not respond as well to high frequencies. If you listen to a lot of loud music your hearing will degrade more rapidly than normal. Although human hearing does not extend above 20 kHz, dogs can hear up to nearly 30 kHz. For this reason, dog whistles emit a frequency of around 27 kHz that dogs can hear but humans can't.

LOUDNESS OF SOUND

We measure the loudness of sound waves using Sound Pressure Level (SPL), which is measured in decibels (dB). In this course we will not do any calculations involving decibels, but you might be interested in some common reference points for the dB scale listed in Table 7-2. The decibel values shown are approximate. I guess those of you who are into heavy metal (I am not) will enjoy the term for a loudness of 120 dB.

0 dB	The level is called the *threshold of human hearing*, because it is the quietest sound human ears are capable of hearing. You will probably never be in a place this quiet in your entire life. (If you were, you would have the sensation of being completely deaf.)
45 dB	This would be about the loudness of faint rustling leaves on a quiet day.
65 dB	Quiet conversation in a quiet room.
85 dB	A moderately noisy environment. Legally, OSHA requires that if the noise level in a workplace (like a factory) is consistently above 85 dB the workers must be supplied with hearing protection. This is because studies have shown that being exposed to noise levels of 85 dB or higher for 8 hours a day over many years can result in permanent hearing loss. MP3 player listeners beware!
95 dB	Loud music in a movie theater.
106 dB	The upper limit of a good stereo system.
110 dB	An excruciatingly loud rock concert.
120 dB	This is usually called the *threshold of pain*, for obvious reasons.
140 dB	Standing next to a commercial jet when it revs up its engines for take off. This is why the ground crew wear hearing protection.
150 dB	A gun being fired right next to your head. This is why sensible people also use hearing protection when at the rifle range. It is also why movies showing people whispering to one another right after firing a bunch of guns at bad guys are not realistic. Temporary hearing loss after a lot of gun fire would make it hard to hear whispers.

Table 7-2. Reference loudnesses of well-known sounds.

CONNECTIONS BETWEEN SCIENTIFIC AND MUSICAL TERMS

It is really handy to make the connections between what we are studying in this chapter and other terms you probably often hear. If you play an instrument, you should know that what musicians refer to as the **pitch** of a musical note is what scientists, and this chapter, call the frequency. If you listen to music (Ha, ha! That was a joke. I know that *everyone* listens to music.) you will appreciate knowing that what music lovers refer to as **volume** or **loudness** corresponds to the amplitude of the sound wave.

ELECTROMAGNETIC WAVES AND LIGHT

It is difficult to understand what exactly electromagnetic waves are, but don't let this bother you. Scientists have been puzzling over electromagnetic radiation in one way or another for many centuries. We now understand that light, and all other electromagnetic radiation, can be described as oscillating electric and magnetic fields in space. (This probably doesn't sound much more comprehensible, but we will be studying fields briefly later on when we get to magnetism, so hang in there!) When we speak of wavelengths and frequencies of electromagnetic radiation, what we are talking about are the wavelengths and frequencies of the electric and magnetic field waves. These oscillating fields are pretty hard to visualize, but Figure 7-13 depicts an attempt at a 3-D representation of a propagating electromagnetic wave. The heights of the waves represent the strengths of the electric and magnetic fields at any point in space as the wave train passes by. (Just in passing, I will mention that the direction of the electric field is what is being referred to when people speak of the *polarization* of light. When light is polarized, the electric field of all the light waves points in the same direction.)

Visible light is only one small piece of the electromagnetic spectrum, the entire frequency range of electromagnetic radiation. Different frequencies of visible light are

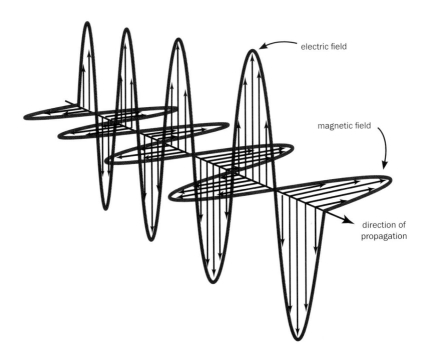

Figure 7-13. Three-dimensional depiction of an electromagnetic wave.

perceived as different colors by our eyes. Red light has the lowest frequency in the visible spectrum. Violet light has the highest frequency in the visible spectrum. As every school child knows, the sequence of colors from low frequency to high frequency is red – orange – yellow – green – blue – violet, as shown in Figure 7-14.

When we speak of electromagnetic radiation sometimes we speak in terms of frequencies, as in the previous paragraph. In other cases we speak in terms of wavelengths. As we saw earlier in this chapter, frequency and wavelength are inversely proportional, so we could just as easily state the difference between red light and violet light in terms of wavelengths this way: Red light has the longest wavelength in the visible spectrum. Violet light has the shortest wavelength in the visible spectrum. In fact, it is pretty easy to remember the wavelength range of the visible spectrum, which is about **700 nm to 400 nm** (red to violet). (Some reference sources have the visible spectrum starting at 750 or 800 nm.) And as you can see from the approximate scale in Figure 7-14, orange light has a wavelength of around 600 to 610 nm, yellow is in the 570 to 580 nm range, green is in the 520 to 550 nm range, and blue light is in the mid to upper 400 nm range.

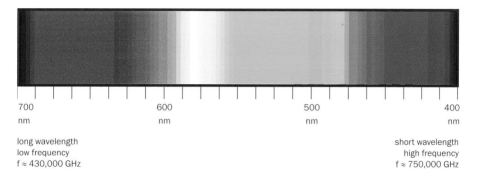

Figure 7-14. The visible light portion of the electromagnetic spectrum.

Electromagnetic radiation with frequencies too low for humans to see is called **infrared radiation**, since red light has the lowest frequency of the visible light spectrum. Electromagnetic radiation with frequencies too high for us to see is called **ultraviolet radiation**, since violet light is the highest frequency in the spectrum of visible light.

The major bands of the electromagnetic spectrum and their approximate frequencies are shown in Table 7-3. You do not need to memorize all of the frequencies, but you do need to know the names for these different regions in the electromagnetic spectrum, and you need to be able to write them in order from lowest frequency to highest frequency.

As additional examples, the electric and magnetic waves of green light with a wavelength of 532 nm oscillate 566,000,000,000,000 times each second! The microwaves your microwave oven uses complete 2,450,000,000 cycles each second.

It is interesting to consider how the wavelengths of these different types of electromagnetic radiation relate to phenomena we are familiar with. You probably all know that the reception of an AM radio in a car blanks out when the car drives under a bridge. If you calculate the wavelength of the carrier frequency of an AM radio station you

Band	Approximate frequency of electromagnetic radiation	Approximate wavelength in a vacuum
AM Radio	1,000 kHz, or 10^6 Hz	300 m
FM Radio	100 MHz, or 10^8 Hz	3 m
Microwaves	10 GHz, or 10^{10} Hz	3 cm
Infrared Radiation	10^{13} Hz	30 μm
Visible Light	10^{14} Hz	700 – 400 nm
Ultraviolet Radiation	10^{15} Hz	300 nm
X-Rays	10^{18} Hz	0.3 nm
Gamma Rays	10^{20} Hz	0.003 nm
Cosmic Rays	10^{22} Hz	0.00003 nm

Table 7-3. Regions in the electromagnetic spectrum.

will see why. Radio station KLBJ here in Austin, Texas broadcasts at 590 kHz. From the wave equation we can calculate the wavelength of the radio waves propagating out from the KLBJ tower, and it turns out to be 508 m, which is nearly a third of a mile. Imagine a bunch of waves this long reflecting around every which way in the air. Very few of them will have just the right orientation to fit under a bridge and pass through, so the signal under there is so weak that the car radio loses it. The wavelengths in the FM range are much smaller, only about 3 m or so, so they fit under bridges much more easily and our FM radios don't blank out when we drive under one. The digital radio channels launched about ten years ago operate in the range of 2.3 GHz! These wavelengths are small enough that the signal should be usable just about everywhere within the broadcast region of the satellites.

How small are they? Well, as another example, consider the metal screen with holes in it on the front door of a microwave oven. With a frequency of 2.45 GHz (not far from where the digital radio signals are), the wavelength of this radiation will be 12.2 cm. The holes in the door screen are about forty times smaller than this, at 3 mm or so, so the waves reflect around inside the oven where they can do their job heating up food, instead of escaping out into the room and harming us.

HARMONICS AND TIMBRE

Let's say you want to play the note on a flute known as "middle C." To do this the air in the flute must resonate at the frequency of middle C, also called C4, which is 278.4375 Hz. Now let's have a clarinet play this same note. To do this the clarinet must also resonate at 278.4375 Hz. How can you tell the difference between the flute and the clarinet if they are both resonating at the same frequency? It's no good saying they sound different. What *is it* about the sound waves these instruments produce that makes them sound different?

Well, one of the reasons they sound different is because while both instruments are resonating at the frequency of middle C, they are also both simultaneously resonating at many other frequencies that are multiples of middle C. The main resonant frequency is called the **fundamental** (middle C in this case) and the other frequencies resonating in the instrument are called **harmonics**. Sometimes the fundamental is called the first harmonic. The second harmonic is usually a frequency that is twice the frequency of the fundamental. The third harmonic is three times the fundamental, and so on. In some instruments the harmonics might go up by odd or fractional multiples, but in general harmonics are multiples of the fundamental frequency. When writing about the different harmonic frequencies that are present in a sound wave the fundamental is denoted as f_1, and the second, third, and higher harmonics are denoted f_2, f_3, and so on.

Figure 7-15 is a sketch of the first three harmonics that would resonate on a string. I drew them on a string because they are easy to visualize (exactly like the standing waves shown in the photo of Figure 7-8), but the same basic idea would apply to waves resonating in any medium, like sound waves in a room or in a musical instrument. The horizontal line in the middle of the figure represents the string as it would be when it was not vibrating. As you can see in the figure, the string length is equal to one half-wavelength of the fundamental, f_1, shown in blue. Two half-wavelengths, or one full wave, of the second harmonic will fit on the same string. Three half-wavelengths of the third harmonic fit on the string. This continues for the higher harmonics, f_4, f_5, and higher.

The different harmonic frequencies will resonate with different intensities (that is, different amounts of energy) in different instruments. Generally, the fundamental has the greatest amplitude and the amplitudes diminish for higher and higher harmonics. The amplitudes of the various harmonics of every violin are similar, which is an important reason every violin sounds like a violin. The harmonic content of every flute is similar, so they all sound like flutes. But even among all of the flutes, there are subtle differences in the harmonic content of the sound resonating in the flute, depending on the materials from which it is made and the details of its construction. So we have a little more energy at this harmonic, a little less energy at that harmonic. This means one can tell one flute from another. A really good instrument is one that has a very desirable blend of harmonics. Cheap instruments have a less desirable blend of harmonics. (The cheapos are probably also more difficult to play and not as loud, but these are separate issues.) The particular combination of harmonics (and some other factors we will not go into) that makes an instrument identifiable as itself, its own unique sound quality, is called the instrument's **timbre** (pronounced "tamber").

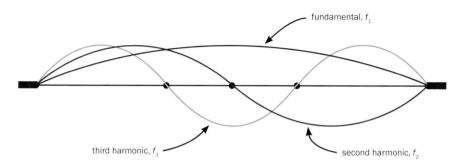

Figure 7-15. Wavelengths of different harmonics that will vibrate in a given medium.

A graphical representation of the relative loudnesses (or intensities) of the fundamental and other harmonics in a particular instrument is called the **harmonic spectrum** of the instrument. Figures 7-16 through 7-18 show the actual harmonic spectra for three different wind instruments playing the same note, D4, the D right above middle C. The frequency of this pitch is 293.665 Hz.

I captured these images from an iPad with a fascinating free app called "n-track tuner" while a friend played the different instruments. On the graphs the vertical scale is the loudness in decibels. The horizontal scale is the frequency. The vertical line just to the right of the capital D is 100 Hz, and the lines to the right go up by 100 Hz each until 1000 Hz. They go up by 1000 Hz each to the right of that until 10 kHz, and the right edge of the graph is at 20 kHz. (If you are a real math geek like me you might like to know that a scale like this where different orders of magnitude are equally spaced – 100, 1000, 10,000 and so on – is called a *logarithmic* scale.)

You can easily see the fundamental in each graph, represented by the left-most large peak in the graph, right around 300 Hz. The second harmonic at about 600 Hz and the third harmonic at around 900 Hz are clearly visible, as are other higher harmonics at multiples of 300 Hz. The jagged curve across the bottom of all the spectra is simply the background noise in the room where these images were captured. (We were in a very quiet room, but there is always background noise. We don't usually notice it, but sensitive instruments like the built-in microphone in the iPad easily pick it up.)

A couple of details should stand out as we examine these spectra. First, in the flute we immediately notice that the intensity of the first and second harmonics (which are the fundamental D pitch and the D one octave up, D5) are at the same intensity. This is unique to the flute spectrum. Second, we notice the very rich harmonics in the saxophone spectrum. The higher-order harmonics in the sax are significantly more intense than the comparable harmonics in the other two spectra, including pronounced harmonics at f_8 and higher. These strong higher harmonics help to explain the "reedy" sound quality the sax is known for.

Figure 7-16. Harmonic spectrum of a flute playing 293.665 Hz (D4).

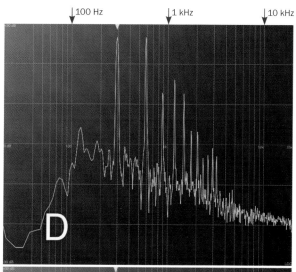

Figure 7-17. Harmonic spectrum of a clarinet playing 293.665 Hz (D4).

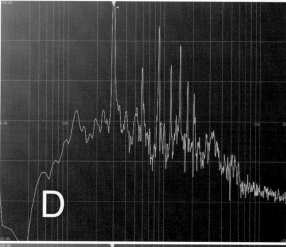

Figure 7-18. Harmonic spectrum of a saxophone playing 293.665 Hz (D4).

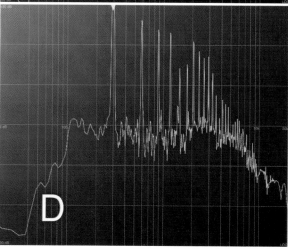

CHAPTER VII EXERCISES

For problems involving sound waves in air, use $v = 342$ m/s.

1. What is the period of a 60.0 Hz tone?

2. What is the frequency of a sine wave with a period of 2.155×10^{-5} s?

3. If a dog whistle emits an ultrasonic tone at 26.0 kHz, what is the period of the wave?

4. For problem 3, what is the wavelength of the wave? State your answer in cm.

5. The Austin FM radio station KMFA broadcasts an electromagnetic radio wave at 89.5 MHz. Determine the wavelength and period in µs of this wave.

6. The San Antonio AM radio station KAHL broadcasts an electromagnetic wave at 1310 kHz. Determine the wavelength and period in µs of this wave.

7. Look at your answers for the previous two problems, and try to explain the fact that when a car drives under a bridge FM radio signals come through fine while AM radio signals do not.

8. Determine the period and frequency of the light emitted by a 542-nm green laser. State the period in ns and the frequency in GHz.

9. A carbon dioxide laser used for cutting steel has a wavelength of 10.6 µm. Determine the period and frequency of the wave. State the period in µs and the frequency in MHz.

10. A consumer can purchase an ultrasonic device to drive pests out of his home that emits a sound wave of 33 kHz. Calculate the period and wavelength of the sound wave this device would emit. State the period in ms and the wavelength in mm.

11. Certain rooms where electronic equipment is tested are completely enclosed by copper screening to keep out electromagnetic waves. This works because electromagnetic waves cannot travel through electrically conductive materials such as copper. If a room is surrounded with screening that has holes 2.00 mm across, what is the approximate frequency range of waves that will be kept out by the screen or that will pass through the screen?

12. In the spectrum of visible light, blue light has a wavelength of 470 nm, green light has a wavelength of 550 nm, and red light has a wavelength of 680 nm. Calculate the frequencies for each of these colors of light.

13. The lowest frequency humans can hear is about 20 Hz. Let's call it 20.00 Hz.

What is the wavelength of this wave in air? Assume sound propagates at 342.0 m/s in air.

14. Calculate the wavelength of a 1.00 kHz tone propagating as a sound wave in each of these different media.

medium	speed of sound
air	342 m/s
water	1,402 m/s
steel	5,130 m/s
helium	965 m/s

15. Calculate the wavelengths in nm for the following types of "rays."

   a. gamma rays, $f = 4.67 \times 10^{20}$ Hz
   b. x-rays, $f = 9.9876 \times 10^{18}$ Hz
   c. microwaves, $f = 2.555 \times 10^{10}$ Hz
   d. ultraviolet light, $f = 1.172 \times 10^{15}$ Hz
   e. infrared light, $f = 2.83 \times 10^{13}$ Hz

16. Consider the wavelength you calculated for microwaves in the previous problem. How does this compare to the 3-mm diameter holes in the screen of the door of a microwave oven? Why are the holes the size they are? (When you are asked to compare two values in science and math it means you should determine their ratio. That is, determine how many times bigger one is than the other, or what percentage the smaller one is of the larger.)

Answers
1. 0.0167 s
2. 46.40 kHz
3. 3.85×10^{-5} s
4. 1.32 cm
5. $\lambda = 3.35$ m, $T = 0.0112$ μs
6. $\lambda = 229$ m, $T = 0.763$ μs
7. -
8. $f = 5.54 \times 10^5$ GHz, $T = 1.81 \times 10^{-6}$ ns
9. $f = 2.83 \times 10^7$ MHz, $T = 3.53 \times 10^{-8}$ μs
10. 0.030 ms, 1.0×10^1 mm
11. Waves with $f > 1.50 \times 10^{11}$ Hz will pass through the screen.
12. blue 6.4×10^{14} Hz, green 5.5×10^{14} Hz, red 4.4×10^{14} Hz
13. 17.10 m
14. air 0.342 m, water 1.40 m, steel 5.13 m, helium 0.965 m
15. gamma 0.000642 nm, x-ray 0.0300 nm, microwave 11,700,000 nm, UV 256 nm, IR 10,600 nm
16. $\lambda/D = 10$

CHAPTER VIII
ELECTRICITY AND DC CIRCUITS

> **OBJECTIVES**
>
> Memorize and learn how to use these equations:
>
> $V = IR$ $\qquad P = VI \qquad P = I^2 R \qquad P = \dfrac{V^2}{R}$
>
> After studying this chapter and completing the exercises, students will be able to do each of the following tasks, using supporting terms and principles as necessary:
>
> 1. Explain what static electricity and static discharges are.
> 2. Describe three ways for static electricity to form, and apply them to explain the operation of the Van de Graaff generator and the electroscope.
> 3. Using the analogy of water being pumped through a filter, give definitions by analogy for voltage, current, resistance, and potential difference.
> 4. Explain what electric current is and what produces it.
> 5. Explain why electric current flows so easily in metals.
> 6. Describe the roles of Alessandro Volta and James Clerk Maxwell in the development of our knowledge of electricity.
> 7. State Kirchhoff's two circuit laws.
> 8. Calculate the equivalent resistance of resistors connected in series, in parallel, or in combination.
> 9. Use Ohm's Law and Kirchhoff's Laws to calculate voltages, currents, and powers in DC circuits with up to six resistors.
> 10. Draw graphs of power vs. current, power vs. voltage, and power vs. resistance, and explain the shape of the graphs in terms of the relationships between variables in the equations derived from Ohm's Law.

This chapter is by far the longest chapter in this textbook. Electricity and circuits are all around us and our lives are massively affected by electrical appliances and gadgets. For this reason, it is important and valuable for you to learn the basics of electrical theory and calculations. However, it is very tedious to put these matters into words. To help, I have included lots of tables, sketches and examples. The result is a very long chapter.

The calculations you will perform in this chapter will be the most difficult ones in the course. Every ASPC student can do these kinds of problems, but it takes lots of practice and patience with the examples. Solving circuit problems is a strictly logical activity, which is another reason learning this material is valuable. But once the light bulb comes on in your head, you will easily be able to solve electric circuits with confidence.

THE AMAZING HISTORY OF ELECTRICITY

Being an electrical engineer, I like to tell the story of electricity and magnetism. Included here is a short account of this interesting saga. As always, the things you need to know and remember are listed on the Objectives List and the Scientists List in Appendix D. This pretty much boils down to just two of the names in this history, and a couple of gadgets you need to know about. But even though you won't be tested on most of the information in this section, I hope you will enjoy reading it.

The earliest recorded observations of electrical phenomena were by the ancient Greeks around 600 BC. They observed that amber rubbed with fur would attract certain materials such as feathers. This is an effect of static electricity, as we will discuss in some detail in this chapter.

Figure 8-1. William Gilbert.

The modern story of our scientific knowledge about electricity and magnetism begins with William Gilbert, Queen Elizabeth's personal physician, who spent a lot of time investigating static electricity and magnetism (Figure 8-1). In 1600 (the same year Kepler moved to Prague to work with Tycho Brahe) Gilbert proposed that the earth acted like a large magnet. Gilbert's work was the first major scientific work published in England. His new attitude toward science emphasized experiment and scientific observations (remember Kepler's keen interest in comparing theory with *data*). He was one of the first scientists to embrace this new attitude, which, of course, was to become the very bedrock principle of science. Gilbert discovered many new materials with electrical properties, and coined the word "electric" from the Greek and Latin words for amber. He also invented an early form of the electroscope, described below.

About 140 years later the Leyden jar was invented, named after the city in the Netherlands where one of the inventors lived. About 10 years after that a new, more sensitive version of the electroscope (which we will discuss soon) was invented in England. Both of these gadgets can store electrical charge, meaning an excess of negatively charged electrons. Of course, we all know that in 1752 Benjamin Franklin proved that lightning was electrical by charging up a Leyden jar connected to his kite flying in an electrical storm. (You know better than to try this yourself, right? If not, have you considered the possibilities of what might happen if the kite string you are holding in your hand gets struck by lightning?) But Franklin also started the convention of denoting the two types of electric charge as positive and negative, and he established the law of conservation of charge. Before Franklin, "positive" and "negative" were called "vitreous" and "resinous," because electricity was believed to consist of two kinds of "fluid." (Back then lots of things were believed to be fluids – fire, electricity, heat, even an "ether" in space that was believed to be necessary for light to travel in. All of these notions were found to be incorrect, causing the facts to change yet again!)

In 1767 Joseph Priestly (1733-1804, English, Figure 8-2) found that the force of attraction between two electric charges varies inversely as the square of the distance between them, just like the gravitational force between masses in Newton's Law of Universal Gravitation we saw back at the end of Chapter II. The exact mathematical expression for the electrical attraction between two charges was discovered in 1785 by French physicist Charles-Augustin de Coulomb, and is known as Coulomb's Law.

Figure 8-2. Joseph Priestly.

"HOLD IT!,"

you are thinking. "Did the author just write that Coulomb's Law for forces of electrical attraction was just like the Law of Universal Gravitation? What does that mean and why didn't he show us what he was talking about? Doesn't he love talking about the mathematical structure of nature?"

Of course! I am so glad you said something. Even though we aren't going to do any calculations with Coulomb's Law, it is still fascinating to consider it, especially because of its similarity to the Law of Universal Gravitation. Let's look at both laws side by side, Newton's on the left, Coulomb's on the right:

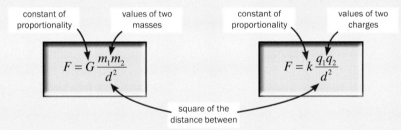

These two equations are both about forces. Newton's equation gives the force of attraction between two masses. Coulomb's equation gives the force between two charges, which could be attractive or repulsive, depending on whether the charges are both positive, both negative or one of each. The thing I want you to notice here is that *these two equations look just alike*. They both have the square of the distance between the objects in the denominator. Newton's law of force between two masses has the two masses multiplied in the numerator, and Coulomb's law of force between charges has the two charges multiplied together in the numerator. Why is the equation that describes gravitational force just like the equation that describes electrical force? Why can these forces be described with such simple equations in the first place? Remember what I wrote when we were looking at Kepler's Third Law of Planetary Motion? The creative hand of a wise Creator is definitely in evidence here!

In 1791 Luigi Galvani (1737-1798), professor of anatomy in Bologna (Italy), was studying muscle contractions in frog legs (Figure 8-3). He noticed that frog legs hanging on copper hooks from an iron balustrade (or railing) twitched as if alive when they touched the iron. He made a fork with iron and copper prongs and reproduced this result in his lab, discovering that electric charge could cause the muscles in a frog's leg to contract. When Galvani's friend, **Count Alessandro Volta** (also Italian, Figure 8-4) heard about this, he was fascinated. After nine years of thought and experimentation he invented the "**Voltaic Pile**" in 1800, the predecessor to the modern battery.

Figure 8-3. Luigi Galvani.

Figure 8-4. Alessandro Volta.

A sketch of the Voltaic Pile is shown in Figure 8-5. In Volta's original design the cells were made of metal disks about four inches in diameter and half an inch thick. The stack of cells was about one and a half feet tall and was encased in a tall, protective glass jar. The pile consisted of a stack of many individual cells. Each cell consisted of a layer of copper, an electrolyte layer, and a layer of zinc. You will have to wait until chemistry to understand what an electrolyte is, but in Volta's pile the electrolyte layer consisted of cloth or cardboard soaked in brine. The invention of the Voltaic Pile was huge, because for the first time scientists had a reliable source of electricity that they could use in the lab so that electricity could be studied further.

By 1820 it was known that electric current and static electricity were different manifestations of the same physical phenomenon. Then a funny thing happened one day

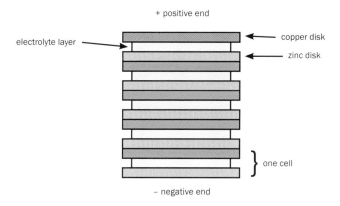

Figure 8-5. The cells in the Voltaic Pile.

Figure 8-6. Hans Ørsted.

in 1820 to Hans Ørsted (1777-1851, Denmark), professor at the University of Copenhagen (Figure 8-6). While walking to the college one morning he realized the connection between electricity and magnetism. He went into his lecture hall and right in front of his students he connected a circuit of wire to a Voltaic Pile and placed a magnetic compass near the wire. When he switched on the current the compass needle deflected. This was the first demonstration of the connection between electricity and magnetism. His students did not realize how huge this was and were unimpressed. (I love the irony of this.) Ørsted discovered that a compass near a conductor would deflect and deduced that electric current produces its own magnetic field. He called this electromagnetism, and immediately published his results. The physics world went crazy!

When André-Marie Ampère (1775-1836, France, Figure 8-7) heard that news, he began experimenting and within a few weeks showed that two parallel current-carrying wires would attract or repel each other just like magnets. He then figured out the equation for the strength of the magnetic field as a function of the current, the law now known as **Ampère's Law**. Ampère's experiments with magnetic fields in coils of current-carrying wire led him to suggest that natural magnetism was the result of tiny circulating currents inside the atoms in magnetic materials. This stunning theoretical leap is now considered to be correct!

Figure 8-7. André Ampère.

Now consider Michael Faraday (1791-1867) in England (Figure 8-8). He had been apprenticed for seven years as a bookbinder and got into physics by reading the books he was binding. He contacted Sir Humphry Davy, one of the most well known and highly respected scientists in London, and asked for a position in his lab. At the advice of one of his cronies, Davy hired Faraday as a bottle washer! So Faraday was given his chance, and boy did he make good on it. In 1831 Faraday demonstrated the opposite effect from Ampère's Law: that a magnetic field can induce an electrical current in a nearby conductor. This is now known as **Faraday's Law of Magnetic Induction**. This discovery led to the

Figure 8-8. Michael Faraday.

development of generators, motors and transformers. Your life would be utterly different, and much more difficult, if we did not have these devices. Faraday had almost no mathematical training, yet he brilliantly theorized the existence of "fields" to explain magnetic phenomena. We will learn some of the details of Ampère's and Faraday's Laws in the next chapter.

This story reaches its conclusion with the amazing Scotsman **James Clerk Maxwell** (1831-1879). In 1864 Maxwell (Figure 8-9) published what is perhaps the greatest achievement in theoretical physics of all time, the four equations now known as Maxwell's Equations. These four equations constitute a complete description of all known electric and magnetic phenomena in the universe! In addition to his supreme achievement with electromagnetic theory, Maxwell also contributed to the pioneering work on the kinetic theory of gases. His theoretical work is ranked at the same level as that of Newton and Einstein.

Figure 8-9. James Clerk Maxwell.

Maxwell's Equations are so elegant and so simple (at least to those who know vector calculus) that I must show them to you. You are years away from learning what these equations and their symbols mean (if you ever do) but just have a *look* at them and consider how so much can be said with so little. The four equations can be written several different ways, but Figure 8-10 shows them in so-called differential form.

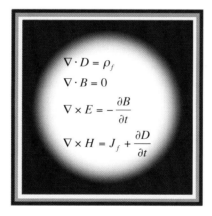

$$\nabla \cdot D = \rho_f$$
$$\nabla \cdot B = 0$$
$$\nabla \times E = -\frac{\partial B}{\partial t}$$
$$\nabla \times H = J_f + \frac{\partial D}{\partial t}$$

Figure 8-10. Maxwell's four beautiful equations.

Back in Chapter II we looked with amazement at how a simple equation like Kepler's Third Law of Planetary Motion could model the real physical world so elegantly. Maxwell's equations, universally regarded as one of the greatest and most beautiful discoveries in the history of physics, are even more astonishing. These equations first proved that light was an electromagnetic wave. From the study of these equations first came the idea that we could create radio waves and transmit information with them, which is without doubt one of the most influential ideas in the history of human technology. And while you are pausing here to consider Maxwell's discovery, consider the profound truth that the universe God made, a universe of apparently infinite complexity, can be modeled with four short equations like these.

CHARGE AND STATIC ELECTRICITY

In the context of physics and chemistry, when we refer to *charge* we mean charged subatomic particles. As you probably already know, there are two kinds of subatomic particles that have charge, protons with positive charge, and electrons with negative charge. All electrical phenomena are due to these charges. All matter is made of atoms and all atoms have protons and electrons in them, so charged particles, or charges for short, are everywhere. (Atoms also contain neutrons, which have no charge.)

Most of the time the colossal numbers of positive and negative charges that matter is made of are so evenly distributed in a substance or object they essentially cancel each other out, so there is no net electrical charge and we aren't even aware of the charges all around us (and in us). But there are two special circumstances when charges do things that make us notice them. These two special circumstances are when charges are in motion, and when charges accumulate somewhere.

As we will discuss more later in the course, electrons in metals are free to move around, which is why metals conduct electricity so well. The protons in metals are locked in the nuclei (plural of nucleus) of the atoms and are not free to move around. For this reason when we talk about charges in motion or charges accumulating, we almost always mean electrons, not protons.

When electric charge is moving or flowing it is called an electric current. We'll get to that soon. The word physicists use to describe something that is not moving is "static," which comes from the Greek word "stasis," meaning "standing still." **Static electricity** is an accumulation of charge that is stationary. An easy way to accumulate static electricity is by combing your hair with a rubber comb. As we will see in the next section, the friction between the comb and the hair can cause static electricity to build up. Then when the comb is held near some small bits of paper the paper will jump onto the comb from the electrical attraction, as shown in Figure 8-11.

Isn't it funny how so many of the things we have learned come in threes? Well, you will not be surprised to learn that there are three ways static electricity can form. In the description that follows I will refer to a device called an **electroscope** to illustrate each of these three ways. Figure 8-12 shows an electroscope when it has no static electricity on it (left) and when it is charged with static electricity (right). An electroscope consists of a metal rod with a metal sphere on top and a pair of metal hooks on the bottom. From the hooks are suspended two strips of aluminum foil called "leaves." The flask and red rubber stopper aren't part of the action; they are only there to hold the metal parts up.

Figure 8-11. Static electricity causes small bits of paper to be attracted to a rubber comb.

Figure 8-12. The electroscope uncharged (left) and charged with static electricity (right).

Consider what would happen if an accumulation of electrons, static electricity, were to be placed on the metal sphere. Since the electrons all have the same negative charge, they would repel each other, so they would try to spread themselves out on the metal parts, attempting to push as far away from each other as possible. (Remember, electrons move easily in metals, because metals are good conductors of electricity, and electricity *is* moving electrons.) So the electrons will spread out up and down the metal parts, including down into the leaves. Since the leaves are free to move, they will swing out due to the electrons' repulsion. The photograph on the right in Figure 8-12 shows the electroscope with its leaves forced outward due to the static electricity on the metal parts of the electroscope.

HOW STATIC ELECTRICITY CAN FORM

The first way static electricity can form is by **friction**. When you walk across a nylon carpet in the winter your feet are actually scraping loose some of the electrons from the atoms in the carpet and they accumulate on and spread out all over your body. There is an accumulation of static charge built up on you! Then when you reach out to touch a door knob (or your sister) the charge that has accumulated on you jumps to the door knob (or your sister) causing the exciting arc and snapping sound that we are all familiar with. The same thing happens when you slide your backside into a car with nylon upholstery. The friction causes electrons to get rubbed onto your body and stay there. Then when you are ready to open the door, the static electricity on you will **discharge** when you touch the metal door handle to get out of the car.

The term discharge is used to denote the arc or spark that occurs when a large accumulation of static electricity is suddenly released. Electrons are constantly trying to get away from each other, because they all have the same negative charge. This is especially true when many of them have been forced to accumulate. So when an object comes close enough for them to jump to, they do! They literally jump through the air, and their violent path through the air releases energy in the form of light and sound, as illustrated in Figure 8-13. The biggest arc any of us will ever see is a lightning bolt, which is a whopping big discharge of the static electricity that builds up in clouds and discharges to the ground.

Figure 8-13. The violet streak of light is an arc from a static discharge as electrons on the large metal sphere jump through the air to the small sphere.

Interestingly, the friction involved when gasoline is flowing in a rubber hose can cause quite a bit of static electricity to build up on a car during fueling, which would obviously be quite dangerous. So gasoline hoses have special designs inside to provide a way to "drain off" the charge on the car rather than causing a sparking hazard with gasoline vapors around.

Let's go back to the electroscope in Figure 8-12. Friction comes into the electroscope demonstration in a couple of ways. First, we always start this demonstration off by rubbing to use friction to create an accumulation of charge which we can then use for the rest of the experiment. The way I usually do it is to rub a Styrofoam cup on my hair, because this friction causes electrons from my hair to accumulate on the cup. (You have probably done this yourself at some point with a balloon, which you can then stick to the wall.) Another way to do it is to rub a glass rod with some silk, which will cause electrons to build up on the glass rod. Or you can do it the way the Greeks did by rubbing a piece of amber with some fur. However you do it, the friction from rubbing is what causes the electrical charge to accumulate.

In my electroscope demonstration friction is used again to get the electric charges off the Styrofoam cup and onto the sphere on the top of the electroscope. Just touching the cup to the metal sphere won't do it, because Styrofoam doesn't conduct electricity so the electrons on the cup cannot make their way to the metal sphere; they are stranded on the cup. But if I rub the metal sphere with the cup the electrons on the cup are literally scraped off onto the metal. You cannot see any of this happening, of course. All you see is that the leaves of the electroscope swing out because the charge that is now on the electroscope is trying to spread out as much as it can (because like charges repel).

The second way static electricity can form is by **conduction**. Now, before we go any further I would like for you to pause and notice that you saw this exact word before when we were talking about heat transfer. The word means something completely different here, so please take note so you don't get these two different uses of the word confused.

Everyone knows that certain substances, like metals, conduct electricity. That is, electrons flow easily in these substances. More on that a bit later, but for now note that if you create a static build-up by friction, and then touch the object that has the static electricity on it with a metal rod, the electric charge will flow, that is *conduct*, in the metal rod and will drain off onto whatever the rod is touching. This effect is demonstrated in the electroscope when the charged object (Styrofoam cup, glass rod, or whatever) touches the metal sphere. When this happens the electric charge conducts down the metal rod in the electroscope and into the metal foil leaves. And now that all of this extra electrical charge is in the leaves, they try to push away from each other. Conduction occurs again when I touch the electroscope with my hand, allowing the extra charge to drain off the electroscope by flowing onto me.

The third way static electricity can form is by a process called **induction**. (Warning: This term will also appear in the next chapter in another context which you will have to keep from getting confused with this one. Sorry.) Let's pause for a brief aside on the word induce. This word basically means "make it happen." If you induce someone to confess a crime, you put them in a position where they figure that is their only option. When labor is induced in pregnant women it is brought about by drugs that make it happen. So, back to electricity. When static electricity, an accumulation of electric charges, is formed by induction it is somehow forced to happen.

Now, consider a block of metal, as illustrated in Figure 8-14. It is chock full of electrons and they are completely free to move around inside this metal, a circumstance we will consider more closely when we get to chemistry. What will all of those electrons do if an object charged up with electrons is brought near to the metal block? As shown in the figure, they will move away from the charged object, because negative charges always try to move away from each other if they can. Even if the charged object never touches the metal block the electrons in the metal block will sort of crowd up together on the opposite side of the block from where the charged object is. In this case, the static electricity accumulation is temporary. It will only stay there as long as the charged object that is causing it, that is, inducing it, is present. Pull the charged object away, and all those electrons in the metal block will relax and spread back out in the block.

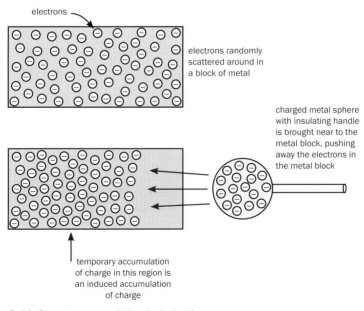

Figure 8-14. Charge accumulation by induction.

This effect is demonstrated in the electroscope when the charged Styrofoam cup is brought near the metal sphere, but is not allowed to touch it. The leaves move apart because of the charge induced in them, or forced down into them, by the nearby presence of the charged Styrofoam cup. But the induced accumulation is temporary, and as soon as the cup is pulled away the charges in the metal parts of the electroscope all relax and spread back out again.

Sometimes when I demonstrate induction with the electroscope students get the mistaken notion that electrical charges jump from the cup to the electroscope. Well, they don't. No charges actually transfer from the cup to the metal. Induction occurs without any charges transferring from the cup to the metal sphere, and it is a temporary effect that goes away as soon as the cup is withdrawn.

A well-known device that produces static electricity is the Van de Graaff generator. These machines are loads of fun, as you can see in Figure 8-15. Figure 8-16 shows photos of the Van de Graaff generator outside and inside. The generator uses an electric motor to run a rubber belt in a loop from the base of the machine up through the plastic tube to the dome at the top. At the bottom there is a wire screen, called a comb, that deposits electrons onto the belt as it runs. In the dome there is another comb that collects the electrons off the belt and allows them to conduct out onto the metal dome where they can cause all kinds of mischief for students who like getting shocked. The photo on the right in Figure

Figure 8-15. *A physics teacher having fun with the Van de Graaff generator.*

Figure 8-16. *The Van de Graaff generator, and a close up of the pulley and comb on the inside of the dome at the top.*

8-16 shows a close-up of the upper comb and how it is positioned. In the photo you can see the amber-colored roller on top of the vertical plastic tube. The rubber belt, which is missing in this photo, wraps over this roller. The metal strands on the lower edge of the comb are very close to the belt without touching it. As the belt turns, the electrons hop off the belt and onto the comb, which is attached with metal legs to the dome. Thus the electrons accumulate on the dome in large quantities. This machine produces quite a large accumulation of static charge on the dome, so much in fact that the dome can produce an impressive arc in the air to a nearby metal object, or to the body of a nearby student!

ELECTRIC CURRENT

We have been discussing static charges. Now we are going to discuss charges in motion. This is what electric **current** is, moving electric charge. Since there are two types of charged particles, positive protons and negative electrons, we would have an electric current if a group of either of them were in motion. However, as before with static electricity, in all practical electric circuits made with wires or other metal conductors, the protons are fixed in place in the metal and it is the electrons that are flowing in the wires. As I said above, in a metal the electrons are completely free to move around. An electric circuit is just an arrangement that forces them to flow in a certain direction so they can do some valuable work for us.

THE WATER ANALOGY

The best way to understand how electricity works is to consider it as analogous to water flowing in a pipe. If we pump water in a closed circuit of piping through, say, a water filter, we have a system that is analogous to an electric circuit in nearly every respect. The sketch in Figure 8-17 shows the water circuit. Imagine that we have a little water pump, like what we would see in an aquarium, and we are pumping the water through the filter to keep it clean. It will be helpful in the discussion that follows for you to know that engineers call the end of the pump where the water goes in the suction end of the pump. The end of the pump where the water comes out is called the discharge.

Let's think for a minute about what causes water to flow in a pipe. Imagine you and a friend are going to play a water game. You are each holding one end of a garden hose, and the hose is filled to the brim with water. Now you both put the end of the hose in your mouth and blow hard. Who is going to lose? That is, whose mouth gets full of water? As we can all easily guess, the person who blows with the lowest pressure will get the mouth full of water. The reason is that water always flows from high pressure to low pressure. In fact, it is the *difference* in pressure that makes it flow at all. If you both put the hoses in your mouths but did not blow, the water would not go anywhere. But when a difference in pressure is created, the water will always flow toward the lower pressure. So in summary, no pressure difference, no flow, and when there is a pressure difference the water flows toward the low pressure.

Now that we know that water always flows toward lower pressure, let's think more about the water circuit and what the pressures must be like at different points around the circuit in order for it to work. In Figure 8-18 is another sketch of the water circuit. This time I have added pressure gauges to the pipes at seven places around the pipe.

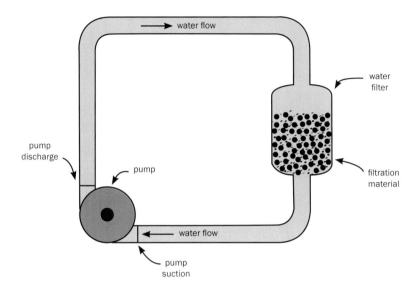

Figure 8-17. A pump circulating water in pipes through a filter.

If we make a graph of what the "pressure profile" around the circuit would look like based on the pressures all these gauges would be reading, we would find that the pressures around the circuit would vary as shown in Figure 8-19.

Notice that the highest pressure in the circuit is at the discharge of the pump, and the lowest pressure in the circuit is at the suction of the pump. It has to be this way, because outside of the pump water always flows toward low pressure. (Inside the pump are blades like those on a fan that are pushing the water toward the discharge, forcing

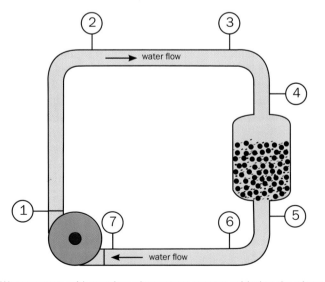

Figure 8-18. Water system with numbered pressure gauges added to the pipes.

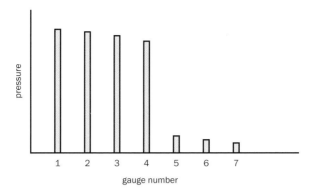

Figure 8-19. The pressure profile around the water system.

the water to go where the higher pressure is. Outside the pump the water flows by itself, and always toward the lower pressure.)

Notice also that the pressure decreases steadily from place to place around the circuit. Again, this has to be the case, because if the pressure ever went up it would mean the water had to flow the other direction, which we all know it does not do. Finally, notice that the largest pressure drop in the system is across the water filter, between gauges 4 and 5. This is because the pipes are unobstructed and only a small difference in pressure between two points will cause the water to flow easily. But the water filter is packed with sand and charcoal and what not, so to make the water flow through it the pressure drop across it has to be very large, just as you would have to blow hard to push water through a clogged hose. In fact, almost all of the pressure drop from the pump discharge to the pump suction will occur across the water filter. All of this behavior is exactly analogous to the way electricity flows in wires.

As a last step of analysis about the water circuit before we make the analogy to electricity, read through Table 8-1, which summarizes the different components in the water system and the roles they each play.

Component	Role
Pump	Makes the water flow. It does this by creating a difference in pressure. Outside the pump water will always and only flow toward lower pressure.
Pipe	Provides a contained pathway in which the water can flow.
Water	Flows in the pipes. Flowing water actually consists of many individual molecules moving along in the pipe.
Filter	Provides resistance to the flow of water. The more material the filter has packed into it the harder it is for the pump to make the water flow. The filter is doing a practical job (cleaning the water) and is the reason we have the water circuit in the first place.

Table 8-1. Components in the water circuit.

THE BASIC DC ELECTRIC CIRCUIT

Electricity flowing in a wire works exactly the same way as water flowing in a pipe. We have the same four basic components and the way each of them works is exactly analogous. Before we proceed you should know the difference between *DC* and *AC* circuits. You have probably heard of these. "DC" means *direct current*. It is a bit of a strange term, but it basically means the currents and voltages in the circuit, which you will understand soon, are constant, steady values. This is what we have with any circuit powered by a battery. The internal electronics inside virtually every electronic gadget are also DC circuits. (The AC from the power outlet is converted to DC right inside the device, or by an adapter built into the plug.) The math for dealing with DC circuits is what we are going to learn in this chapter. "AC" means *alternating current*. We are not going to study AC circuits very much in this course, except for a bit in the next chapter. But in an AC circuit the current and voltage are constantly changing, increasing and decreasing back and forth, going from positive to negative and back again. This kind of electric system is used for the entire electrical power distribution system that carries electricity to all the homes, offices and factories. So the electricity that comes from a wall outlet is AC.

Let's now consider a DC electric circuit consisting of a battery, some wire, and a resistor, as shown in Figure 8-20. This type of drawing is called a *schematic diagram*. Schematic diagrams use symbols to show the components in the circuit and how they are connected, but they do not show what the circuit actually looks like physically. The wires may be a mile long, bending along on power poles along a roadside, or they may be tiny copper connections inside a computer. The schematic only shows the components and how they connect to each other.

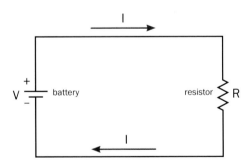

Figure 8-20. The basic DC circuit.

Before we begin analyzing this circuit a few words about the symbols are necessary. The flowing current is indicated by the arrows labeled with a capital I. The letter I is used as the symbol for electric current. (I suppose this is because the letter C is used for another electrical variable, capacitance, which we will not get into in this course.) The symbol for the battery in the circuit in Figure 8-20 is actually the symbol for a single cell battery, like a flashlight battery. The long bar on the end of the battery symbol indicates the positive end of the battery. The current comes out of this end of the battery and flows in the wire towards the resistor. Often the + and − symbols are not shown. The student is supposed to know that the long bar is the positive end. As you know, all batteries have a certain voltage, which I will explain further below. But the +/– symbols indicate that the voltage at the + end of the battery is higher than the voltage at the – end, the difference being the voltage of the battery. Note that the – symbol does not mean the voltage is actually *negative*, just as the suction of a pump does not have a negative pressure. The +/– symbols simply tell you which end of the battery has the highest voltage (+).

The small batteries you are familiar with for flashlights and other gadgets are single cell batteries. The voltage a battery can produce is fixed by its cell chemistry, so if we want a higher voltage we have to stack cells on top of one another, just as Volta's pile was made out of a stack of individual cells. A car battery has six cells, which would be depicted as shown in Figure 8-21. However, it is common when studying DC circuits to just use the battery symbol for a single cell battery, regardless of the battery voltage, and this is what I will do in my diagrams.

Figure 8-21. The electrical symbol for a six-cell battery.

Now we are ready to study the electric circuit the same way we studied the water circuit. Just like the water circuit, there are four components in the circuit. The roles they play are nearly identical to those played by the components in the water circuit, and I have summarized them in Table 8-2. Compare Table 8-2 carefully to Table 8-1.

Component	Role
Battery	Makes the current flow. It does this by creating a difference in voltage. Outside the battery current will always and only flow toward lower voltage.
Wire	Provides a contained pathway in which the current can flow.
Current	Flows in the wires. Flowing current actually consists of many individual electrons moving along in the wire.
Resistor	Provides resistance to the flow of current. The more resistance the resistor has the harder it is for the battery to make the current flow. The resistor is doing a practical job and is the reason we have the electric circuit in the first place. The resistor can be an actual resistor, or some other device represented by a resistor.

Table 8-2. Components in the electrical circuit.

TWO SECRETS

I don't want to confuse you, but there are two little important details from all of this explanation and analogy that need to be said. You won't really need to think about them at all when doing your circuit calculations, but I need to tell you about them so you won't get a wrong impression.

Everything we said above about current flowing from positive to negative and all that is valid only within the convention adopted a hundred years ago that we do the math with circuits by assuming the flow of positive charge. (As you know, protons have positive charge.) This convention has the advantage of eliminating a ton of negative signs from our calculations that don't do anything but get in the way, so this is a good convention. In reality, as I explained above, it is not positive charge that flows in the wires of the circuit but negatively charged electrons. So here is secret number one: Do believe everything I told you above about the direction the current is flowing and all that, and do the calculations calculations the way I will explain below. Just remember: We define current as the flow of fictitious positive charges. In reality, the electrons are actually flowing in the wire in the opposite direction!

One other thing. Remember how we said that for current to flow there must be a voltage difference, just like there must be a pressure difference for water to flow? Well, strictly speaking this is absolutely correct. However, in practical circuit analysis the voltage difference between the end of the battery and the near end of the resistor or light bulb or whatever in the circuit is typically so small that we can completely ignore it in our circuit calculations. So we will! This means we can assume that any two points connected together by solid wire are at the same voltage. So I have drawn the electric circuit again in Figure 8-22 and indicated the voltage at a few places with this simplification in mind. I drew the circuit with a 9 V battery.

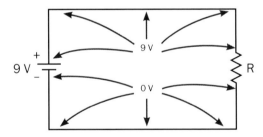

Figure 8-22. Circuit showing voltages (for practical purposes) on the wires.

Remember, there actually is an extremely minute voltage drop along the wire from the end of the battery to the resistor. If there weren't, current would not flow. But this is our second secret: The voltage difference is so small we are going to just forget about it and pretend that along any continuously connected wire the voltage is the same.

ELECTRICAL VARIABLES AND UNITS

I don't know why, but textbooks rarely point out certain confusing things about the variables and units we use when we are studying circuits. For example, the variable we use to describe the force with which the battery is pushing the current is voltage, with symbol V, and the units we use are volts, symbol V. So we measure voltage with volts. These words kind of look alike and their symbols are the same. But the other variables don't look at all like their units of measure, and their symbols are not the same. The easiest way to lay it out is in another nice table, as I have done in Table 8-3.

Variable	Symbol	Units of Measure	Symbol for Units
voltage	V	volts	V
current	I	amperes, or amps	A
resistance	R	ohms	Ω
power	P	watts	W

Table 8-3. Electric variables and units of measure, and their symbols.

I will mention a couple of points here about units. First, because the terms amp and ohm begin with vowels there are some spelling exceptions when using the metric prefixes with these electrical units of measure. Take note of how to spell these particular units:

kilamps   megamps
kilohms   megohms   microhms

Some others that you might think would be spelling exceptions are not, such as:

milliamps   microamps
milliohms

Second, resistor values in common electrical systems or electronic systems are usually in ohms, kilohms or megohms. Voltages for common circuits in computers or powered by batteries are just a few volts. Now, from Ohm's Law (which we will get to very shortly), current (designated with a capital I) is equal to I = V/R. If the voltage is just a few volts, the current units will come out to amps, milliamps, or microamps, depending on the what the resistor values are. For example, if V = 6 V and R = 5.1 kΩ,

$$I = \frac{V}{R} = \frac{6 \text{ V}}{5100 \text{ }\Omega} = 0.0011765 \text{ A} = 1.1765 \text{ mA}$$

(I put all those decimals in there deliberately, despite what we have learned about significant digits. I will explain why farther down.) This happens so often that you can save yourself a lot of time with circuit calculations by just remembering that when the resistance is in kilohms the current is most easily expressed in milliamps. If the resistance is in megohms the current is typically best expressed in microamps.

Finally, as you know by now, I usually like to work out the MKS units we are using when we encounter new derived units. I assure you that the units of measure in the table above are all MKS units and that they all boil down to combinations of fundamental units. (Actually, current, in amperes, *is* one of the seven fundamental units. Of course, nobody really says amperes anymore, we all just say amps now for short.) But working out the units for volts and ohms is beyond what we need to get into. I will only explain one of them, namely, power.

Power is the rate at which energy is delivered or used. (We will get to the equation for power in a few pages.) A watt (W) of power is an energy rate of one joule per second, or

$$1\ W = 1\frac{J}{s}$$

For example, a traditional 60 W light bulb, such as we all had in our houses before we replaced the bulbs with compact fluorescents, draws 60 joules of energy every second from the city electrical distribution system. When your parents pay the electric bill they pay for those joules. (There is a meter on the side of your house that measures how many joules you use, although the units they use aren't the nice joules (J) we use, they are the weird kilowatt-hours (KWH). It's still just a unit for energy.) When I present the equations below for calculating power in terms of voltage, current, and resistance, just take my word for it that all those combinations of units all amount to the same thing, joules per second (J/s) or watts (W).

OHM'S LAW

The basic equation relating voltage, current, and resistance is called Ohm's Law and is

$$V = IR$$

This is one of the most famous equations in physics. One handy way to remember it is that it spells the word *man* in Latin (*vir*). As I mentioned above, we will consider the voltage on a single piece of wire to be the same everywhere on it. If we apply this rule and Ohm's Law to a simple circuit with one resistor like the one in Figure 8-20, this would mean that the voltage of the battery is equal to the voltage drop across the resistor. So if you know the value of the resistance, you can find out how much current is flowing through the resistor. You simply solve Ohm's Law for the current I, giving

$$I = \frac{V}{R}$$

Then put in the voltage and resistor values and calculate the current that will be flowing in that circuit. A calculation like this for a circuit with a single resistor is quite simple. However, in the real electrical world circuits usually have many resistors, not just one. So very soon we are going to learn the analytical techniques for determining the voltages and currents that will be present in a circuit with several resistors.

WHAT EXACTLY ARE RESISTORS AND WHY DO WE HAVE THEM?

There are little devices called resistors that are used in electronic circuits to regulate the voltages and currents throughout the circuit. A complicated electronic gadget like a computer has hundreds of resistors in its circuitry. Figure 8-23 shows an image of

Chapter VIII Electricity and DC Circuits

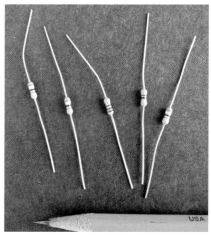

Figure 8-23. Photographs of resistors, new (left) and installed (right).

some low power resistors, the way they look just out of the package, and the way they look when installed on a circuit board in an electronic gadget. (The gadget pictured is built using technology that is nearly obsolete. Resistors connected in new electronic devices these days are quite a bit smaller and look a bit different.) The colored bands are coded and represent the resistance, in ohms, of the resistor. You will learn more about the resistor color code when you do the DC circuits experiment.

Now you may or may not end up being an electrical engineer designing circuits with actual resistors. You may relate more immediately to the electronic devices in your house than you do to electronic parts. But at a basic level, most of the electrical devices we use – electric driers, lighting, hedge trimmers, computers, MP3 player chargers, and on and on – act like resistors in the electric circuits that power everything around us. (AC circuits are more complicated than that, but this is an adequate simplification for our purposes here.) Light bulbs and electric heaters and toasters are actually quite similar to actual resistors, so we can use resistors to model those devices in our circuit calculations. In summary, when we do electric circuit calculations we use the generic symbol for a resistor. But the resistors in our drawing could represent just about any device that needs electric power to operate.

THROUGH? ACROSS? IN?

The prepositions and adverbs we use to describe what is going on in an electric circuit are important, so I want to take a moment here to make these clear. Current flows *in* a wire, just like water flows *in* a pipe. The voltage drops *across* a resistor, just like the pressure drops *across* the water filter. What we mean by this is that the voltage "upstream" of the resistor is higher than the voltage "downstream" of the resistor, so the voltage drops *across* it. We never speak of the voltage *through* a resistor because the voltage is not what is going through it. The current is what is going through it. So currents flow *in* wires and through *resistors*. Voltages drop *across* resistors.

While we are discussing word issues, you should also know that for historical reasons voltage is also often called *potential*, and a voltage difference is sometimes called a *potential difference*. The phrase "potential difference" does not mean, "there is a possible difference." It means "difference in voltage" or "voltage drop." Finally, another term for voltage from the old days that you still see from time to time is *electromotive force*, or EMF. I actually like this term, because it reminds me of how the battery is providing the force that moves the current, just like a pump providing the force that moves the water.

VOLTAGES ARE RELATIVE

Consider for a moment our discussion back in Chapter V about energy. The gravitational potential energy in an object depends on how high up it is. But how high an object is depends on where the zero reference is for height. Where is zero height? The table top? The floor? The ground outside the building? Sea level? The answer is that it doesn't matter. The only reason we ever calculated the E_G in an object was so we could predict how much work it would take to get it up there or how fast it would be going if it fell down. All these calculations really only depend not on the absolute height, which is hard to even define, but on the *difference* in height from some reference point like the floor or the table top to where the object is. It is really only the difference that matters.

The same situation holds for voltages. Defining voltage in an absolute sense will have to wait for a future physics course. For now all we will really concern ourselves with are voltage differences, or voltage drops. We will simply use the lowest voltage in the circuit as our reference point. Remember where this is? The lowest pressure in the water circuit is at the suction of the pump. Analogously, the lowest voltage in an electric circuit will be at the negative end of the battery. Since the only thing that matters about a voltage is how high it is relative to the reference, we will just say that our reference voltage is the wire connected to the negative end of the battery. Further, we might as well call this voltage zero volts.

In fact, we might as well point out that in the great electrical systems of our noble nation the earth itself is used as the zero voltage reference, and in our power distribution systems we realize this by physically bolting part of the circuit to the ground. This is done by driving a long copper rod into the ground (at your house this rod is about eight feet long) and connecting the power system to it with a hefty copper wire. This is the electrical "ground" that you have probably heard of plenty of times but didn't understand. The electrical people call it ground because the reference point for zero volts in the circuit is the ground, and the circuit is connected to the ground, literally.

In an electric circuit diagram an electrical ground connection is indicated by the symbol shown in Figure 8-24. Wherever ground is in an electric circuit, that is where the zero volt reference is. Now, I am not going to draw the ground symbol in on every

Figure 8-24. The symbol for electrical ground.

circuit example in this book. This is mainly because a lot of common electric circuits, such as flashlights, MP3 players, cell phones and so on aren't actually connected to the ground unless they are being charged up. So we usually don't show a ground symbol on DC circuits. But I will show it once just to illustrate.

Figure 8-25 shows our one-resistor circuit with the ground symbol shown. I have also labeled the voltage, current and resistance using the standard symbols. V_B stands for the voltage of the battery. From now on, we will simply say that the negative end of the battery is where zero volts is. All of the voltages in the circuit go up from there.

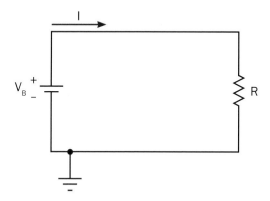

Figure 8-25. An electric circuit with ground connection shown.

POWER IN ELECTRICAL CIRCUITS

In an electric circuit the battery, or whatever the power supply actually is, even if it is not a battery, supplies the energy to the devices that need it. The energy is supplied at a certain rate, a certain number of joules per second. This energy rate is called the **power** and we measure it in watts (W). The basic equation for power is

$$P = VI$$

where P is the power (W, that is, J/s).

There are two more equations for power that we can derive if we substitute for V or I in this equation using Ohm's Law. If we simply substitute for V, we get

$$P = I^2 R$$

and if we solve Ohm's Law for

$$I = \frac{V}{R}$$

and substitute this into the power equation for I we get

161

ASPC

$$P = \frac{V^2}{R}$$

Each of these three equations for power is useful, depending on which quantities one has to work with, so you should know and be able to use all three of them.

While we are on the subject of power, let's consider how the conservation of energy works in electrical circuits. The battery supplies energy to the circuit, and this energy is "consumed" (as we say) by the devices in the circuit. If we are talking about the energy supplied by the battery in one second, then this is the power supplied to the circuit, and this power supplied to the circuit must equal the power consumed by all of the devices in the circuit. For resistors, the power is *dissipated* (that is, given off) as heat. For other devices such as motors the power is used to do mechanical work ($W = Fd$), as with a fan pushing air or a garage door opener lifting the door.

For the circuit shown in Figure 8-20, assume that the battery voltage is 5.60 V and the resistor value is 7.7 kΩ. Determine the current flowing in the circuit, the power supplied by the battery, and the power consumed by the resistor. State your results in mA and mW.

Back in the section entitled "Electrical Variables and Units" I mentioned that with battery voltages in volts, resistor values in kilohms will result in currents in milliamps. In addition, if the current is in milliamps the power will be in milliwatts. We can save the work of the unit conversions in this problem by just using kilohms for the resistor value. This will automatically give us the units required for the answers.

As usual, begin by writing the givens.

$V = 5.60$ V

$R = 7.7$ kΩ

$I = ?$

$P = ?$

To solve for the current flowing in the circuit we use Ohm's Law.

$V = IR$

$$I = \frac{V}{R} = \frac{5.60 \text{ V}}{7.7 \text{ k}\Omega} = 0.7273 \text{ mA}$$

I included an extra significant digit because I am going to use this result later to calculate the power values. But we must also state the current as one of our answers, so rounding to three significant digits we have

$I = 0.727$ mA

To compute the power supplied by the battery we will use the basic power equation. We have the battery voltage, and we have just computed the current flowing through the battery, so we can dive right in.

$$P = VI = 5.60 \text{ V} \cdot 0.7273 \text{ mA} = 0.407 \text{ mW}$$

We could apply the same power equation to compute the power consumed by the resistor. If we did, we would use the same voltage and the same current, so we would obviously get the same answer. This makes sense, of course, because of the conservation of energy. The power produced by the battery must equal the power consumed by the resistor. But since we calculated the current flowing through the resistor, let's compute the power consumed by the resistor using the power equation that has the current in it. This gives

$$P = I^2 R = (0.7273 \text{ mA})^2 \cdot 7.7 \text{ k}\Omega = 0.407 \text{ mW}$$

This is the power consumed by the resistor, and is the same as the value we computed for the power produced by the battery.

TWO-RESISTOR NETWORKS

So far you have seen a circuit with one battery and one resistor. We have now covered all of the basics for such circuits and are ready to learn how to handle circuits with more than one resistor. The bundle of resistors in a circuit, all wired together but not connected to the battery, is called a resistor *network*. Our first task is to learn the different ways resistors can be connected together in a network.

Resistors in circuits can be connected together in two main ways. The first kind of connection is shown in Figure 8-26. The key feature here is that all of the current that enters resistor R_1 must also enter resistor R_2. There are no other branches or pathways in the circuit, so the current, represented by the red arrows, has no option but to go through both resistors. This kind of connection is called a **series** connection and we say that the two resistors are connected "in series." We could also wire up three or four resistors in series, or as many as we like. It is important to note that in a series connection there is nothing between the resistors, no other devices, no connections, nothing except the single wire connecting the series resistors together. If there is anything else there the resistors are not connected in series. An important principle to note for series connections is this: *When resistors are connected in series, the same current passes through each of them.*

Figure 8-26. Two resistors connected in series.

The second way of wiring resistors together is shown in Figure 8-27. The connections shown in the three sketches are the same. I drew it three different ways to show different ways this type of resistor network can be drawn. This kind of connection is called a **parallel** connection, and we say that the two resistors are connected "in parallel." Just as with series connections, we can connect as many resistors in parallel as we like. The three networks shown in Figure 8-27 are identical. A key feature to note about a parallel connection like this is that if two resistors are connected together in parallel, then they are connected together at both ends. If they are not connected together at both ends, or if there is more than one resistor in one of the branches, they are not connected in parallel.

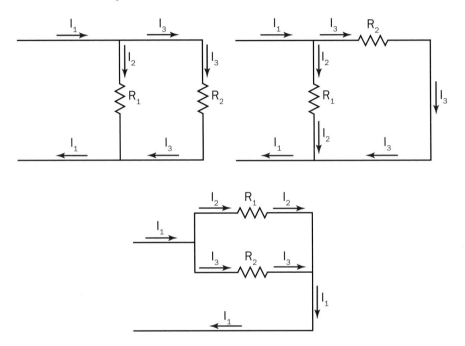

Figure 8-27. Three identical resistor networks showing two resistors connected in parallel.

Notice that in a parallel network the current heading towards the resistors, I_1, will divide at the *junction*, or *node*, in the wires. Some of the current, which I show as I_2, goes through resistor R_1, and the rest of the current, which I have labeled I_3, goes through the other resistor, R_2. The amount of current that will go through each branch depends on the resistance in each branch. Keep thinking about the water analogy. If water is flowing in a pipe and comes to a place where the pipe has two branches the water will divide, some going one way and some the other.

Now let's see what these two basic types of networks look like when they are connected to a battery. The sketches in Figure 8-28 each show circuits with two resistors in series. These two circuits are identical. The sketches in Figure 8-29 each show circuits with two resistors in parallel. As with the series circuits in the previous figure, these two circuits are identical. They may look different, but they are not.

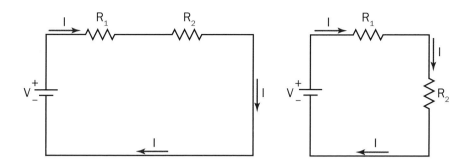

Figure 8-28. Identical DC circuits with two resistors connected in series.

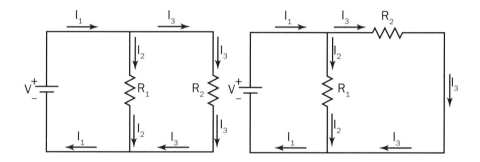

Figure 8-29. Identical DC circuits with two resistors connected in parallel.

EQUIVALENT RESISTANCE

When we perform calculations with circuits our ultimate goal is to determine the voltages and currents in every part of the circuit. To do this, we must first find out how much current is flowing into the resistor network from the battery. Until we do this, we generally won't be able to figure out much else.

Here is an important first principle to keep in mind: The battery cannot tell how many resistors there are or how they are wired together. As far as the battery is concerned, there may be one resistor or a hundred. To the battery, regardless how many resistors there are, the network *seems* like just one resistor. The value of the apparent resistance that the battery perceives is called the **equivalent resistance**, for which we use the symbol R_{EQ}.

Consider this analogy. A horse pulling a cart doesn't know the difference between having 20 bales of hay on the cart each weighing 50 pounds, and 40 bales of hay, each weighing 25 pounds. To the horse, it feels like a 1,000 pound load. This is because it *is* a 1,000 pound load. We could even remove all the hay bales and put a single machine weighing 1,000 pounds on the cart and the horse would not know the difference. Similarly, to the battery, the entire resistor network feels like one "load." (The term *load* is actually used when discussing electric circuits to describe the resistor network that the power source is connected to.) Regardless of how the resistors are connected, and how many of them there are, the battery responds as if there were only a single resistor connected with a resistance value equal to the R_{EQ} of the resistance network. I will now discuss how one determines the value of R_{EQ} for different types of resistor networks. This presentation is lengthy, not because this topic is particularly difficult, but because there are a number of details to review and I want to explain them very carefully. So read on.

We will take this one step at a time. First, the equation for calculating the equivalent resistance for resistors in series is

$$R_{EQ} = R_1 + R_2 + R_3 + ...$$

This equation works for any number of resistors, as long as they are all connected in series. For resistors in parallel, the easiest equation to use for calculating R_{EQ} works only for two resistors. The equation for calculating the equivalent resistance for two resistors in parallel, including the special symbol we use to indicate parallel resistances, is

$$R_{EQ} = R_1 \| R_2 = \frac{R_1 R_2}{R_1 + R_2}$$

There is an equation that can handle more than two resistors in parallel, but we will not use it for our calculations. (This equation has an interesting mathematical symmetry with the series equation. Because of this, you might want to at least see the equation, so here it is: $\frac{1}{R_{EQ}} = \frac{1}{R_1} + \frac{1}{R_2} + \frac{1}{R_3} + ...$) When using the equation above you will want to remember two tips to help check yourself as you go. First, the resistance of two resistors in parallel is always lower than either of the single values that go into the calculation. So your R_{EQ} calculation for two resistors in parallel should always give a result that is less than the values of R_1 and R_2. Second, when two identical resistors are in parallel the equivalent resistance is half of the value of one of the resistors. For example, two 8 kΩ resistor in parallel will have an R_{EQ} of 4 kΩ. This is a time-saving tip.

As examples, the sketches in Figure 8-30 show one series network and one parallel network. I drew each network inside an imaginary box with two terminals (the little circles) to show how the battery would be connected (as the red arrows show) to the network. The black arrow is there to show that the R_{EQ} value is the single apparent resistor value that the battery will think is out there when connected to the network. Remember, the battery does not know what is in the imaginary box.

The R_{EQ} value for the series network is

$$R_{EQ} = 40 \text{ Ω} + 530 \text{ Ω} = 570 \text{ Ω}$$

Chapter VIII   Electricity and DC Circuits

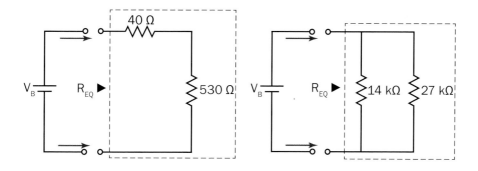

Figure 8-30. Series and parallel resistor networks connecting to a battery.

To the battery, this resistor network will seem like a single 570 Ω resistor. The resistor values in the parallel network are all in kilohms. This is fine, as long as all of them are in the same units. Since both resistors are in kilohms, we will just do the calculation in kilohms. The R_{EQ} value for the parallel network is

$$R_{EQ} = 14 \text{ k}\Omega \parallel 27 \text{ k}\Omega = \frac{14 \text{ k}\Omega \cdot 27 \text{ k}\Omega}{14 \text{ k}\Omega + 27 \text{ k}\Omega} = 9.2195 \text{ k}\Omega$$

To the battery, this parallel resistor network will seem like a single resistor with a resistance of 9.2195 kΩ. We'll discuss significant digits in these calculations next.

SIGNIFICANT DIGITS IN CIRCUIT CALCULATIONS

As you probably noticed immediately from the parallel resistor example, my calculation of the equivalent resistance has several more digits in it than we would normally permit according to the significant digit rules we have been using since the beginning of this book. You are right! Here is what is going on.

In electric circuit calculations for circuits with more than one resistor there is often more than one way to do the calculations, and there are usually many steps in the calculations. The ordinary significant digit rules do apply to these calculations, as always. But if we use them it makes it difficult for you, as the student, to make sure you are doing the steps correctly. If you make a small error somewhere, your answer can look correct but be incorrect, because the rounding can cover up the error. On the teacher's end, scoring papers can be a nightmare if the teacher cannot tell whether an answer is correct or not, and it can easily happen that a student's answer was calculated correctly, but looks different from the answer key because of rounding repeatedly over several steps of calculations. By the same logic, incorrect answers can look correct.

In this course we are going to handle this problem by using a special digits rule for circuit calculations: *For circuits with more than one resistor, ignore the usual significant digit rules and show four decimal places in every value for every calculation.* This includes equivalent resistances, voltages, currents and powers. Four decimal places is enough to tell whether your calculation has been done correctly or not. When comparing your answers to the answer key the first three decimal places should agree nearly every time. The last place may differ from the key, but if the first three agree you probably did the problem correctly.

LARGER RESISTOR NETWORKS

We have seen how to compute R_{EQ} for series and parallel networks with two resistors. The next step is to apply these techniques to larger networks. Consider the three-resistor network shown in Figure 8-31. For resistor networks with more than two resistors we always begin the process of calculating R_{EQ} by *starting at the right side of the network*. We work our way from right to left, identifying small groups of resistors that are connected either in series or parallel. Each time we identify a little group, we calculate a sort of sub-R_{EQ} for that group. Then replace the group with the R_{EQ} for the group. Do this repeatedly, working from right to left until every resistor in the network has been combined into a single resistor value, the R_{EQ} for the entire network. Hopefully, working the following two examples will make this clear.

To begin calculating R_{EQ} for the three-resistor network in Figure 8-31, first note that one of the resistors is in kilohms and the other two are in megohms. This will not do. All resistors need to be in the same units before we begin combining their values together. In these calculations you should not trouble yourself to convert all of the resistor values to MKS units, which would be ohms. We don't want to have to write down dozens of zeros all the time. Instead, pick the metric prefix that occurs most frequently in the resistor values in the circuit, and rewrite all of the other resistor values so they have the same prefix. We will rewrite 830 kΩ as 0.83 MΩ and do the R_{EQ} calculation in MΩ.

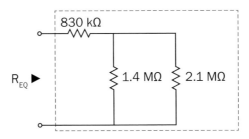

Figure 8-31. Example three-resistor network.

Beginning on the right hand side, we see that the 1.4 MΩ and 2.1 MΩ resistors are in parallel. So we will calculate an equivalent value of resistance that we can substitute for them.

$$1.4\text{ M} \parallel 2.1\text{ M} = \frac{1.4\text{ M} \cdot 2.1\text{ M}}{1.4\text{ M} + 2.1\text{ M}} = 0.8400\text{ M}\Omega$$

Imagine at this point that in the network the 1.4 MΩ and 2.1 MΩ resistors have been replaced by their 0.8400 MΩ equivalent. The network may now be represented as shown in Figure 8-32.

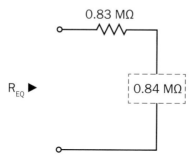

Figure 8-32. Three-resistor network with two of the resistors replaced by their equivalent resistance.

Now we can see that the 0.83 MΩ resistor is in series with the 0.84 MΩ group, so we can use the series resistance calculation to compute R_{EQ} for the entire network as

$R_{EQ} = 0.8300 \text{ M}\Omega + 0.8400 \text{ M}\Omega = 1.6700 \text{ M}\Omega$

Now it's time for an example of one of the most difficult equivalent resistance calculations. In this course, circuits with five or six resistors will be as difficult as they get. With that in mind, let's tackle the network shown in Figure 8-33. You'll notice I didn't put the Ω symbol by the resistor values. It is common to see circuits that just show the numerical value and the metric prefix, and since all resistor values are in ohms or -ohms with a prefix, the Ω symbol is assumed. As before, we check the units and notice that four of the six resistors are in kilohms, so that's what we'll use for the R_{EQ} calculation.

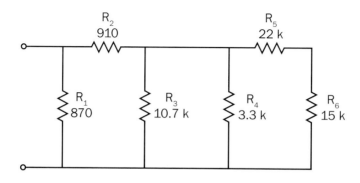

Figure 8-33. Example six-resistor network.

As we calculate R_{EQ} for this circuit, I will show you a helpful technique for keeping track of where you are in your calculation. This technique involves marking the circuit with lines through it as your R_{EQ} calculation includes more and more resistors.

As before, we begin by working from right to left. We notice that R_5 and R_6 are in series, so

$$R_5 + R_6 = 22\text{ k} + 15\text{ k} = 37\text{ k}\Omega$$

Now I suggest you mark the circuit as shown in Figure 8-34, indicating that the resistance of everything to the right of the green line is 37 kΩ.

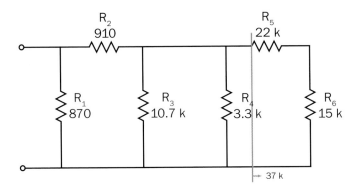

Figure 8-34. First mark on the network, indicating we know the equivalent resistance of everything to the right of the line.

Moving to the left, we observe that R_4 is in parallel with the group we just finished with. So we execute a parallel calculation between the 37 kΩ and R_4.

$$R_4 \parallel 37\text{ k} = \frac{3.3\text{ k} \cdot 37\text{ k}}{3.3\text{ k} + 37\text{ k}} = 3.0298\text{ k}\Omega$$

Now we mark the circuit as before, as shown in Figure 8-35. (Notice I used four decimal places.) Next, R_3 is in parallel with what we have done so far, so we execute that calculation.

$$R_3 \parallel 3.0298\text{ k}\Omega = \frac{10.7\text{ k} \cdot 3.0298\text{ k}}{10.7\text{ k} + 3.0298\text{ k}} = 2.3612\text{ k}\Omega$$

Now we mark the circuit as shown in Figure 8-36. Next, everything to the right of the third green line is in series with R_2, so we add these two values. Don't forget that we are using kilohms for all resistor values.

$$R_2 + 2.3612\text{ k} = 0.910\text{ k} + 2.3612\text{ k} = 3.2712\text{ k}\Omega$$

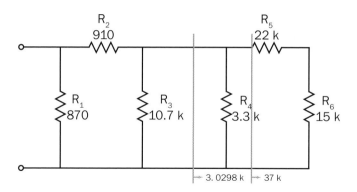

Figure 8-35. Second mark on the network.

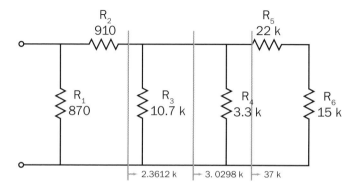

Figure 8-36. Third mark on the network.

Now we draw in draw in one last green line as in Figure 8-37. Finally, R_1 is in parallel with everything we have so far. Writing the value of R_1 in kilohms, we have the value for R_{EQ}.

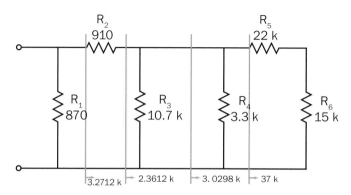

Figure 8-37. Fourth mark on the network.

171

$$R_{EQ} = R_1 \| 3.2712 \text{ k}\Omega = \frac{0.870 \text{ k}\Omega \cdot 3.2712 \text{ k}\Omega}{0.870 \text{ k}\Omega + 3.2712 \text{ k}\Omega} = 0.6872 \text{ k}\Omega$$

In summary, here are the steps of our procedure for calculating R_{EQ}.

- Make sure all of the resistor values in the network use the same units.
- Work from right to left, identifying small groups of resistors to combine into a single value.
- As you work your way to the left keep track of where you are by drawing a line through the circuit each time you take in another resistor or group of resistors.
- When you reach the left side and have included every resistor in the calculation you have your value for R_{EQ}.

KIRCHHOFF'S LAWS

We are almost ready to put a battery on a resistor network and start calculating the voltages and currents. In addition to the equations for combining resistors in series or parallel circuits, there are three equations we will use to do the circuit calculations. You already know one of them, Ohm's Law, or $V = IR$. The other two are called Kirchhoff's Laws.

Kirchhoff's Junction Law says that at any junction, or node, in a circuit, the sum of all the currents entering the node must equal the sum of all the currents exiting the node. This makes perfect sense. Think again about the water analogy. If three pipes are feeding water into some point in the plumbing network, and two pipes are where the water exits, then the sum of the three water flows coming into the junction must equal the sum of the two water flows going out. The water has no where else to go.

As a quick illustration, assume you have a section of an electrical circuit as shown in Figure 8-38.

If you know I_1 and I_2, then I_3 is easy to determine, because $I_1 = I_2 + I_3$, so I_3 is just $I_1 - I_2$. Using the values shown in the sketch, I_3 has to be

$$I_3 = I_1 - I_2 = 14.5 \text{ A} - 6.9 \text{ A} = 7.6 \text{ A}$$

The other law is **Kirchhoff's Voltage Drop Law**. This law states that if you scan around any loop in a circuit the sum of the voltage rises must equal the sum of the voltage drops. Sometimes it takes students a while to get their minds around the terminology

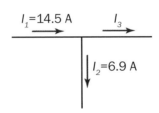

Figure 8-38. A junction, or node, in a circuit.

in this law, so I am going to expand on this with an example and with a new analogy. Hopefully this will enable you to grasp this concept, which like everything else in the chapter is relatively simple but tricky to put into words. First, the example.

Figure 8-39 shows a sketch of a circuit in which I have shown the voltage drops. I have shown three of the voltage drops as unknowns, A, B and C. The other values you may take as givens, although soon you will be able to calculate them for yourself.

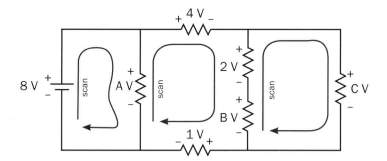

Figure 8-39. Scanning around the loops in a circuit.

Pick any loop you like in this circuit. Scan around the loop, marking down the voltage rises and drops. Kirchhoff's Voltage Drop Law says they must be equal. Begin in the first loop next to the battery, starting in the bottom left corner of the loop. This is the zero volt point in the circuit. As we move around the loop in a clockwise circle we first see a rise of 8 volts as we cross the battery and then a drop of A volts to complete the loop. Applying the voltage drop law we get

8 V = A V

A = 8 volts

This first loop illustrates a general principle you should make note of. *Devices in parallel always have the same voltage drop.* Here, the resistor is in parallel with the battery, so the voltage across them must be the same. You will use this most often with resistors, or, as in this case, with a resistor in parallel with the battery.

Now scan the middle loop. Beginning in the lower left corner and scanning around the circuit in a clockwise direction we first have a rise of 8 volts, then a drop of 4 volts, a drop of 2 volts, a drop of B volts, and a drop of 1 volt. Using Kirchhoff's law we have

8 V = 4 V + 2 V + B + 1 V

B = 8 V – 4 V – 2 V – 1 V = 1 volt.

Finally, scanning the third loop we begin with two rises, 1 volt and 2 volts, followed by one drop of C volts. Using Kirchhoff's law we have

$1\,V + 2\,V = C$

$C = 3$ volts

Note that we could also solve for C by scanning around the large outer loop of the circuit, where we have a rise of 8 volts followed by drops of 4 volts, C volts, and 1 volt, giving

$8\,V = 4\,V + C + 1\,V$

$C = 8\,V - 4\,V - 1\,V = 3$ volts

I also promised a new analogy to aid in understanding the voltage drop law. Imagine a huge old mansion with many different staircases. It has been converted into a sort of crazy fun house so that there is one main staircase for going up to the top floor, like the voltage of a battery, and many little ones for going down, like the voltages across resistors. These directions correspond to the direction current is flowing in a circuit. The different routes one can take to get to the top floor and then back down to the ground are like the loops we scanned in the previous example. The number of steps in each staircase corresponds to the value of a voltage rise or voltage drop. The different floors correspond to different voltages above the zero voltage wire of a circuit.

So consider the sketch in Figure 8-40. The horizontal lines represent the different floors in the house. Notice that no matter what route you choose to take to get down, you have to go down 16 steps, which is equal to the number of steps in the up staircase. Also, between any two floors in the house there are the same number of steps, no matter which staircase you take. This means that this house is like the voltage drop law. You can scan around any loop you want involving any staircases you like, but the total steps you go up has to equal the total steps you go down to get back to where you started. And you may also notice that those two staircases at the bottom that have four steps are like resistors in parallel. As I mentioned before, any two resistors in parallel have the

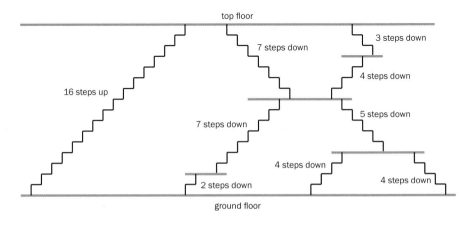

Figure 8-40. Steps in a house as an analogy for voltage rises and drops in a circuit.

same voltage drop, just as these two staircases have to have the same number of steps, since they start and end at the same floors.

PUTTING IT ALL TOGETHER TO SOLVE DC CIRCUITS

The final step in this study is to put everything to use to solve DC circuits. Let's begin with a procedure that outlines how to proceed. The steps to solve a circuit are as follows:

1. Calculate the equivalent resistance, R_{EQ}. Work from right to left in the circuit.
2. Number all of the resistors if they are not already numbered on the circuit diagram. Label the current leaving the positive end of the battery as I_1. Label the voltage drop across each resistor using the same number as the resistor number. Label the current flowing in each of the other branches of the circuit.
3. Calculate the current I_1 using the battery voltage, the equivalent resistance, and Ohm's Law, $V = IR$. Solving Ohm's Law for I we have $I = V/R$. Applying this to the battery voltage and equivalent resistance of a circuit we have

$$I_1 = \frac{V_B}{R_{EQ}}$$

4. Working left to right, use Ohm's Law and Kirchhoff's two laws to calculate the individual currents and voltage drops for the resistors. You will always apply one of these three laws. For any resistor, if you know two out of the three variables in Ohm's Law you can calculate the third. In any loop, if you know all of the voltage drops but one, you can use Kirchhoff's Voltage Drop Law to determine the unknown voltage drop. If at any junction you know the values of all of the currents entering and leaving the junction but one, you can use Kirchhoff's Junction Law to determine the one unknown current.
5. When required, calculate the powers consumed by the individual resistors using their individual voltage, current and/or resistance values.
6. Assuming that the battery voltage is a few volts (very common), use this rule of thumb to expedite dealing with the metric prefixes on currents and powers: If all the resistors are in Ω then the currents will be in A and powers in W. If all the resistors are in kΩ then the currents will be in mA and powers in mW. If all resistors are in MΩ then the currents will be in μA and the powers in μW.

As examples I will solve each of the four circuits given as examples above for calculating R_{EQ}. For each of these the first step of calculating the equivalent resistance is already complete. For each circuit I will add a battery with a certain voltage value, and label all of the currents and voltage drops. Then I will solve for all the voltages and currents. Finally, I will compute the power being delivered into the circuit by the battery, and the power being consumed by each of the resistors. The total power consumption of the resistors should match the power the battery is producing, except for possible small differences due to rounding.

Our first example will be to compute the voltages, currents and powers for the circuit shown in Figure 8-41.

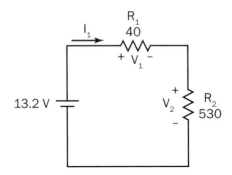

Figure 8-41. Example two-resistor series circuit.

From the R_{EQ} calculation above, we have $R_{EQ} = 570\ \Omega$. Next we calculate the current I_1 as

$$I_1 = \frac{V_B}{R_{EQ}} = \frac{13.2\ \text{V}}{570\ \Omega} = 0.0232\ \text{A}$$

With this current we solve for the two voltage drops using Ohm's Law:

$$V_1 = I_1 R_1 = 0.0232\ \text{A} \cdot 40\ \Omega = 0.9280\ \text{V}$$

and

$$V_2 = I_1 R_2 = 0.0232\ \text{A} \cdot 530\ \Omega = 12.2960\ \text{V}$$

According to Kirchhoff's Voltage Drop Law, voltages V_1 and V_2 should add up to the battery voltage, 13.2 V. They actually add up to a teeny bit more because of the rounding in our I_1 calculation.

Now for the powers. The power produced by the battery is

$$P_B = V_B I_1 = 13.2\ \text{V} \cdot 0.0232\ \text{A} = 0.3062\ \text{W}$$

For the resistor power consumption we have three equations we can use, and they will all give the same powers, except for small differences due to rounding. I will use two different equations for the powers for no reason except just to illustrate. The power consumed by R_1 is

$$P_{R1} = I_1^2 R_1 = (0.0232\ \text{A})^2 \cdot 40\ \Omega = 0.0215\ \text{W}$$

and the power consumed by R_2 is

$$P_{R2} = \frac{V_2^2}{R_2} = \frac{(12.2960 \text{ V})^2}{530 \text{ }\Omega} = 0.2853 \text{ W}$$

Adding the two resistor powers together gives 0.3068 watts, which is very close to the 0.3062 watts we calculated for the power the battery is delivering to the circuit. The difference is due to rounding.

The next example is for the circuit in Figure 8-42.

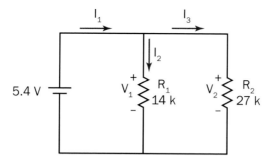

Figure 8-42. Example two-resistor parallel circuit.

The R_{EQ} value we calculated previously for this circuit was 9.2195 kΩ. However, this circuit is an example of the *one time* that R_{EQ} is not needed for computing the voltages and currents of the resistors. As I have mentioned a couple of times before, devices in parallel always have the same voltage across them. This means that *in any circuit in which every resistor is in parallel with the battery, the voltage drop across every resistor is the battery voltage*. Notice here that both of the resistors are in parallel with the battery, so

$$V_1 = V_2 = 5.4 \text{ V}$$

With these voltages we can calculate the currents using Ohm's Law as

$$I_2 = \frac{V_1}{R_1} = \frac{5.4 \text{ V}}{14 \text{ k}\Omega} = 0.3857 \text{ mA}$$

and

$$I_3 = \frac{V_2}{R_2} = \frac{5.4 \text{ V}}{27 \text{ k}\Omega} = 0.2000 \text{ mA}$$

Notice that I left the resistances in kilohms, which means the current units are in milliamps, as previously discussed. For the powers this time I will just use the $P = VI$ equation for each of them. Because this was a parallel circuit I never did calculate I_1, so I will need to do it now because I need it to compute the power produced by the battery. I can calculate it the regular way with V_B and R_{EQ}, or I can use Kirchhoff's Junction Law. I will use the Junction Law this time.

$$I_1 = I_2 + I_3 = 0.3857 \text{ mA} + 0.2000 \text{ mA} = 0.5857 \text{ mA}$$

Now I can calculate the power produced by the battery, which is

$$P_B = V_B I_1 = 5.4 \text{ V} \cdot 0.5857 \text{ mA} = 3.1628 \text{ mW}$$

Notice that since the current was in milliamps the power is in milliwatts. Calculating the power for the resistors using the same equation, we have

$$P_{R1} = V_1 I_2 = 5.4 \text{ V} \cdot 0.3857 \text{ mA} = 2.0828 \text{ mW}$$

and

$$P_{R2} = V_1 I_3 = 5.4 \text{ V} \cdot 0.2000 \text{ mA} = 1.0800 \text{ mW}$$

These two powers add up to equal the power produced by the battery calculated just above.

The third example uses the three-resistor circuit in Figure 8-43.

Our R_{EQ} value from above was 1.6700 MΩ. As before, we begin by calculating the current coming out of the battery.

$$I_1 = \frac{V_B}{R_{EQ}} = \frac{12.0 \text{ V}}{1.67 \text{ MΩ}} = 7.1856 \text{ μA}$$

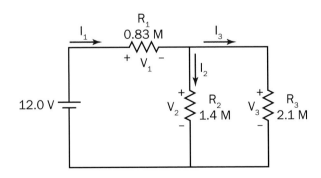

Figure 8-43. Example three-resistor circuit.

And as previously mentioned, since the resistor value is in megohms the current is in microamps. Now that we know this, we have to ask what we can calculate next using one of the three laws (remembering that we are working from left to right). The answer is that since we know the current going through R_1 we can calculate V_1 using Ohm's Law. This gives

$$V_1 = I_1 R_1 = 7.1856 \ \mu A \cdot 0.83 \ M\Omega = 5.9640 \ V$$

Now again we ask what we can calculate next. We are working our way from left to right. As we move to the right of R_1 we have a junction, and we also have R_2 nearby. Can we use Kirchhoff's Junction Law at the node? No, because although we know I_1, we don't know either of the other two currents. Can we make use of Kirchhoff's Voltage Drop Law? Yes. In the first loop we know every voltage except V_2. Writing the voltage drop equation for this loop gives

12.0 V = 5.9640 V + V_2

V_2 = 12.0 V − 5.9640 V = 6.0360 V

Now that we know V_2 we can calculate current I_2 using Ohm's Law.

$$I_2 = \frac{V_2}{R_2} = \frac{6.0360 \ V}{1.4 \ M\Omega} = 4.3114 \ \mu A$$

Still working our way to the right, we can either look at the node to the right of R_1, or we can simply note that R_3 is in parallel with R_2 and thus has the same voltage drop. The simplest thing to do is the latter approach, but just for its instructional value, let's look at the node. We now know two out of the three currents at the node at the top of the circuit. Applying the junction law here we have

$I_1 = I_2 + I_3$

$I_3 = I_1 − I_2 = 7.1856 \ \mu A − 4.3114 \ \mu A = 2.8742 \ \mu A$

Finally, knowing I_3 we can use Ohm's Law to determine V_3.

$$V_3 = I_3 R_3 = 2.8742 \ \mu A \cdot 2.1 \ M\Omega = 6.0358 \ V$$

You can see that this value differs only a tiny bit from the value we already had for V_2. And as you know by now, this is because of rounding and is the reason we are using four decimal places for these calculations.

There are three equations I could use to calculate the powers in this circuit. This time I will use $P = VI$ for the battery and $P = I^2R$ for the resistors.

$P_B = V_B I_1 = 12.0 \text{ V} \cdot 7.1856 \text{ μA} = 86.2272 \text{ μW}$

$P_{R1} = I_1^2 R_1 = (7.1856 \text{ μA})^2 \cdot 0.83 \text{ MΩ} = 42.8553 \text{ μW}$

$P_{R2} = I_2^2 R_2 = (4.3114 \text{ μA})^2 \cdot 1.4 \text{ MΩ} = 26.0234 \text{ μW}$

$P_{R3} = I_3^2 R_3 = (2.8742 \text{ μA})^2 \cdot 2.1 \text{ MΩ} = 17.3482 \text{ μW}$

As before, if we add up all the powers consumed by the three resistors we get a total that is extremely close to the power we calculated that the battery is producing.

And now for the last and most complex example. I think it is safe to say that this is the most complicated problem in this entire text. Well, the hardest one was bound to show up somewhere, and here it is!

Before we start, let me just say that the calculation that follows is a nice workout, but that is all it is. You will not see any circuits this complicated on quizzes, and you certainly won't have to calculate every voltage, current and power in the circuit all on the same problem. We are solving this circuit here because it is good practice, it is fun, and we're crazy. So take a deep breath, keep your head, and let's go!

The circuit for this example is shown in Figure 8-44. Solving this circuit involves nothing but the repeated use of the three laws we have to work with.

The R_{EQ} value from above is 0.6872 kΩ. Calculating I_1 we get

$$I_1 = \frac{V_B}{R_{EQ}} = \frac{9.0 \text{ V}}{0.6872 \text{ kΩ}} = 13.0966 \text{ mA}$$

Now moving to the right into the resistor network we ask what we are able to calculate next. We do not have enough information to use the junction law at the node above R_1. But we note that R_1 is in parallel with the battery, so

$V_1 = V_B = 9.0 \text{ V}$

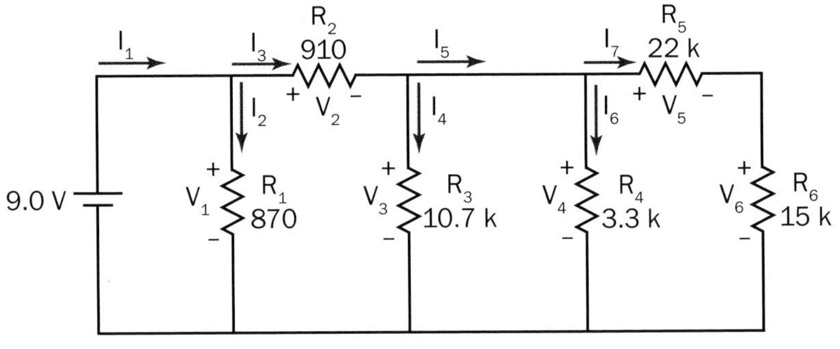

Figure 8-44. Example six-resistor circuit.

Knowing V_1 we can use Ohm's Law to calculate I_2 (making sure to write R_1 in kilohms):

$$I_2 = \frac{V_1}{R_1} = \frac{9.0 \text{ V}}{0.87 \text{ k}\Omega} = 10.3448 \text{ mA}$$

Knowing this current, we can use the junction law now at the node above R_1 to solve for I_3.

$$I_3 = I_1 - I_2 = 13.0966 \text{ mA} - 10.3448 \text{ mA} = 2.7518 \text{ mA}$$

Knowing I_3 we can now use Ohm's Law to determine the voltage drop V_2 across R_2, which is

$$V_2 = I_3 R_2 = 2.7518 \text{ mA} \cdot 0.91 \text{ k}\Omega = 2.5041 \text{ V}$$

We have completed work on R_1 and R_2. Now what? Notice that once again we do not have enough information to use the junction law on the next node (to the right of R_2). However, we can use the voltage drop law on the big loop that includes the battery, R_2 and R_3. Doing so gives us

$$9.0 \text{ V} = V_2 + V_3$$
$$V_3 = 9.0 \text{ V} - V_2 = 9.0 \text{ V} - 2.5041 \text{ V} = 6.4959 \text{ V}$$

With V_3 we now use Ohm's Law to get I_4.

$$I_4 = \frac{V_3}{R_3} = \frac{6.4959 \text{ V}}{10.7 \text{ k}\Omega} = 0.6071 \text{ mA}$$

Next, notice that R_4 is in parallel with R_3, so

$$V_4 = V_3 = 6.4959 \text{ V}$$

Now we can calculate several currents. First, use the junction law on the node above R_3 to get

$$I_5 = I_3 - I_4 = 2.7518 \text{ mA} - 0.6071 \text{ mA} = 2.1447 \text{ mA}$$

and Ohm's Law to get I_6,

$$I_6 = \frac{V_4}{R_4} = \frac{6.4959 \text{ V}}{3.3 \text{ k}\Omega} = 1.9685 \text{ mA}$$

Hang on, we're almost done! We now compute I_7 by using the junction law on the node above R_6.

$$I_7 = I_5 - I_6 = 2.1447 \text{ mA} - 1.9685 \text{ mA} = 0.1762 \text{ mA}$$

This is the last current, and we can use it with Ohm's Law to get the last two voltages:

$V_5 = I_7 R_5 = 0.1762$ mA $\cdot$ 22 kΩ = 3.8764 V

$V_6 = I_7 R_6 = 0.1762$ mA $\cdot$ 15 kΩ = 2.6430 V

Now we have solved the hardest circuit, and the hardest problem in the book. Let's quickly calculate the powers using $P = VI$ for everything.

$P_B = V_B I_1 = 9.0$ V $\cdot$ 13.0966 mA = 117.8694 mW

$P_{R1} = V_1 I_2 = 9.0$ V $\cdot$ 10.3448 mA = 93.1032 mW

$P_{R2} = V_2 I_3 = 2.5041$ V $\cdot$ 2.7518 mA = 6.8908 mW

$P_{R3} = V_3 I_4 = 6.4959$ V $\cdot$ 0.6071 mA = 3.9437 mW

$P_{R4} = V_4 I_6 = 6.4959$ V $\cdot$ 1.9685 mA = 12.7872 mW

$P_{R5} = V_5 I_7 = 3.8764$ V $\cdot$ 0.1762 mA = 0.6830 mW

$P_{R6} = V_6 I_7 = 2.6430$ V $\cdot$ 0.1762 mA = 0.4657 mW

Whoa! Adding up the resistor powers we get 117.8736 mW, which is, again, quite close to the power we calculated the battery is producing. The reason there is a bit more difference this time is because we kept rounding and rounding to get through all those resistors. Now you see why we use four decimal places! If we didn't, the rounding errors would add up so much that our result wouldn't even be that close. If you followed this example all the way through to the end you should be ready for anything!

CHAPTER VIII EXERCISES

Introductory Circuit Calculations

For this set of exercises use the basic single-resistor circuit shown in Figure 8-20. Use our ordinary significant digits rules in your answers for this set (not the special rule for circuits).

1. The current in a circuit is 13.00 A. The voltage powering the circuit is 25.00 V. Calculate the resistance in the circuit.

2. Determine the current that will flow from a 24-V battery into a 250-Ω resistance. Express your answer in mA.

3. The resistance in a circuit is 12.20 kΩ. If the supply voltage is 4.500 V, calculate the current in the circuit. Express your answer in mA.

4. Calculate the supply voltage in a circuit if 0.0300 mA of current is flowing through 33.3 MΩ of resistance.

5. Calculate the power supplied by the power supply in problems 1, 2, 3 and 4.

6. A standard light bulb is powered by a voltage of 120 V and consumes 60.00 W of power. Calculate the resistance of a standard light bulb and the current that will flow in it.

7. Household electrical appliances operate at a voltage of 120 V, and the maximum current that is allowed to flow in a household circuit is 12 A. Determine the maximum power that can be consumed by a household appliance (such as a hair dryer) and express your answer in kW.

8. A circuit in a pocket calculator draws 13.5 µA from a 6.0 V battery. Determine the power the calculator is consuming and express your answer in µW.

9. A generator at a power station produces 155 MW of power at a voltage of 762 V. Determine the current the generator is producing.

Answers
1. 1.923 Ω
2. 96 mA
3. 0.3689 mA
4. 999 V
5. a) 325.0 W b) 2.3 W c) 1.660 mW d) 30.0 mW
6. R = 240 Ω, I = 0.50 A
7. 1.4 kW
8. 81 µW
9. 203,000 A

Relationships Between Variables in Electric Circuits

For this exercise you will again refer to the basic DC circuit shown in Figure 8-20, with one change. Instead of a fixed-voltage battery, assume the voltage source is adjustable. For this circuit make two graphs. First, make a graph of I vs. V for the circuit assuming a resistance value of 3.3 kΩ. In your table of values for this graph use voltages ranging from 0 to 15 V. Compute a table of values with at least five points, and show your table of values with your graph.

Next, make a graph of the power consumed by the resistor (P_R) as a function of the current I in the resistor, or P_R vs. I. To create your data set you will need to use one of the versions of the power equation, such as $P = I^2 R$. In your table of values for this graph use currents ranging from 0 to 5 mA. Compute a table of values with at least five points, and show your table of values with your graph.

Equivalent Resistance Calculations

Determine R_{EQ} for these four networks. Use 4 decimal places in every calculation.

1.

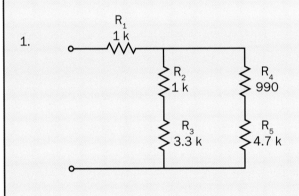

2.

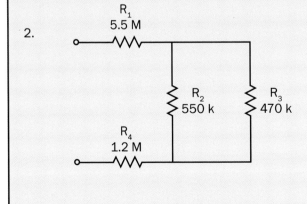

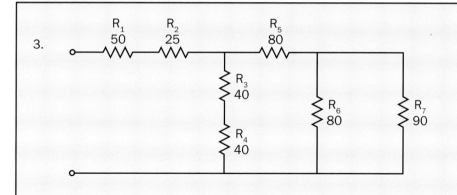

Answers
1. 3.4491 kΩ
2. 6.9534 MΩ
3. 123.3721 Ω
4. 3.2184 Ω

Multi-Resistor Circuit Calculations I

Calculate the voltage, current, or power as indicated. It is necessary to calculate R_{EQ} first for every problem except the second one. Use four decimal places.

1. Calculate the voltage across resistor R_1.

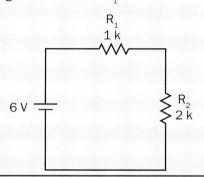

2. Calculate the current in resistor R_2.

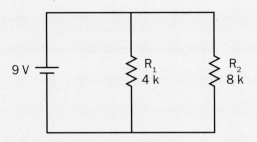

3. Calculate the voltage drop across resistor R_4.

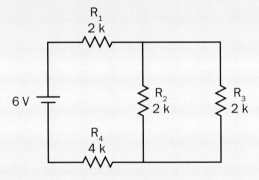

4. Calculate the power consumed by resistor R_3.

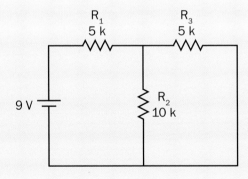

5. Calculate the the power consumed by each individual resistor and add these powers together to get the total power consumed by resistors R_1, R_2, R_3, and R_4. Then compute the power produced by the battery and see if it matches the power consumed by the resistors.

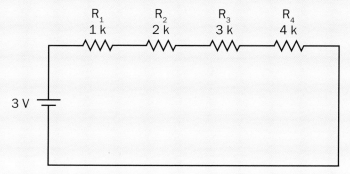

Answers
1. $V = 2.0000$ V
2. $I = 1.1250$ mA
3. $V = 3.4284$ V
4. $P = 2.5920$ mW
5. Total Power $= 0.9000$ mW

Multi-Resistor Circuit Calculations II

Use four decimal places in all of your calculations. Your answer should match the given answer in the first three decimal places.

1. Compute the voltage and current for R_2.

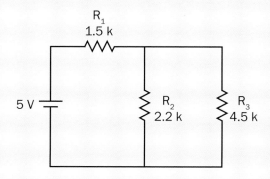

2. Compute the voltage across R_3.

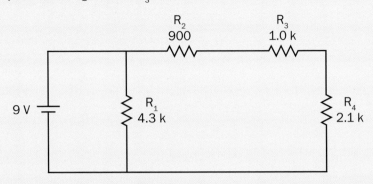

3. Compute the power consumed by R_4.

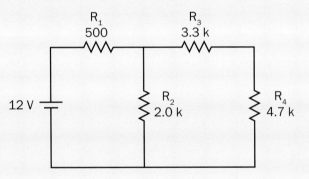

4. Compute the voltages and currents for all 6 resistors.

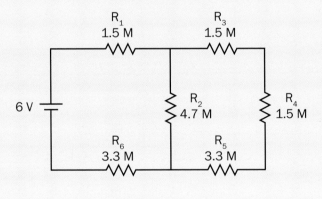

Chapter VIII — Electricity and DC Circuits

Answers
1. $R_{EQ} = 2.9776\ k\Omega$; For R_2, $V = 2.4812\ V$ and $I = 1.1278\ mA$
2. $R_{EQ} = 2.0723\ k\Omega$; For R_3, $V = 2.2500\ V$
3. $R_{EQ} = 2.1000\ k\Omega$; For R_4, $P = 6.1383\ mW$ to $6.1392\ mW$ (depending on which power equation you use)
4. $R_{EQ} = 7.4918\ M\Omega$
   R_1: $V = 1.2013\ V$ $I = 0.8009\ \mu A$
   R_2: $V = 2.1557\ V$ $I = 0.4587\ \mu A$
   R_3: $V = 0.5133\ V$ $I = 0.3422\ \mu A$
   R_4: $V = 0.5133\ V$ $I = 0.3422\ \mu A$
   R_5: $V = 1.1293\ V$ $I = 0.3422\ \mu A$
   R_6: $V = 2.6430\ V$ $I = 0.8009\ \mu A$

Multi-Resistor Circuit Calculations III

Use four decimal places in all of your calculations. Your answer should match the given answer in the first three decimal places.

1. Compute the voltage and current for R_3.

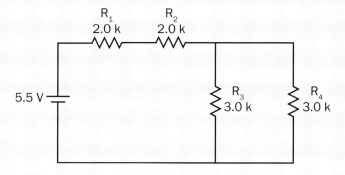

2. Compute the voltage and current for R_2.

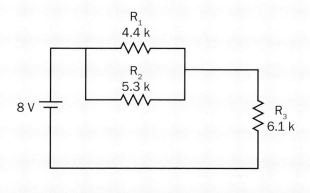

3. Compute the voltage, current, and power for R_2 and R_3.

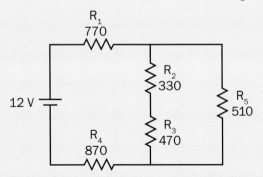

4. Compute the voltage, current, and power for R_5.

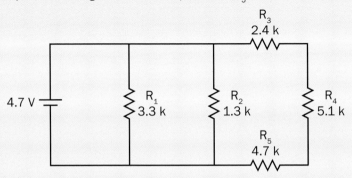

Answers
1. R_{EQ} = 5.5 kΩ; For R_3, V = 1.5 V and I = 0.5 mA
2. R_{EQ} = 8.5041 kΩ; For R_2, V = 2.2617 V and I = 0.4267 mA
3. R_{EQ} = 1.9515 kΩ
   R_2 and R_3 have the same current, I = 2.3932 mA
   For R_2, V = 0.7898 V, P = 1.8901 mW
   For R_3, V = 1.1248 V, P = 2.6919 mW
4. R_{EQ} = 0.8664 kΩ
   For R_5, V = 1.8104 V, I = 0.3852 mA, and P = 0.6974 mW

Multi-Resistor Circuit Calculations IV

Put away your calculations for Multi-Resistor Circuit Calculations II and do the entire set over again.

Multi-Resistor Circuit Calculations V

Put away your calculations for Multi-Resistor Circuit Calculations III and do the entire set over again.

CHAPTER IX
FIELDS AND MAGNETISM

> **OBJECTIVES**
>
> After studying this chapter and completing the exercises, students will be able to do each of the following tasks, using supporting terms and principles as necessary:
>
> 1. Explain what a field is.
> 2. Describe three major types of fields, the types of objects that cause each one, and the objects or phenomena that are affected by each one.
> 3. State Ampère's Law.
> 4. State Faraday's Law of Magnetic Induction.
> 5. Explain the difference between the theories of gravitational attraction of Einstein and Newton.
> 6. Apply Ampère's Law and Faraday's Law of Magnetic Induction to given physical situations to determine what will happen.
> 7. Explain the general principles behind the operation of solenoids, generators and transformers.
> 8. Use the right-hand rule to determine the direction of the magnetic field around a wire or through a solenoid.
> 9. Explain why transformers work with AC but not with DC.

TYPES OF FIELDS

My definition of a **field** consists of two parts. First, a field is a mathematical abstraction that describes a region in space that will influence specific kinds of matter or radiation. Second, this influence, the result of a field being present, is a force. Different kinds of fields affect different kinds of things, but the effect is always a force on the thing.

You have been hearing about fields all of your life: the **gravitational field** around the earth, a **magnetic field** around a magnet, and maybe even **electric fields**. The gravitational field around the earth is what pulls you toward the earth's center. The force on you is your weight. You have felt the affects of a magnetic field when holding the ends of two magnets near each other. You feel the resulting forces when the magnetic poles attract or repel each other. And you have probably seen the results of the electric field present when static electricity is around. Dry, clean hair that stands up when you brush it, synthetic clothes clinging together when removed from the dryer, a balloon that sticks to the wall after being rubbed, the leaves swinging out in the electroscope – these are all evidences of the forces present due to the electric fields caused by the build up of electric charge we call static electricity.

The diagrams below help in visualizing what is going on when a field is present. The arrows in these diagrams are called "field lines," and represent the direction of the force on an object placed in the field, assuming that the object in question is the type of object that could be affected by that type of field.

The gravitational field caused by an isolated mass is depicted in Figure 9-1, and is spherically shaped around the massive object. Remember that in both Newton's and Einstein's theories of gravity everything with mass causes a gravitational field, but gravity is such a weak force that the force isn't usually noticeable unless the object causing the field is very massive, like the earth, moon, or sun. As far as we know right now, there is only one kind of mass, so there is only one way to draw the gravitational field diagram. (Scientists are now thinking there may be more than one kind of mass. If so, and we confirm it experimentally, that would be yet another fact that changed!)

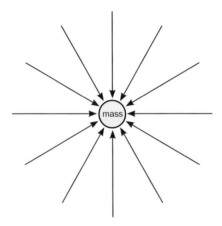

Figure 9-1. Gravitational field around a mass.

There is an important difference between the gravitational theories of Einstein and Newton. In Newton's Theory of Universal Gravitation only an object with mass can be affected by the gravitational field that exists around another object with mass. However, you will hopefully recall from Chapter II that Einstein's general theory of relativity treats gravity as a curvature in "space-time," rather than as a mysterious attraction between objects. As a result, Einstein was able to predict that starlight would bend in its path as it passed near another star such as our sun, even though light does not have mass. As we saw, this prediction was confirmed by Eddington's photographs during the solar eclipse of 1919. The fact that gravitational fields affect electromagnetic radiation is reflected in the summary table I will present below.

Unlike mass, there are two kinds of electric charge. Electric charge is what causes electric fields, depicted in Figure 9-2. In this case, our convention is that the arrows represent the direction of the force on a positive charge placed in the field. Since like charges repel, the arrows point away from the positive charge in the diagram on the left. Since opposite charges attract, the arrows point toward the negative charge on the right.

With both gravitational and electric fields, the thing causing the field, a particle of mass or charge, can exist by itself. But with the magnetic field, illustrated in Figure 9-3, this is not the case (as far as we know). Magnetic fields are caused by magnets. A magnet always has two poles, which we call north and south. If you cut a magnet in

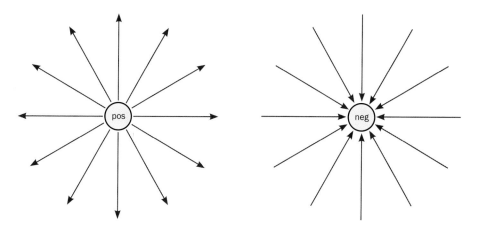

Figure 9-2. Electric fields around two kinds of charge.

half you won't have a north or south pole by itself. You will instead have two magnets, each with north and south poles. Physicists like to express this by saying, "there is no magnetic monopole." The result of this is that the field lines in a magnetic field don't just go off into space. Instead, they start on a north pole and land on a south pole of the same magnet.

When we think about these three different types of fields it is very helpful to know what can cause each one and what kinds of things can be affected by each one. Table 9-1 summarizes this information. This table is not exhaustive, but it covers the basics. There are some other strange substances affected by magnetism (liquid oxygen, for example), but those I have listed will do for us. The term *ferrous* in the table means "made of iron." This term comes from the Latin word for iron, *ferrum*. This same Latin word gives iron its chemical symbol, Fe, which we will see again later.

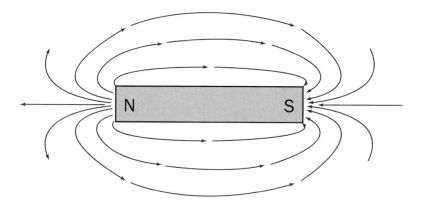

Figure 9-3. Magnetic field around a magnet.

193

	Gravitational Field	Electric Field	Magnetic Field
What can cause this type of field?	mass	charge	magnets, current-carrying wires
What can be affected by this type of field?	mass, electromagnetic radiation	charges	magnets, current-carrying wires, ferrous metals

Table 9-1. Causes of fields and what they affect.

LAWS OF MAGNETISM

There are four important magnetic laws commonly studied in physics courses. We are going to learn two of them. These two laws happen to be intimately related to the operation of the electrical devices that surround us in the modern world, so it is particularly interesting and helpful to know these two laws. They were named after their discoverers, André Ampère and Michael Faraday, whom we met in the last chapter. I will present the two laws in this section, and in the next section we will explore their important applications.

Ampère's Law states that when current flows in a wire a magnetic field is created around the wire, and the strength of the magnetic field is directly proportional to the current. Another aspect to this law relates to what happens if the current-carrying wire is wound into a coil. As I will explain below, the magnetic field around such a coil of wire is magnified over what it would be if the wire were not wound into a coil. In a coil, the magnetic field is proportional not only to the current, but also to the number of loops, or turns (as we say) in the coil.

Before discussing the next law we have to get a grip on a difficult and abstract concept, **magnetic flux**. Recall that graphically we can show the presence of fields in space by drawing field lines indicating the forces that would be present on an object placed in the field. In diagrams of this type, the closer together the field lines are, the stronger the field is. Consider Figure 9-4, which depicts a close-up of the north end of a magnet and the field lines in its vicinity. Near the magnet the field lines are closer together, and as we all know, the magnetic field is stronger near the magnet.

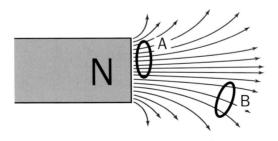

Figure 9-4. Magnetic field lines near the end of a magnet.

Imagine that we place a coil of wire near the end of this magnet at position A and just hold it there. Does it seem reasonable to say that something is passing through this coil of wire? Well, whatever it is that appears to be passing through the coil, our word for it is magnetic flux. In fact, there is nothing physically passing through the coil, and despite the suggestive sound of the word flux, nothing is flowing anywhere. What we do have passing through the coil is part of this mathematically abstract thing called the magnetic field, as indicated by the field lines. Now consider our little coil of wire being at position B instead of A. The difference, as suggested by the figure, is that there is less flux passing through the coil at position B than there is at position A. You can easily imagine that if our little coil of wire were very far away from the magnet there would be essentially no flux passing through it at all, because the magnetic field out there would be so weak.

Now we can state the next law, which is **Faraday's Law of Magnetic Induction**, or Faraday's Law, for short. This law states that if a changing magnetic flux is passing through a coil of wire a current will be induced in the coil. Recall from the previous chapter that induction comes from the word induce, which means "make it happen." Here the magnetic field interacting with this coil of wire is making a current "happen" in the coil. This principle is wonderful – we are able to create electrical current in wires without even touching them or connecting a power supply to them! Because of this principle we can generate electrical power and transport it to our homes and factories. When this became possible it utterly transformed society. As with Ampère's Law, the amount of current induced according to Faraday's Law depends on the number of turns in the coil.

It is critical for you to notice that current will only be induced in the coil if the flux passing through the coil is *changing*, that is, increasing or decreasing. Holding the coil steady at location A in the figure does nothing, even though a lot of flux is passing through the coil. But if you rotate the coil, for example, like a coin spinning on a table top, the flux passing through the coil will be highest when the coil is facing the magnet, and essentially zero when the coil is on edge to the magnet (because the flux will then be passing by the coil without passing through it). Anything else you can imagine that would cause the amount of flux passing through the coil to increase or decrease will also induce current in the coil, but only as long as the change in flux is happening. These possibilities would include moving the coil back and forth, closer to and farther away from the magnet, moving the magnet away from the coil, or even somehow making the diameter of the coil increase or decrease.

THE RIGHT-HAND RULE

Our convention for determining the direction of the magnetic field lines around a current-carrying wire is called the **right-hand rule**, illustrated in Figure 9-5. According to the rule, if you grasp the wire with your right hand, with your thumb pointing in the direction of the flowing current, the direction your fingers point as they wrap around the wire is the circular direction the magnetic field lines point around the wire.

Now, an interesting thing happens if we form this wire into a coil as shown in Figure 9-6. If you apply the right-hand rule to the wire at various places you can establish the direction of the magnetic field at different places, as I have shown in the figure. Now consider two different regions relative to this coil of wire, the region outside

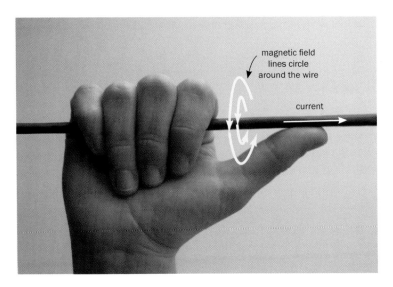

Figure 9-5. The right-hand rule.

the coil (where we are), and the cylindrical region down through the center of the coil (where the brown rod in the figure is). Notice, from the little circular arrows, that are outside the coil, both in front of it and behind it, the field lines point to the left. Inside the coil, where the brown rod is, the field always points to the right.

To get another view of this, imagine that we slice through the coil, cutting every through every loop simultaneously. Then imagine we throw away one half and look at the cut ends of the wires in the other half. In Figure 9-7 the black dots represent the cut ends of the wires. Shaded in the background you can see the remaining pieces of the coils connecting the black dots. (In this diagram I did not include the rod down the center of the coil.) Now examine the shape of the magnetic field going around and

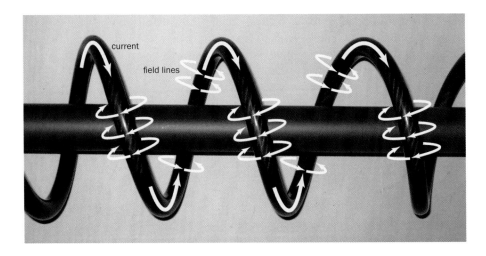

Figure 9-6. Magnetic field lines around a wire wrapped into coils.

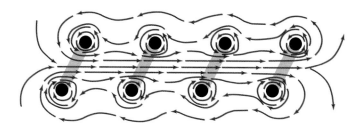

Figure 9-7. Cross-section of the coils showing the magnetic field lines.

through this coil. Down through the center of the loops the field points to the right. The field lines emerge from out of the coils on the right end and circle around the outside of the coils, heading back to the left, and enter back into the coils from the left end. For the purposes of this sketch I have greatly simplified the magnetic field lines in order to show the primary routes through the coil and back around to the other end. In actuality, if the coils are spaced apart the way they are in the sketch a lot of the field lines that could be marching together down the center of the coil will be pointlessly circulating around the individual coil wires. Nevertheless, what we have done with this coiled wire geometry is create an electric magnet, or an **electromagnet**. This electromagnet has its noth pole on the right end of the coil where the field lines come out, and its south pole on the left end where the field lines enter the coil.

SOLENOIDS, GENERATORS AND TRANSFORMERS

In the coil of wire we were just discussing, the tighter the coils are wound together, the better the electromagnet will be because no flux can leak out between the coils and circulate around the coil wires. In a tightly wound coil almost all of the flux will stay inside the coil as it travels straight down the center. (Remember, the word "travels" is a metaphor here; nothing is really moving or flowing anywhere.) This is shown in Figure 9-8, which depicts another cutaway view, this time of a long, tightly wound coil. A long, tightly wound coil of wire like this is called a **solenoid**. If a solenoid is wound onto a steel rod, called a *core*, the strength of the magnetic flux is increased considerably. So a solenoid is a powerful electromagnetic that can be easily switched on and off.

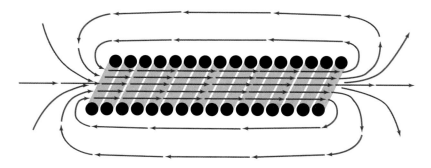

Figure 9-8. Cross-section of a solenoid showing the magnetic field lines.

By placing small moving parts made of ferrous metal near the end of the solenoid, the parts can be made magnetically to move back and forth as the current to the solenoid is turned on and off. Thus, solenoids are devices that can take an on or off electric current and use it to cause mechanical back and forth motion. It turns out that solenoids are hugely practical and hundreds of devices have been invented that use solenoids.

A common example of a solenoid consumers use every day is in the starting circuit in automobiles. When the driver turns the key to start the car, an electric motor begins spinning that turns the car engine just long enough for the ignition system to start up and begin running the car. The starter motor has a gear on it to turn the engine, but once the engine starts the starter motor with its little starter gear needs to get out of the way of the running engine or it will be torn to shreds. This is neatly managed by a solenoid that quickly moves the starter gear in one direction to engage the engine while starting the car and moves in the opposite direction when the key is released to get the starter gear out of the way of the car engine that is now running on its own.

Solenoids are especially useful in the world of industrial electric circuits, because they can function as electrically operated switches in many different applications. These switches are called *relays*, and are used extensively in industrial control systems. A small control relay is shown in Figure 9-9. In the close-up on the right I have indicated the coil (covered in white tape), the solenoid core, the moving arm that is pulled back and forth as the coil is energized or de-energized, and the electrical contacts at the bottom of the moving arm that are opened and closed as the relay switches on and off.

Solenoids are little gadgets that a lot of people don't even know about. Electric **generators**, by contrast, are massive devices that everyone knows about. Everyone knows that power stations are located all over the country generating electrical power to run the electrical devices in houses, factories, offices, and industrial plants. What

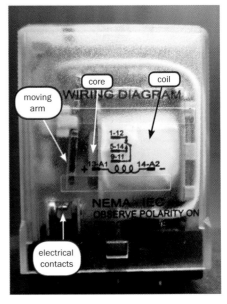

Figure 9-9. A relay (left) and a close-up showing the major internal parts.

you should know is that Faraday's Law of Magnetic Induction is the simple principle that allows an electric generator to generate hundreds of megawatts of electrical power.

Before I explain how generators work I will just mention that electric **motors** are basically generators running backwards. Whereas a generator uses mechanical power from an engine to make electric current, a motor uses electric current to run a mechanical machine. Since these two devices are so similar, they are basically the same on the inside. I will describe how a generator works, but you can keep in mind that an electric motor is essentially the same thing.

Consider Figure 9-10, which depicts the poles of a C-shaped magnet with a coil of wire in between them. There is a vertical axle attached to the coil of wire which allows it to rotate. This axle must be connected to an engine that can make the coil rotate. Because the north and south poles of the magnet are close together and facing each other on each side of the coil, there will be a strong magnetic flux passing from the north pole to the south pole, right through the space where the coil is mounted. In the figure I have shown the field lines pointing from the north end of the magnet to the south end of the magnet. Faraday's Law of Induction says that when the magnetic flux passing

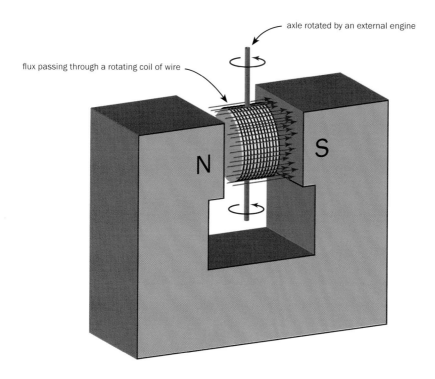

Figure 9-10. Rotating coil and magnetic poles inside a basic generator.

through a coil of wire is changing a current will be induced in the coil of wire. Also, as mentioned above, the amount of current will depend on the number of turns of wire there are in the coil, so the coil I have drawn in the figure has lots of turns. The position in which the coil is shown in the figure is the position where a maximum amount of flux will pass through the coil. As the coil rotates from here, the flux passing through it will decrease from this maximum value. The flux will fall to zero, rise to maximum again but in the reverse direction (because the coil has rotated half way around), fall to zero again, and so on.

I have sketched what this rising and falling flux looks like in Figure 9-11. This curve has what we call a "sinusoidal" shape, and this sinusoidal variation of the flux through the coil produces a sinusoidally shaped current in the coil, which is exactly the shape of the voltage and current curves for the AC power distribution system. The curves look exactly like the wave curves we explored back in Chapter VII, and they can be examined with the same mathematics. In America the frequency of the AC curves is 60 Hz. To produce this frequency in the oscillating flux, the engine must turn the coil so that it completes 60 rotations per second. In Europe the frequency is 50 Hz. Since the frequency in America is 60 Hz (which I will write as 60.00 Hz), this means the period is

$$\tau = \frac{1}{f} = \frac{1}{60.00 \text{ Hz}} = 0.01667 \text{ s}$$

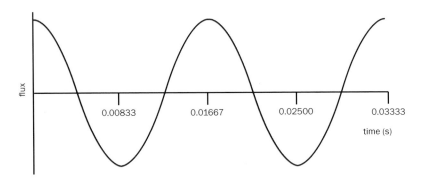

Figure 9-11. Flux through the rotating coil vs. time, at a 60 Hz rotation rate.

I have shown the time values on the diagram for the changing flux to complete one half cycle, one cycle, one and one half cycles, and two cycles.

In my sketch in Figure 9-10 I drew the magnet as if it were an ordinary permanent magnet. In practice, this would work for a tiny demonstration generator, but using a permanent magnet large enough for an industrial generator would be very impractical. Instead, large coils of wire operating according to Ampère's Law can be used to create the magnetic field for the generator. These coils are called *field coils*. Part of the generator's own output is used to power these coils. Figure 9-12 is a photograph of a device that is designed in just this way. The photo is actually of a small motor, not of a generator, but as I said earlier, a motor is just a generator running backwards. Both devices depend on coils of wire to produce the stationary magnetic flux, and have rotating coils on a shaft inside the magnetic field. The entire winding system that is fixed in place (the field coils) is called the *stator*. All of the rotating coils and the frames they are wound on is called the *rotor*. In the photograph you can see that on the rotor there

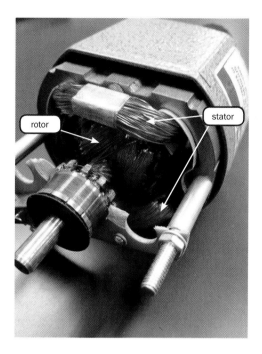

Figure 9-12. Inside look at a small motor showing the stator and rotor coils.

is more than just a single coil of wire. In fact, the motor shown has six coils, which is pretty common for a small motor like this one.

Before we move on, aren't you curious how the electric current gets in or out of the rotor while the rotor is spinning? I assumed you would be, so I took one more photo to show you a common way this is done. In Figure 9-13 you can see a close-up of a round device on the end of the rotor shaft called the *commutator*. Pressing against the commutator is a chunk of carbon called the brush, or since there are usually more than one of them, the *brushes*. The wiring from outside the motor attaches to the brushes and a spring inside the brush holder keeps the brushes pressed against the commutator. The copper segments on the commutator are connected to the different coils on the rotor, so as the rotor turns different coils are connected one after the other through the brushes to the outside and are interacting each in turn with the magnetic field. Of course, the carbon bushes wear down because they are constantly pressing against the spinning commutator, so the brushes eventually have to be replaced. If you own a car long enough you may eventually have to replace the brushes inside the starter motor, as I once did.

There is one more type of device that is ubiquitous in our power distribution system that you should know about and understand. These devices are called **transformers**, and they have nothing whatsoever to do with the popular movies. But before I tell you about them I want to describe the background that makes them necessary. The example that follows contains a lot of math, but we will not do any calculations with transformers. I am including the computations in the example to help you understand the problem that transformers solve for us. So try to follow the calculations, which are based on the power equations from the previous chapter, but don't panic.

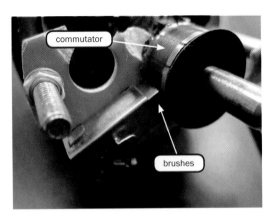

Figure 9-13. The brushes and commutator allow the current to get in or out of the spinning rotor coils.

Using a very simple electric circuit model for getting power from the power station to a nearby city, let's take a quick look at how this system will work in practice. In the circuits we studied in Chapter VIII we neglected the voltage drops in the wires. But in a large distribution system the long wires will have some significant resistance, and thus there will be a voltage drop in the wires. This voltage drop will mean that power is lost from the system due to heating in the wires caused by the wires' own resistance.

Let's assume we have a typical, power generator at a power station that can produce 100 MW of power at 700 V. The current coming out of the generator will be $I = P/V =$ 100,000,000/700, which is about 143,000 A. For reasons we will not go into, the real current in this situation would be smaller than this by a factor of $\sqrt{3}$ (about 1.73), so we would need to transport 82,500 A of current from the power station to the city.

Our first very practical problem is that conductors large enough to carry this much current without simply melting would be as big around as tree trunks. Not exactly ideal for transporting the electricity 10 miles to a city. But let's say we solve that problem by dividing up the current into smaller portions and using multiple transmission lines to carry the power. Each smaller circuit will carry only 300 A, so we will have 275 of these! The resistance of such wire might be around 0.05 Ω per 1,000 ft. For 10 miles, that is 52,800 ft, which is about 2.6 Ω of resistance for just *one* wire going *one* direction. Computing the power lost in this wire we will use $P = I^2R$, which gives 234 kW of lost power for each circuit. For 275 circuits, that's a total of 64 MW of lost power. This is almost 2/3 of the total power we had to start with! If we include the other wiring necessary to allow the current to flow back to the power station, there is no usable power left at all. Clearly, transporting lots of power at 700 V is not practical.

The problem is that the power lost in the wiring varies as the *square* of the current. High currents mean catastrophic power losses. Here's the solution: Since power is $P = VI$, we will reduce the current, and thus the power losses, if we raise the voltage. Let's say we raise the voltage to, say, 200,000 V, which is quite realistic for those cross-country power lines you see up on the big steel trestles. This would mean that 100 MW of power would only require 100,000,000/200,000, or 500 A of current. Again, the actual current will be $\sqrt{3}$ less than this, or 289 A. If we divide this into only 10 circuits, that's 29 A apiece. Total power lost in each circuit is now 2,200 W, or 22,000 W for all 10 circuits. This may still sound like a lot, but it's not really. A dozen or so homes doing the laundry use this much power. If we transport 100 MW of power, we lose 22,000 W due to wire resistance, or 0.022%. Voila! All we have to do is have a device that can easily raise the voltage from 700 V at the power station up to 200 kV for transportation. Then at the city, we use a similar device to drop the voltage back down to a safe 240 V, which is the voltage that comes to our homes.

This is where Ampère's and Faraday's Laws comes in with a nifty invention called the **transformer**. A transformer is a very simple device that uses these laws to raise or lower an AC voltage while simultaneously lowering or raising the current. In Figure 9-14 we have a square ring, called the core, made of steel plates. Coils are wound around the core on two sides. The coil where the electric current comes in is called the primary coil. Electric current leaves the transformer at the secondary coil. Here is the beautiful part: The ratio of the primary voltage to the secondary voltage will be the same as the "turns ratio" of the transformer. The turns ratio is simply the ratio of the number of turns in the primary coil to the number of turns in the secondary coil. (Another beautiful thing about transformers is that they have no moving parts, so they almost never wear out!)

Using the turns ratio rule in our example, if we want to boost the voltage from 700 V to 200,000 V, that is a ratio of 1:286. Let's say we want 10 turns of wire in the primary coil. This means we need 2,860 turns in the secondary coil. That's no problem. When our power arrives at the outskirts of the city we will drop the voltage from 200 kV to say, 50 kV using another transformer, only this time the primary coil will have the larger number of turns. A ratio of 200:50 is 4:1, so if we want 100 turns in the secondary, we will need 400 in the primary. Electrical power is distributed around cities at various voltages, but they tend to be in the range of 50 kV. Then when it gets to a particular neighborhood the voltage is dropped again down to the 10-15 kV range. Finally, at your house it dropped again by one last transformer down to 240 V.

As I said previously, you don't need to be able to do this math in ASPC. I presented these sample calculations merely to show how important transformers are to our power distribution system, which is essential for the conveniences we enjoy.

Now let's take a look at the magnetic principles that make transformers work. Referring again to Figure 9-14, we apply an AC voltage to the primary coil. An AC voltage is sinusoidal, just like the sinusoidal flux we saw in Figure 9-11. A sinusoidal voltage like this will cause a sinusoidal AC current to flow in the primary coil. According to Ampère's Law, a magnetic flux will be created inside the coil, just as with the solenoids we studied before. Here, however, the flux is not steady. Since the current in the primary coil is sinusoidal, the flux in the transformer core will also be sinusoidal, reaching a peak, decreasing to zero, reversing direction to a negative peak, back to zero, over and over. The flux in the core will be oscillating at the same frequency as the current in the primary coil. Now, magnetic flux "flows" very easily inside of iron or steel, like our transformer core. So the flux created by the primary coil goes around the core, passing through the secondary coil. In the figure I have drawn lines in the transformer core representing the magnetic field in the core. I put arrows on the field lines to indicate that the field is oscillating back and forth, clockwise then counter-clockwise, changing direction 60 times per second. So inside the transformer core the flux is continuously changing. This means that in the secondary coil Faraday's Law of Magnetic Induction

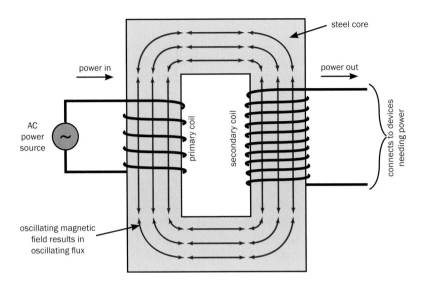

Figure 9-14. Primary and secondary coils in a transformer.

kicks in. Since a changing magnetic flux is passing through the secondary coil, a current will be induced in the secondary coil. This secondary current, which is also sinusoidal, flows out of the transformer to whatever electrical devices are connected to it.

Notice what would happen if we connected a battery (DC voltage) instead of an AC voltage source to the transformer primary. Ampère's Law would still operate in the primary coil, and magnetic flux would still be created in the primary coil and in the transformer core. But the flux would be static or steady, that is, not changing. Thus, no current would be induced in the secondary coil at all, because Faraday's Law requires the flux through a coil to be changing to induce current in the coil. Transformers do not work with DC currents and voltages.

You probably have no idea how many transformers you use and depend on every day. There is a transformer at your house that lowers the power distribution voltage from 10 or 15 kV down to the 240 V that goes to your house. In neighborhoods where the power lines are underground, these house transformers are often inside green metal boxes in the yards of the houses. If the power lines are above ground on poles, the house transformers are on the power poles in gray canisters. Nearly every piece of electronic equipment you use has a transformer in it. For portable devices such as a mobile phone or laptop computer, the device runs on a DC battery. The transformer is in the charger that recharges the battery. (After the transformer reduces the voltage inside a charger down to a level the charger can use, there is another gadget in there that converts the small AC voltage from the transformer into DC for charging the battery, but we won't be talking about that.)

Figure 9-15 shows a couple of small transformers of the type used in electronic devices such as a stereo system or computer. In the close-up on the right, if you look carefully at the slots in the two white plastic pieces you can see the windings of the two coils in the transformer. Wires are attached to the coils in various places and connected

Chapter IX   Fields and Magnetism

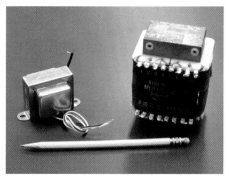

Figure 9-15. Small transformers.

to the silver-colored terminals you see, and from there they would be connected to the rest of the electronics.

We are surrounded all the time by motors, generators, transformers, and relays, so I went into a fair amount of detail in this chapter describing how all of these electrical devices work. One of the things I enjoyed about getting my undergraduate degree in electrical engineering was that I then understood *how* they all work. If this kind of stuff stimulates your curiosity, then maybe you should consider getting an engineering degree, too. It when I was 15 years old in ninth grade, perhaps just like you, when I decided to get an engineering degree myself!

CHAPTER IX EXERCISES

Fields Study Questions

1. What kinds of things cause a gravitational field?

    a. Does the Holy Spirit have a gravitational field around Him?
    b. Does the archangel Gabriel have a gravitational field around him?
    c. Does an atom of nitrogen create a gravitational field?
    d. Does a neutron in an atom of nitrogen create a gravitational field?

2. What kinds of things are affected by a gravitational field?

    a. Is heat affected by a gravitational field?
    b. Is your soul affected by gravity?
    c. Is the broadcast signal from your favorite radio station affected by gravity?
    d. Is Beethoven's Ninth Symphony affected by gravity?
    e. When you are listening to Beethoven's 9th Symphony, is the sound wave affected by gravity?

3. Describe circumstances under which your body would and would not be affected by an electric field.

4. Which of these will create an electric field?

    a. An isolated electron.
    b. An ordinary atom.
    c. A Van de Graaff generator.
    d. A neutron.
    e. A basketball.
    f. Beethoven's Ninth Symphony.
    g. A plasma. (A plasma is a gas made of ions, which are atoms that do not have equal numbers of protons and electrons.)

5. Which of these would be affected by a magnetic field?

    a. A copper wire with no current flowing in it.
    b. A copper wire with current flowing in it.
    c. A block of cast iron.
    d. An aluminum window frame.
    e. A banana.
    f. The stainless steel screws used to repair a person's bones after a bad injury.
    g. A compass.
    h. A compass in a vacuum.
    i. A compass under water.

j. The compass in Jonah's pocket after he was swallowed by the fish.
k. The steel girders in our classroom building.

Magnetism Study Questions

1. Does the current in a solenoid have to be changing in order for a magnetic field to be produced? Why or why not?

2. The figure below shows a solenoid with a current flowing in it due to the battery. Use the right-hand rule to determine the direction of the magnetic field inside of the solenoid. Using this information, identify which end of the solenoid (the left or right) will be the magnetic north pole of this electromagnet.

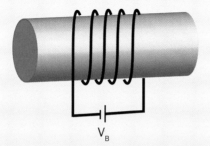

3. The figure below depicts a magnetic field produced by a solenoid and a coil of wire placed in the field. Using Faraday's Law, briefly but clearly describe what will be going on in the coil in each of the following circumstances:

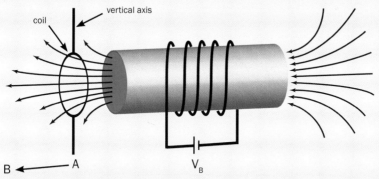

   a. The coil is moving rapidly to the left from point A to point B.
   b. The coil is stationary as shown, but its diameter is rapidly getting smaller and smaller.
   c. The coil and the solenoid are both stationary, but the current in the solenoid is increasing.
   d. The coil is rotating in place about a vertical axis, like a coin spinning on

a table top.

4. In the preceding problem, what difference would it make if the coil were made with three turns (loops) of wire instead of one?

5. Use the laws of magnetism to write a few sentences explaining how a generator works.

6. Use Ampere's and Faraday's Laws to write a few sentences explaining how a power transformer works.

7. Explain why a transformer will work with AC current in the primary coil, but not with DC current.

CHAPTER X
SUBSTANCES

OBJECTIVES

Memorize and learn how to use this equation:

$$\text{solubility} = \frac{\text{mass of solute (g) at saturation}}{\text{volume of solvent (mL)}}$$

After studying this chapter and completing the exercises, students will be able to do each of the following tasks, using supporting terms and principles as necessary:

1. Define and distinguish between these terms pertaining to substances: substance, pure substance, alloy, mixture, compound, heterogeneous mixture, homogeneous mixture, suspension, element and solution.
2. Make a tree chart that relates together in their proper hierarchy the terms in the previous item. In each category of the chart list two or more examples of representative common substances.
3. Use and define the following terms: solution, solute, solvent, solubility, soluble, saturated solution, supersaturated solution and precipitate.
4. Define and perform calculations pertaining to solubility.
5. Define and distinguish between these additional terms: chemical symbol, chemical formula and atom.
6. Given either the name or the chemical symbol of any of the elements listed in Table 10-1, give the corresponding symbol or name. *It is very important that you memorize this list well.*
7. Identify these basic parts of the Periodic Table of the Elements: periods, groups, alkali metals, alkaline earth metals, noble gases, halogens, transition metals, metalloids, inner transition metals (also called rare earth elements) and nonmetals.
8. Define and distinguish between physical properties, chemical properties, physical changes, and chemical changes. Give several examples of each.
9. Define and give examples of the terms malleable and ductile.
10. Write a paragraph explaining the relation between atoms, elements, molecules, chemical reactions and compounds.

The following two objectives are treated in Chapter II and in Appendix C in the experiment on Solubility:

11. Use the terms meniscus and parallax error when discussing experimental error.
12. Describe how accuracy and precision relate to measurements made with a triple-beam balance or graduated cylinder.

With this chapter we begin the study of chemistry that will occupy us for the remainder of the course. Chemistry is basically the study of how the electrons in atoms of substances interact with one another to form other substances.

REVIEW OF SOME BASICS

We will be talking a lot about **atoms** in our study of chemistry, so I will summarize here the basics about atoms. Much of this information you probably already know.

All matter is made of atoms. Atoms are mostly empty space. Each atom has a tiny nucleus in the center which contains all of the atom's **protons** and **neutrons**. Since the protons and the neutrons are in the nucleus, they are collectively called **nucleons**. The masses of protons and neutrons are very nearly the same, although the neutron mass is very slightly greater. Each proton and neutron has nearly 2,000 times the mass of an electron, so the nucleus of an atom contains practically all of the atom's mass. Outside the nucleus is a weird sort of cloud surrounding the nucleus containing the atom's **electrons**. Sometimes people have modeled electrons as orbiting the nucleus like planets orbiting the sun, but we have known for a century that this is not at all an accurate description of what is going on with electrons. It is actually very hard to say what is *really* going on, but we are not really going to get into that in this course.

Neutrons have no electric charge. Protons and electrons each contain exactly the same amount of charge, but the charge on protons is positive and the charge on electrons is negative. If an atom has no net electric charge it must contain equal numbers of protons and electrons.

Another topic I am sure you have studied before is the phases of matter. The three basic phases of a substance are solid, liquid and gas or vapor. If a substance normally exists in nature as a gas, we call it a gas. Oxygen and nitrogen are examples of gases. But if a substance is normally a liquid, then if we boil some we call it a vapor. Steam is water vapor.

This is all you need to know going in. Prepare to learn a lot more.

TYPES OF SUBSTANCES

A **substance** is anything that contains matter. There are many different types of substances, but they all fall into two categories, **pure substances** and **mixtures**. I am going to describe pure substances first, but it will be helpful to know that a **mixture** is made any time different substances are combined together without a chemical reaction occurring. If a chemical reaction does occur when different substances are combined, a pure substance called a **compound** is formed. Thus, these are the two major types of substances that are composed of combinations other substances, compounds and mixtures. The major distinction between mixtures and compounds is that if no chemical reaction occurs when substances are combined, the resulting substance is a mixture. If a chemical reaction does occur when substances are combined, a compound (or perhaps more than one compound) is formed. We will come back to compounds in a moment.

There are two kinds of **pure substances**, one of which is compounds. The other kind of pure substance is a group of substances called **elements**. We will discuss elements first.

In previous science classes you may have seen or studied the Periodic Table of the Elements (PTE), which lists all of the known elements. This famous table will play a major role in our work for the rest of this course, and we will begin studying it in more detail in the next chapter. The Periodic Table is shown in Figure 10-1. There is also an image of the PTE inside the back cover of this text.

The factor that defines each element in the PTE is the number of protons the element has in each of its atoms. For example, carbon is element number six in the PTE. This means that an atom of carbon has six protons. All carbon atoms have six protons. If an atom does not have six protons, it is not a carbon atom, and if an atom does have six protons, it is a carbon atom. An element is therefore a type of atom, classified according to the number of protons the atom has. Elemental carbon is therefore any lump of atoms that contain only six protons apiece. Oxygen is another example of an element. Pure oxygen is a gas (ordinarily) that contains only atoms with eight protons each, because oxygen is element number eight. Other examples of elements you have heard of are iron, gold, silver, neon, copper, nitrogen, lead, and many others.

There are several regions and groups of elements in the PTE that have special names that you need to know. At the most general level, the elements can be divided into **metals**, **nonmetals**, and **metalloids**, as shown in Figure 10-2. As you can see from this figure, the large majority of elements are metals. The term *metalloid* is probably new to you, but if you are very familiar with the inside of a computer you have probably heard of semiconductors. Semiconductors, as their name implies, conduct electricity under certain conditions, but not others, and are at the heart of all contemporary electronic devices. The metalloids are the elements that are most commonly used in the manufacturing of semiconductor materials. You do not need to memorize which elements are metalloids, but you do need to know where these three general categories of elements are in the PTE. The reason elements 113-118 are not colored in is that the properties of these elements are not yet known well enough to classify them. However, if the patterns in the PTE hold true, we would expect elements 113-115 to act like metals and 117-118 to act like nonmetals. Element 116 or 117 or perhaps both might act like metalloids.

There are also a number of smaller clusters of elements with specific names. The 18 columns in the PTE are called **groups**. The seven rows are called **periods**. Figure 10-3 shows the specific names of several more clusters of elements.

As you can see, there are special names for elements in several of the groups. Group 1 elements are called the **alkali metals**. Group 2 elements are called the **alkaline earth metals**. Group 17 elements are called the **halogens**. You may have heard of halogen lamps being used for car headlights or other types of lighting. They are called this because there is a small amount of halogen gas like bromine (element number 35) or iodine (element number 53) in these lamps, along with more gas from an element in Group 18. Group 18 elements are called the **noble gases**. The remarkable characteristic about the noble gases is that when it comes to chemical reactions, these elements are very, shall we say, *reserved*. That is, they don't react much at all with other elements, at least not compared to how the other elements behave. In fact, formerly we thought that the noble gases didn't react at all, so they were called the "inert gases" for a while. Now we know that they can form compounds, but for practical purposes you may consider

Figure 10-1 (next page). The Periodic Table of the Elements.

1																	18
1 H Hydrogen 1.0079	2											13	14	15	16	17	2 He Helium 4.003
3 Li Lithium 6.94	4 Be Beryllium 9.012											5 B Boron 10.811	6 C Carbon 12.011	7 N Nitrogen 14.007	8 O Oxygen 15.999	9 F Fluorine 18.998	10 Ne Neon 20.17
11 Na Sodium 22.990	12 Mg Magnesium 24.305	3	4	5	6	7	8	9	10	11	12	13 Al Aluminum 26.982	14 Si Silicon 28.086	15 P Phosphorus 30.974	16 S Sulfur 32.066	17 Cl Chlorine 35.453	18 Ar Argon 39.948
19 K Potassium 39.098	20 Ca Calcium 40.078	21 Sc Scandium 44.956	22 Ti Titanium 47.88	23 V Vanadium 50.942	24 Cr Chromium 51.996	25 Mn Manganese 54.938	26 Fe Iron 55.847	27 Co Cobalt 58.933	28 Ni Nickel 58.71	29 Cu Copper 63.546	30 Zn Zinc 65.38	31 Ga Gallium 69.723	32 Ge Germanium 72.59	33 As Arsenic 74.922	34 Se Selenium 78.96	35 Br Bromine 79.904	36 Kr Krypton 83.80
37 Rb Rubidium 85.467	38 Sr Strontium 87.62	39 Y Yttrium 88.906	40 Zr Zirconium 91.224	41 Nb Niobium 92.906	42 Mo Molybdenum 95.94	43 Tc Technetium 98.906	44 Ru Ruthenium 101.07	45 Rh Rhodium 102.906	46 Pd Palladium 106.42	47 Ag Silver 107.868	48 Cd Cadmium 112.411	49 In Indium 114.82	50 Sn Tin 118.69	51 Sb Antimony 121.75	52 Te Tellurium 127.60	53 I Iodine 126.905	54 Xe Xenon 131.30
55 Cs Cesium 132.905	56 Ba Barium 137.327	71 Lu Lutetium 174.967	72 Hf Hafnium 178.49	73 Ta Tantalum 180.947	74 W Tungsten 183.85	75 Re Rhenium 186.207	76 Os Osmium 190.23	77 Ir Iridium 192.22	78 Pt Platinum 195.09	79 Au Gold 196.967	80 Hg Mercury 200.59	81 Tl Thallium 204.37	82 Pb Lead 207.2	83 Bi Bismuth 208.980	84 Po Polonium 209	85 At Astatine 210	86 Rn Radon 222
87 Fr Francium 223	88 Ra Radium 226.025	103 Lr Lawrencium 260	104 Rf Rutherfordium 261	105 Db Dubnium 262	106 Sg Seaborgium 263	107 Bh Bohrium 262	108 Hs Hassium 265	109 Mt Meitnerium 266	110 Ds Darmstadtium 281	111 Rg Roentgenium 281	112 Cn Copernicium 285	113 Uut Ununtrium 286	114 Uuq Ununquadium 289	115 Uup Ununpentium 289	116 Uuh Ununhexium 293	117 Uus Ununseptium 294	118 Uuo Ununoctium 294

57 La Lanthanum 138.906	58 Ce Cerium 140.115	59 Pr Praseodymium 140.908	60 Nd Neodymium 144.24	61 Pm Promethium 145	62 Sm Samarium 150.36	63 Eu Europium 151.964	64 Gd Gadolinium 157.25	65 Tb Terbium 158.925	66 Dy Dysprosium 162.50	67 Ho Holmium 164.930	68 Er Erbium 167.26	69 Tm Thulium 168.934	70 Yb Ytterbium 173.04
89 Ac Actinium 227.028	90 Th Thorium 232.038	91 Pa Protactinium 231.036	92 U Uranium 238.029	93 Np Neptunium 237.048	94 Pu Plutonium 244	95 Am Americium 243	96 Cm Curium 247	97 Bk Berkelium 247	98 Cf Californium 251	99 Es Einsteinium 254	100 Fm Fermium 257	101 Md Mendelevium 258	102 No Nobelium 259

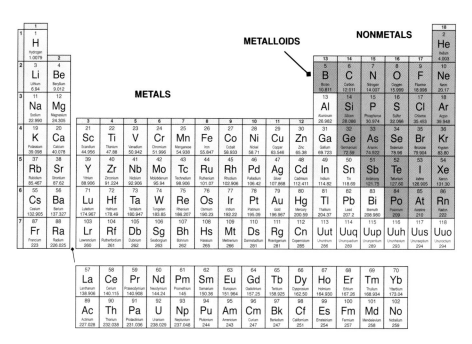

Figure 10-2. Major divisions of elements.

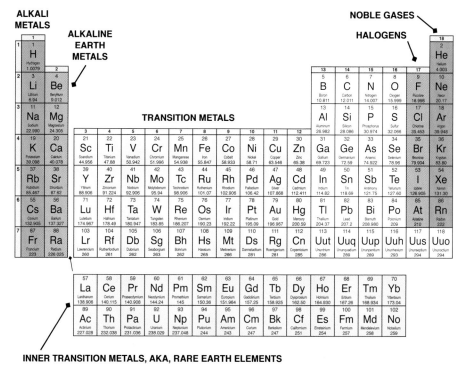

Figure 10-3. Special clusters of elements.

213

them as essentially non-reactive. There are one or two other group names, but you can learn those in a future course if you need to know them.

There is also a big cluster of metals in the middle called the **transition metals**, and two rows of elements down at the bottom called the **inner transition metals**, or **rare earth elements**. Did you notice the gap in the PTE between Groups 2 and 3? That gap is where the inner transition metals go. But if we put them in there, the PTE becomes so long that it is inconvenient to use or publish in charts or in books. So these elements are always pulled out of the main part of the PTE and shown underneath it. You will not need to know anything about the inner transition metals for this course, but you might be interested to learn that a lot of them are radioactive, such as uranium (element number 92) and plutonium (element number 94).

For every element there is a **chemical symbol** which is used in the PTE and in the chemical formulas for compounds (which we will discuss next). For some elements a single upper case letter is used, such as N for nitrogen and C for carbon. For other elements an upper case letter is followed by one lower case letter, such as Na for sodium and Mg for magnesium. (The three-letter symbols at the right side of Period 7 beginning with U are just placeholders until official names and two-letter symbols are selected by the appropriate governing officials.)

Some of the chemical symbols are based on the Latin names, such as Ag for silver, which stands for *argentum*, and Pb for lead, which stands for *plumbum*. There is no need for you to memorize the Periodic Table. That would truly be torture for most us. However, there are many elements that one encounters so frequently in scientific study that you should memorize the chemical symbols for them. I have listed the 30 elements in Table 10-1 that you need to know for this course. For each one, you need to be able to give the symbol from the name and vice versa. Spelling matters, so be sure to notice the unusual spellings in elements like sulfur and fluorine.

Name	Symbol	Name	Symbol	Name	Symbol
Aluminum	Al	Hydrogen	H	Oxygen	O
Bromine	Br	Iodine	I	Phosphorus	P
Calcium	Ca	Iron	Fe	Potassium	K
Carbon	C	Lead	Pb	Silicon	Si
Chlorine	Cl	Lithium	Li	Silver	Ag
Chromium	Cr	Magnesium	Mg	Sodium	Na
Copper	Cu	Mercury	Hg	Sulfur	S
Fluorine	F	Neon	Ne	Tin	Sn
Gold	Au	Nickel	Ni	Uranium	U
Helium	He	Nitrogen	N	Zinc	Zn

Table 10-1. Names and chemical symbols for 30 common elements.

Now that you know what elements are we can describe **compounds** in greater detail. As the name implies, a compound is formed when two or more elements are chemically bonded together, which is always the result of a **chemical reaction**. A chemical reaction is any process in which connecting bonds between atoms are formed or broken. It takes a chemical reaction to bond atoms together, and it takes a different chemical reaction to break them apart.

There are several different types of chemical bonds, which we will study in a later chapter. But for now the point to pick up on is that when chemical bonds between the atoms of different elements form, the result is a compound, a type of pure substance. A compound may consist of only two elements, as in the case of water, which is composed of hydrogen and oxygen, or carbon dioxide, which is composed of carbon and oxygen. All compounds can be represented by a **chemical formula**, which specifies the elements that are in the compound, and in what ratios. For example, the chemical formulas for water and carbon dioxide are written as H_2O and CO_2, respectively. Other compounds are composed of several elements, such as acetic acid, which is denoted by the formula $C_2H_4O_2$, and which consists of hydrogen, carbon, and oxygen. Another example is ammonium phosphate, which is denoted by the formula $(NH_4)_3PO_4$ and is composed of nitrogen, hydrogen, phosphorus, and oxygen.

When atoms bond together to form a compound the atoms in the compound can be arranged in two different basic types of structures. In many cases the atoms join together in small, individual, identical groups called **molecules**. Water is composed of molecules, each molecule consisting of two hydrogen atoms and one oxygen atom, as depicted in Figure 10-4. You should notice the characteristic elbow shape of a water molecule. This shape is responsible for many of water's unusual properties. Other examples of compounds composed of molecules are ammonia (NH_3), sulfur dioxide (SO_2), and methane (CH_4).

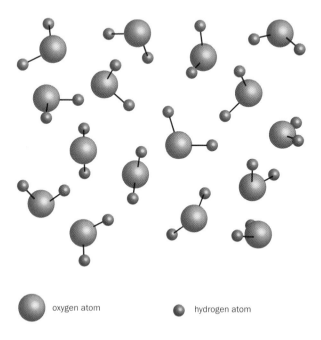

Figure 10-4. Water molecules.

The other common way atoms can combine is to form together in a continuous, geometric arrangement. These compounds are called **crystals**, and the structure the atoms in the compound make when they join together is called a **crystal lattice**. The number of different arrangements atoms can make in a lattice is endless, and these arrangements are responsible for many of the unusual properties crystals possess. But what all lattices have in common is the regular arrangement of the atoms into repeating, geometrical patterns. A sketch of the very simple crystal structure for sodium chloride (NaCl, that is, table salt) is shown in Figure 10-5. I added some outlines to the crystal to help you visualize this 3-D depiction.

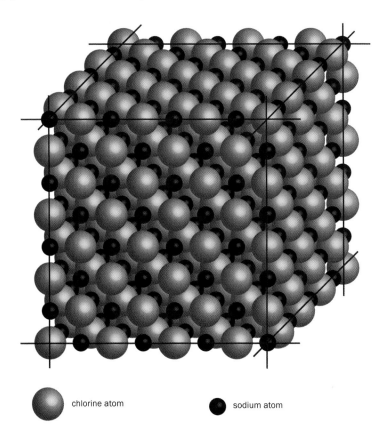

Figure 10-5. Crystal lattice for sodium chloride.

One last point to note about compounds is that the properties of a compound are completely different from the properties of any of the elements in the compound. Consider water and oxygen. Oxygen is an invisible gas that you breathe in the air and that supports combustion. Hydrogen is an invisible, flammable gas. Water is composed of oxygen atoms bonded to hydrogen atoms, but one cannot breathe water and water does not combust nor support combustion. Or consider sodium chloride, which is the table salt you commonly apply to your food. We all have to have salt in our diets, and we find it tasty. But both sodium and chlorine, the two elements of which sodium chloride is composed, are deadly dangerous in their pure, elemental forms.

So far we have been discussing one major category of substances, pure substances, which, again, consists of elements and compounds. The other major category is **mixtures**. Any time substances are mixed together without a chemical reaction occurring a mixture is formed. (Remember – if a chemical reaction occurs compounds are formed, not mixtures.) If you toss vegetables in a salad, you've made a mixture. If you put sugar in your tea or milk in your coffee, you've made a mixture. If you mix up a batch of chocolate chip cookie dough, a bowl of party mix, or the batter for a vanilla cake, you've made mixtures.

There are two basic kinds of mixtures, **heterogeneous mixtures** and **homogeneous mixtures**. The simple difference between these is that in a heterogeneous mixture you can visually tell that there is more than one substance present in the mixture; you can see the different substances in the mixture with your eyes. In a homogeneous mixture the substances are mixed together so finely that you cannot see them and the substance appears uniform. Of the example mixtures in the previous paragraph, salads, chocolate chip cookie dough, and party mix are all heterogeneous mixtures. Sweetened tea, coffee with milk, and vanilla cake batter are all homogeneous mixtures.

There are two special types of homogeneous mixtures that are so common that you should know them by name. These are **solutions** and **suspensions**. A solution is a liquid mixture formed when one substance, called the **solute, dissolves** into another substance, called the **solvent**. When salt is dissolved in water, the salt is the solute and the water is the solvent. As another example, when a packet of Kool-Aid is dissolved in water, the Kool-Aid is the solute and the water is the solvent. The solute does not necessarily have to be a solid. Mixing cherry syrup (solute) into a Coke (solvent) is a solution, as is vodka (solute) in tomato juice (solvent). When two liquids are mixed into a solution like this the one that has the most volume is generally called the solvent, although it doesn't really matter. One could just as easily say that the two liquids are dissolving into each other.

There is a special class of *solid* solutions called **alloys**. An alloy is a solid solution of metals. Usually, to make an alloy the metals must be melted first so that they are liquids. But once the metals are melted they can be thoroughly mixed together and allowed to cool, and then the alloy is formed. There are three particular alloys that you encounter all of the time, so you should know what they are made of. **Steel** is an alloy made of iron with a small amount of carbon mixed in. There are many different steel alloys, including many different alloys called "stainless" steel because they don't rust. Steel is one of the most popular building materials in the world, and some of these alloys include other metals in addition to the iron and carbon.

Other well-known alloys are **brass**, which is an alloy of copper and zinc, and **bronze**, an alloy of copper and tin which is often used as a sculptural medium. Both brass and bronze contain copper, but it can be hard to remember which one is made with zinc, and which one with tin. Here is my helpful way to remember which is which: "Bronze has a z, so it doesn't have the z. Brass doesn't have a z, so it has the z." The z here stands for *zinc*. Clever, huh?

As one more final example of alloys, most cars these days are available with "alloy wheels." These wheels are made mostly of aluminum, but there are small quantities of other metals blended in to give the wheels the desired mechanical properties. There are many different alloys of aluminum, all with slightly different mechanical or electrical properties (physical properties). This is because aluminum is the material of choice for a great variety of applications, so quite a few different aluminum alloys have been developed for different purposes.

In addition to solutions, the other special type of homogeneous mixture is **suspensions**. A suspension is a homogeneous mixture in which there are two different phases of substances involved. Substances typically exist in three different phases, solid, liquid and gas (or vapor). If substances in two different phases are mixed together and they stay in their different phases they make a suspension. A common example of a suspension is clouds, which have water droplets (liquid) mixed in air (gas). Shaving cream is another good example, which has tiny air bubbles (gas) mixed into a liquid. Toothpaste has tiny solid particles mixed into a liquid. Other common examples of suspensions are blood and milk.

The relationships between all these different types of substances are summarized (or will be soon) in Figure 10-6.

DO-IT-YOURSELF GRAPHICS!

Now that we have reviewed all of the types of substances we will be covering in this course, you should make a chart, a sort of "family tree" of the substances. Your chart should include the ten terms listed in the first item of the Objectives List for this chapter. Most chemistry books show this chart in the text. I am not going to because you will remember it much better if you develop it yourself based on what was presented here in this section.

Start at the top with the term SUBSTANCES, and draw in the various substances in a tree underneath, showing which types are subsets of other types. You will have ten boxes in all in your family tree when you finish. While you are at it, write in two examples for each type of substance you list, and don't use any particular example more than once.

Figure 10-6. The family tree of substances.

MORE ON SOLUTIONS

As you know, there is only so much sugar one can stir into a glass of iced tea before the sugar will not dissolve any longer. If one continues to put sugar into the liquid after this point the sugar will simply pile up on the bottom of the glass. A solution that has as much solute dissolved in it as it can hold is said to be **saturated**. Sometimes it is possible to create a solution that is **supersaturated**, meaning that solvent is in an unstable condition of actually having more solute dissolved in it than it can hold. You probably remember making a solution like this back in Grammar School by dissolving sugar into water while the water was hot. Cool water cannot hold as much sugar in solution as hot water can, so as the water cools it becomes supersaturated. In the Grammar School experiment students usually put the supersaturated solution in a jar with a pencil across the top and a piece of string hanging down into the sugar water. As the solution cools the sugar **precipitates** out of solution (that is, it is no longer dissolved in the solvent) by forming crystals of solid sugar on the string. Any time a solute precipitates out of solution like this the material that has come out of the solution is called a **precipitate**. As you see, the term *precipitate* is used both as a noun and as a verb.

For different solutions we denote the amount of a particular solute that will dissolve into a particular solvent at saturation as the **solubility**. The solubility of a particular solution is the amount of the solute, in grams or kilograms, that will dissolve in one milliliter or cubic meter of the solvent, so we can calculate it as

$$\text{solubility} = \frac{\text{mass of solute (g) at saturation}}{\text{volume of solvent (mL)}}$$

Usually if we are using grams for the solute mass we use milliliters for the solvent volume. If we use kilograms for the solvent mass, then cubic meters is typical for the solvent volume. And sometimes you will see solubilities stated as grams of solute per 100 mL of solvent. The solubilities for many solutions are temperature dependent, and often more solute will dissolve into the solvent at higher temperatures.

As an example, let's look at the solubility of a compound called sodium nitrate ($NaNO_3$) in water. At room temperature one can dissolve 122 grams of sodium nitrate in 150 mL of water. From this we can calculate the solubility of sodium nitrate in water as

$$\text{solubility} = \frac{\text{mass solute (g) at saturation}}{\text{volume solvent (mL)}} = \frac{122 \text{ g}}{150 \text{ mL}} = 0.81 \frac{\text{g}}{\text{mL}}$$

We can easily express this in other units by performing the appropriate unit conversions. Expressed in kg/m^3 we would have

$$0.81 \frac{\text{g}}{\text{mL}} \cdot \frac{1 \text{ kg}}{1000 \text{ g}} \cdot \frac{1000 \text{ mL}}{\text{L}} \cdot \frac{1000 \text{ L}}{\text{m}^3} = 810 \frac{\text{kg}}{\text{m}^3}$$

To express this solubility in g/100 mL, we would have

$$0.81 \frac{\text{g}}{\text{mL}} \cdot \frac{100}{100} = 810 \frac{\text{g}}{100 \text{ mL}}$$

Now that you know the basics, here are two more example calculations. Solubility problems are just a matter of setting up ratios and solving for the unknown.

Given that the solubility of sodium nitrate in water is 0.81 g/mL at room temperature, how much sodium nitrate would it take to make a saturated solution in 750 mL of water?

Set up two ratios and solve for the unknown quantity.

$$\frac{0.81 \text{ g}}{\text{mL}} = \frac{(x) \text{ g}}{750 \text{ mL}}$$

$$(x) \text{ g} = \frac{0.81 \text{ g}}{\text{mL}} \cdot 750 \text{ mL} = 608 \text{ g}$$

Rounding this value to 2 significant figures we have our final result of 610 g.

A technician measures the solubility of a certain compound in alcohol, and finds that 24.5 g of the compound will dissolve in 175 mL of alcohol. How much alcohol would it take to make a saturated solution that will completely dissolve 145 g of the compound?

$$\frac{24.5 \text{ g}}{175 \text{ mL}} = \frac{145 \text{ g}}{(x) \text{ mL}}$$

$$(x) \text{ mL} \cdot 24.5 \text{ g} = 145 \text{ g} \cdot 175 \text{ mL}$$

$$(x) \text{ mL} = \frac{145 \text{ g}}{24.5 \text{ g}} \cdot 175 \text{ mL} = 1{,}036 \text{ mL}$$

Rounding this values to 3 significant figures we have 1,040 mL.

PHYSICAL AND CHEMICAL PROPERTIES AND CHANGES

All substances have certain properties. We divide up the different properties substances can possess into two broad classes. Some properties have to do with the physical characteristics of the substance, such as color, shape, size, phase, texture, and density. These properties are called **physical properties**. Below is a list of examples of statements about the physical properties of substances. Consider how each one relates to the definition of physical properties just given.

- Iron is gray in color.
- Rust (iron oxide) is dark orange in color.
- Mica is shiny.
- Glass is smooth, but has sharp edges.
- The density of germanium is 5.323 g/cm^3.
- Ethyl alcohol is transparent and colorless.
- Algae is green.
- Helium is a gas at atmospheric pressure and room temperature.
- Concrete is rough in texture.
- At standard pressure water freezes at 0°C.
- Milk is opaque and white in color.

- Oil is slippery.
- Clay brick is ochre in color.
- At 4.0°C the density of water is 1.0 g/cm³.
- Aluminum is malleable and ductile.
- Cast iron is not malleable.
- Glass is not malleable.
- Play-Doh is malleable, but not ductile.
- Jello is not ductile.

You probably noticed a couple of unfamiliar terms in the above list. The terms **malleable** and **ductile** are used to describe two important properties possessed by many metals. A substance is malleable if it can be hammered into different shapes, or hammered flat into sheets. A substance is ductile if it can be "drawn" into a wire. Wire drawing is a process of making wire by pulling the metal through a small hole in a metal block called a *die*. Usually the metal is already formed into a wire of larger diameter. The end of this larger diameter wire is hammered down or filed to get it through the hole in the die, and then a machine pulls the wire through the die to make the new, smaller diameter wire. Substances that can be drawn through a die like this without simply snapping are said to be ductile.

Notice from the above examples that a good student of science needs to be careful when describing physical properties. We need to make sure our statements are accurate in cases where temperature or pressure affect the property in question. For example, it is inaccurate to say that H_2O is a liquid. A more accurate statement would be to say that one of the physical properties of water is that it is a liquid between 0°C and 100°C. An even more accurate statement would be to specify that the preceding sentence is correct at standard pressure, because at very low pressures the boiling and freezing points of water are different.

The second broad class of properties has to do with the kinds of chemical bonds a substance will form, that is, the chemical reactions a substance will or will not participate in. These properties are called **chemical properties**. We have not yet studied chemical reactions, so you may not know that much about them. However, there are two common chemical reactions that you are quite familiar with, burning, and rusting. Both of these are chemical reactions in which a substance combines with oxygen. Fiery explosions are simply **combustions** that happen very rapidly. But whether the combustion happens slowly, as with a log on a fire, or rapidly, as with a firecracker, combustion is a chemical reaction with oxygen. Substances that will react with oxygen in this way are said to be *flammable* or *combustible*. (Oddly, *inflammable* also means the same thing! Go figure. English is a strange language.)

The more general term for rusting is **oxidation**. Like combustion, oxidation is also a chemical reaction with oxygen, it just happens to proceed very slowly and generally affects different types of materials. When iron oxidizes it forms iron oxide, which we commonly call rust. There are several different forms of iron oxide, colored red, yellow, brown and black. Other metals oxidize as well. When copper oxidizes it can form two different oxides, one red and one black. This is why copper objects exposed to the air will turn dark brown or black. (Over a longer period of time the copper oxide forms other compounds, such as copper carbonate, which give the copper its pretty blue-green color. The Statue of Liberty is made of copper, and has been there for a long time. It is essentially covered with a layer of copper carbonate.) Aluminum also oxidizes. Aluminum oxide is dark gray, and anyone who has done a lot of hand work with aluminum parts will have noticed his or her hands blackened by the particles of aluminum oxide building up. Figure 10.6 shows a few different oxides.

Figure 10-6. Oxides: red iron oxide (left), yellow iron oxide (center), and black copper oxide (right).

Here are some examples of how one would describe chemical properties of substances:

- Hydrogen is combustible.
- Aluminum oxidizes to form aluminum oxide.
- Water is not flammable.
- Gold does not oxidize. (This is why it is so valuable. It stays shiny and clean.)
- Baking soda reacts with vinegar.
- Iron oxidizes to form iron oxide, or rust.
- Sodium reacts violently with water.
- Hydrogen reacts with a number of different polyatomic ions to form acids.
- Dynamite is explosive.
- Chewing gum is not explosive. (Except in movies like *Mission Impossible*.)
- Sodium hydroxide reacts with aluminum.
- Sulfuric acid reacts with many metals.

The two broad classes of properties we have been discussing, physical properties and chemical properties, are related to two broad classes of changes that substances can undergo. If a substance experiences a change with respect to one of its physical properties, we call this a **physical change**. When a physical change occurs the substance is still the same substance, it just looks different. If a chemical reaction occurs to a substance, this is called a **chemical change**. Chemical properties basically describe the kinds of chemical changes a substance can make. When a chemical change occurs, the original substances that went into the reaction are converted into new substances with totally different physical and chemical properties. When asked to describe a given change as physical or chemical, ask yourself if the substance is still the same substance, or if it has actually gone through a chemical reaction and has become a different substance.

Table 10-2 contains a number of examples of physical and chemical changes, with reasons given for identifying the type of change as physical or chemical.

Change	Type	Reason
Glass breaking	physical	The broken glass is still glass, it just changed shape.
Firecracker exploding	chemical	This explosion is a combustion. All combustions are chemical reactions. The substances in the firecracker have reacted to form new substances such as ash and various gases.
Mercury boiling	physical	The mercury is still mercury, it has just changed from the liquid phase to the vapor phase.
Copper turning dark brown or black	chemical	This occurs because the copper is oxidizing and forming copper oxide, a new substance. This is a chemical reaction.
Iron pipes corroding	chemical	Corrosion is a chemical reaction. In this case, the iron reacts with whatever was in the pipes to form a new substance.
Water evaporating	physical	The substance is still H_2O, it has simply changed phase from liquid to vapor.
Mixing cake batter	physical	The eggs and flour and so on have formed a mixture, but no chemical change (reaction) has occurred.
Baking cookies	chemical	The heat caused a chemical reaction to occur in the dough. The substance is no longer dough. It is cookie.
Spilled pancake batter drying out	physical	No chemical reaction occurred. Dried batter is still batter. (If you want a pancake you have to cook it, which would be a chemical change.)
Molten lead hardening	physical	The lead is still lead, it just changed phase from liquid to solid.
Balloon popping	physical	The balloon material is still the same material, it is just in shreds now. The air inside the balloon is at a lower pressure and is not contained in the balloon any longer, but it is still air.

Table 10-2. Examples of physical and chemical changes.

CHAPTER X EXERCISES

Solubility Calculations

1. The solubility of potassium chloride (KCl) in water is 0.32 g/mL. How much water would it take completely to dissolve 3.50 kg of KCl? State your answer in liters.

2. The solubility of lithium chloride (LiCl) in water is 1.01 g/mL at 100°C. How much LiCl will dissolve in 5.00 m^3 of water? State your answer in kilograms.

3. A student finds that a maximum of 1.750 kg of a certain substance will dissolve in 3.66 L of a certain solvent. Determine the solubility of this substance in this solvent, and express your result in g/mL.

4. Suppose that a certain industrial cleaning solution is manufactured by preparing a saturated solution consisting of 3,610 kg of solute dissolved in 45,550 L of solvent. Determine the solubility of this solute in the solvent and express your result in kg/m^3.

5. Let's say the factory that makes the cleaning solution in the previous question purchases 46,800,000 kg of solute each year. How much solvent do they need to buy to make cleaning solution out of all the solute they purchase? State your answer in ML.

Answers
1. 11 L
2. 5,050 kg
3. 0.478 g/mL
4. 79.3 kg/m^3
5. 5.90 x 10^2 ML

Physical and Chemical Changes

Identify each of the following as a physical change or a chemical change.

1. An avalanche	9. Boiling mercury
2. A cigar burning	10. Welding steel
3. Spilling a glass of milk	11. Allowing molten iron to harden
4. Digesting your food	12. Filling a helium balloon
5. Swatting a fly	13. A car exhaust pipe rusting
6. Stirring cream into coffee	14. Frying chicken
7. Firing a popgun	15. Snow melting
8. Firing a real gun	16. Paint drying

CHAPTER XI
ATOMIC MODELS

> **OBJECTIVES**
>
> Memorize and learn how to use this equation:
>
> $$\rho = \frac{m}{V}$$
>
> After studying this chapter and completing the exercises, students will be able to do each of the following tasks, using supporting terms and principles as necessary:
>
> 1. State the key features of the atomic models envisioned by Democritus, Dalton, Thomson and Rutherford. State the key atomic discoveries each of these men made, and identify the aspects of their atomic models that we now regard to be correct or partially correct.
> 2. Describe the key experiments and contributions of Lavoisier, Mendeleev, Thomson, Millikan and Rutherford.
> 3. State the law of conservation of mass in chemical reactions.
> 4. Use the density equation to compute the density, volume or mass of a substance.
> 5. Calculate the volume of right rectangular solids and right cylinders.

THE HISTORY OF ATOMIC MODELS

The story of atomic theory starts back with the ancient Greeks. As we look at how the contemporary model of the atom developed we will hit on some of the great milestones in the history of chemistry along the way.

In the fifth century BC the Greek philosopher **Democritus** proposed that everything was made of tiny, indivisible particles. Our word atom comes from the Greek word *atomos*, meaning "indivisible." Democritus' idea was that the properties of substances were due to characteristics of the atoms they are made from. So atoms of metals were supposedly hard and strong, atoms of water were assumed to be wet and slippery, and so on. At this same time there were various views about what the basic elements were. One of the most common views was that there were four elements, earth, air, water, and fire, and that everything was composed of these.

Not much real chemistry went on for a very long time. During the medieval period, of course, there were the alchemists who sought to transform lead and other materials into gold. This cannot be done by the methods available to them and their efforts were not successful.

But in the seventeenth century things started changing as scientists became interested in experimental research. The goal of the scientists described here was to figure out what the fundamental constituents of matter were. This meant figuring out how atoms were put together, what the basic elements were, and understanding what was going on when various chemical reactions took place. The nature of earth, air, fire and water were under intense scrutiny over the next 200 years.

Figure 11-1. Robert Boyle

An early contributor was English scientist Robert Boyle, who is remembered now for Boyle's Law relating the volume and pressure of gases (Figure 11-1). In 1661 Boyle proposed that substances were not elements if they were formed of two or more "components."

French scientist **Antoine Lavoisier** is considered the "father of modern chemistry" because of his reliance on experimental research, as well as his important discoveries (Figure 11-2). His wife Marie Anne Lavoisier worked with him, illustrating his experiments and assisting with the publication of his work. Unfortunately, Lavoisier was beheaded in 1794 at the age of 51 during the Reign of Terror that took place in France right after the French Revolution.

Lavoisier proposed the existence of compounds composed of two or more elements. He performed two very famous and ingenious experiments with the element tin that led him to some very important discoveries. Lavoisier was trying to understand what was going on when tin became tin oxide. Left to itself, the tin will gain weight while forming the tin oxide, and Lavoisier was trying to determine where this extra weight was coming from. He decided to seal the tin inside a container and see if it gained weight then. In Lavoisier's first experiment, illustrated in Figure 11-3, he placed a sample of tin inside a sealed glass container. He heated the material with sunlight focused on the tin through a magnifying glass. He found that after the tin had formed the tin oxide the container weighed exactly the same as it had before. This led him to formulate the "**law of conservation of mass in chemical reactions**." In contemporary language this law states that the mass of the reactants, the substances going into the reaction, is equal to the mass of the products, the substances resulting from the reaction. In other words, no mass is gained or lost in a chemical reaction.

Figure 11-2. Antoine Lavoisier and his wife.

Since the glass container didn't gain any weight in the first tin experiment, Lavoisier took the tin experiment one step further. This time he placed the tin under a glass cover that was open on the bottom. The bottom of the cover was immersed in water, as illustrated in Figure 11-4.

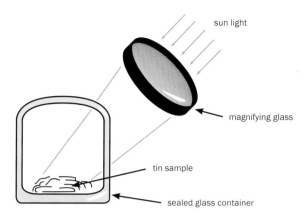

Figure 11-3. Lavoisier's first tin experiment.

As the tin formed the tin oxide in this second experiment, Lavoisier observed that the water came up into the glass cover where the oxide was forming. Lavoisier concluded from this that a portion of the air was combining with the tin to form the oxide. This was correct, but Lavoisier went further. He noted that right at 20% of the air space under the glass had been filled in with water. He concluded from this that air

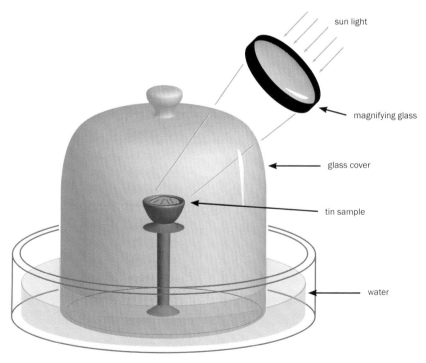

Figure 11-4. Lavoisier's second tin experiment.

was composed of at least two components, that one of them was 20% of the air, and that this particular component was combining with the tin to form the oxide. The water was coming into the glass to fill in the space left by this component as it left the air and combined with the tin. All of this was correct, but keep in mind that back then *no one even knew what air was made of*. We now know that air is indeed composed of approximately 20% oxygen, and that it is the oxygen reacting with the tin that forms tin oxide. Lavoisier's experiment was brilliant and his scientific reasoning was amazing!

Around this same time the English chemist Joseph Priestly heated some mercuric oxide and produced a gas that made things burn brighter and hotter (Figure 11-5). Hearing of this, Lavoisier realized that Priestly's gas was the same substance as the one reacting with tin in his oxide experiments. Lavoisier named this gas "oxygen," which means "acid former." You will learn a lot about acids when you get to chemistry.

Figure 11-5. Joseph Priestly.

In 1766 English scientist Henry Cavendish conducted an experiment that produced hydrogen, which Cavendish discovered was flammable. Once again, Lavoisier put two and two together and concluded that this new hydrogen gas combined with oxygen to produce water. In the 1790s the French chemist Joseph Proust proposed the "law of definite proportions," which states that elements combine in definite mass ratios to form compounds. This new law was one of the important clues confirming the existence of atoms.

Then we come to another major milestone: In 1803 English scientist John Dalton produced the first scientific model of the atom (Figure 11-6). **Dalton's atomic model** was based on five main points, which you need to know:

1. All substances are composed of tiny, indivisible particles called atoms.
2. All atoms of the same element are identical.
3. Atoms of different elements have different weights.
4. Atoms combine in whole number ratios to form compounds.
5. Atoms are neither created nor destroyed in chemical reactions.

Figure 11-6. John Dalton.

The impressive thing about Dalton's atomic theory is that even today the last three of these points are regarded as correct, and the first two are least partially correct. On the first point, it is still scientifically factual that all substances are made of atoms, but we now know that atoms are not indivisible. This should be obvious, since you know that atoms themselves are composed of protons, neutrons and electrons. The second point is correct in every respect but one. Except for the number of neutrons in the nucleus, every atom of the same element is identical. However, we now know that there are different **isotopes** for each element. Isotopes, as we will study in more detail later, are varieties of atoms of the same element that differ in the number of neutrons they have.

A lot of work went on during the 19th century, much of it centering on gases. French chemist Joseph Gay-Lussac discovered that 2 L of hydrogen combine with 1 L of oxygen to form 2 L of water. Gay-Lussac inferred from this that the correct formula for water should be H_2O. (Oddly, Dalton did not share this view, holding that the correct formula for water should be HO.) In the early 1800s Italian physicist Amedeo Avogadro hypothesized that hydrogen and oxygen gases were composed of molecules, and that these molecules were in turn composed of atoms. This was the first time anyone had proposed such a thing, and the scientific community pretty much ignored Avogadro's work for about 50 years. But Avogadro was right.

Moving on toward the later part of the nineteenth century, many individual elements had been discovered by this time. Another huge milestone was the publication of the first Periodic Table of the Elements by Russian chemist **Dmitri Mendeleev** in 1869 (Figure 11-7). Mendeleev had discovered that there were patterns in the properties of the elements that he could use to order the elements into rows and columns. When he arranged the elements this way he noted that there appeared to be gaps between some of the elements. This led him to predict the properties of several elements which had not yet been discovered. One of these was the element germanium, which Mendeleev called *ekasilicon*. Germanium is right below silicon in the Periodic Table of the Elements. Mendeleev predicted the color, density and atomic weight of this unknown element. His predictions turned out to be quite accurate once the element was discovered 15 years later. Table 11-1 shows a comparison between Mendeleev's ekasilicon prediction and the actual properties of germanium.

Figure 11-7. Dmitri Mendeleev.

Our attention now shifts from the chemists to the physicists as we take a look at the most famous developments that led to our contemporary understanding of how atoms are structured.

"Ekasilicon" Prediction		Germanium Discovered	
Date Predicted	1871	Date Discovered	1886
Atomic Mass	72	Atomic Mass	72.6
Density	5.5 g/cm³	Density	5.32 g/cm³
Color	dark gray	Color	light gray

Table 11-1. Predicted and actual characteristics of germanium.

English scientist **J. J. Thomson** worked at the Cavendish Laboratory in Cambridge, England (Figure 11-8). In 1897 he conducted a series of landmark experiments that revealed the existence of electrons. Because of his work he won the Nobel Prize in Physics in 1906 and was knighted in 1908. Thomson's ingenious set-up is sketched in Figure 11-9.

Thomson placed electrodes from a high-voltage source inside of a very elegantly made, sealed-glass vacuum tube. This apparatus can generate a so-called *cathode ray* from the negative electrode, called the *cathode*, to the positive one, called the *anode*. A cathode ray is simply a beam of electrons, but this was not known at the time. (In the era prior to flat-screen displays, television sets and computer monitors used cathode rays to hit the screen from behind and create the picture on the screen.) The anode inside Thomson's vacuum tube had a hole in it for some of the electrons to escape through, which created a beam of "cathode rays" heading toward the other end of the tube.

Figure 11-8. J. J. Thomson.

Thomson placed the electrodes of another voltage source inside the tube, above and below the cathode ray and discovered that the beam of electrons deflected when this voltage was turned on. The dashed line in the figure shows the direction of the beam without the deflection voltage turned on. The deflection of the beam toward the positive electrode led Thomson to theorize that the beam was composed of negatively charged particles, which he called "corpuscles." (The name *electron* was first used a few years later by a different scientist.) By trying out many different arrangements of cathode ray tubes, including the use of a magnetic field to deflect the beam, Thomson confirmed that the ray was negatively charged. Then using the scale on the end of the tube to measure the deflection angle, he was able to determine the charge-to-mass ratio of the individual electrons he had discovered, which is 1.8×10^{11}. (The units of this value, as we would now say, are coulombs/kilogram, or C/kg.) When you think about all the work involved in the hand-blowing of every glass tube, the skill of getting the metal parts inside the

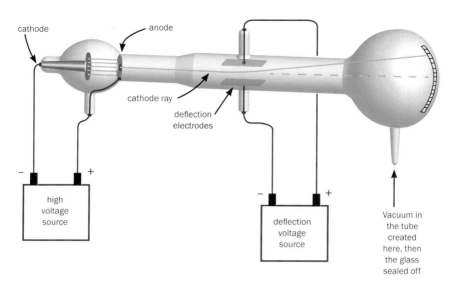

Figure 11-9. Thomson's cathode ray tube experiment.

glass vacuum tube, and the whole idea of this experiment, you have to admit that this is a *very* cool experiment. *What a great scientist!*

Thomson went on to theorize that electrons came from inside atoms. He developed a new atomic model that envisioned atoms as tiny clouds of massless, positive charge sprinkled with thousands of the negatively charged electrons. This model is usually called the "**plum pudding**" model, so let's call it that. But since American students rarely know what plum pudding is, you might think of it as the "watermelon model."

Figure 11-10. Robert Millikan.

The red meat of the watermelon is like the overall cloud of positive charge, and the seeds are like the negatively charged electrons scattered around inside it.

In 1911 American scientist Robert Millikan (Figure 11-10) devised a brilliant experiment that allowed him to determine the charge on individual electrons, 1.6×10^{-19} coulombs (C). Using the value of the charge/mass ratio determined by Thomson, Millikan was then able also to calculate the mass of the electron, 9.1×10^{-31} kg. Millikan's apparatus is sketched in Figure 11-11.

Millikan's famous experiment is called the "**oil drop experiment**". Inside of a metal drum about the size of a large bucket Millikan placed a pair of horizontal metal plates connected to an adjustable high-voltage source. The upper plate

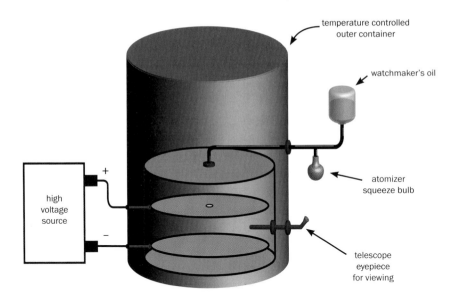

Figure 11-11. Millikan's oil drop experiment.

had a hole in the center and was connected to the positive voltage, the lower plate to the negative. He used an atomizer spray pump (like cartoons have on bottles of perfume) to spray in a fine mist of watchmaker's oil above the positive plate. Some of these droplets would fall through the hole in the upper plate and move into the region between the plates. Connected through the side of the drum between the two plates was a telescope eyepiece and lamp so that Millikan could see the oil droplets between the plates.

The process of squirting in the oil droplets with the atomizer sprayer caused the droplets to acquire a charge of static electricity. As you now know, this means the droplets had excess electrons on them, and they would have picked them up by friction as the droplets squirted through the sprayer tube. As Millikan adjusted the voltage between the plates he could make the charged oil droplets hover when the voltage was just right. Millikan took into account the weight of the droplets, and the viscosity of the air as the droplets fell and was able to determine that every droplet had a charge on it that was a multiple of 1.6×10^{-19} C. From this he deduced that this must be the charge on a single electron, which it is. Millikan won the Nobel Prize in Physics in 1923 for this work.

The last famous experiment in this basic history of atomic models was initiated in 1909 by one of Thomson's students at Cambridge, New Zealander **Lord Ernest Rutherford** (Figure 11-12). Rutherford was already famous when this experiment occurred, having just won the Nobel Prize in Chemistry the previous year. Rutherford's **"gold foil experiment"** resulted in the discovery of the atomic nucleus. To understand this experiment you need to recall from Chapter V that an "alpha particle," or α-particle, using the Greek letter alpha, is a particle composed of two protons and two neutrons. Such a particle is exactly the same as the nucleus of a helium atom. Alpha particles are naturally emitted by some radioactive materials.

Rutherford's experiment is sketched in Figure 11-13. Rutherford created a beam of α-particles by placing some radioactive material (radium bromide) inside a lead box

Chapter XI Atomic Models

Figure 11-12. Ernest Rutherford.

with a hole in one end. The α-particles from the decaying radium atoms streamed out of the hole. Rutherford aimed the α-particles at a thin sheet of gold foil that was only a few hundred atoms thick. Surrounding the gold foil was a ring-shaped screen coated with a material that would glow when hit by an α-particle. Rutherford could then determine where the α-particles went after encountering the gold foil.

Rutherford found that most of them went straight through the foil and struck the screen on the other side. However, occasionally an α-particle (one particle out of every several thousand) would deflect with a large angle. Sometimes these deflected particles would bounce almost straight back. This was astonishing and was not at all what Rutherford had expected.

Thomson had theorized that the positive charge in the atom was dispersed around throughout the atom. Because of this Rutherford expected the massive and positively charged α-particles to blow right through the gold foil. Most of them did, but when some of them ricocheted backward the astonished Rutherford commented that it was like firing a huge artillery shell into a piece of tissue paper and having it bounce back and hit you! Rutherford's work led to his new proposal in 1911 for a model of the atom. Rutherford's model includes these key points, which you should know:

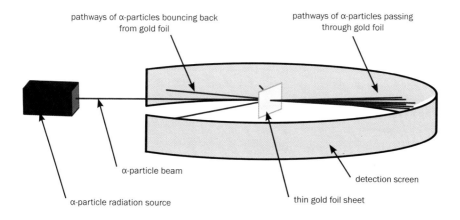

Figure 11-13. Rutherford's gold foil experiment.

- The positive charge in atoms is concentrated in a tiny region in the center of the atom, which Rutherford called the nucleus.
- Atoms are mostly empty space.
- The electrons, which contain the atom's negative charge, are outside the nucleus.

Just what do we mean when we say that atoms are mostly empty space? Well, the diameter of an atom is about 20,000 times as big as the diameter of the nucleus. So imagine that you are standing at center field of a large football stadium holding a nickel in your hand. (Close your eyes and consider the vast space of the empty stadium surrounding you.) If the nucleus of an atom were the size of your nickel, the atom would be as big as the entire stadium! That is a lot of empty space!

We are going to study two more recent atomic models in more detail in the next chapter. But before we get to them let's just finish off this brief history of discovery about atoms by noting when neutrons came into the picture. It took another twenty years before James Chadwick, another Englishman, discovered the neutron (Figure 11-14). Before World War I Chadwick studied under Rutherford (at Cambridge, of course). Then the War began. Not only did the war interrupt the progress of the research in general, but Chadwick was a prisoner of war in Germany. Working back in England after the war he discovered the neutron in 1932 and received the Nobel Prize in Physics for his discovery in 1935. And, of course, ten years later he was knighted, too.

Chadwick's discovery of the neutron enabled physicists to fill in a lot of blanks in their understanding of the basic structure of atoms. But years before Chadwick even made his discovery Rutherford's atomic model was already being taken to another level through the work of Niels Bohr, whom we will meet in the next chapter.

Figure 11-14. James Chadwick.

DENSITY

In the previous chapter I mentioned that density is one of the physical properties substances possess. In our historical review of the 19th century you recall that Mendeleev was able to predict the density of germanium, which at that time had not yet been discovered. It is now time to formally introduce the equation for density. You will be satisfied, I'm sure, to learn that this is the last equation you will have to learn for ASPC!

Density is a measure of how much matter is packed into a given volume for different substances. No doubt you are already familiar with the concept of density. You know that if you hold equally sized balloons in each hand, one filled with water and one filled with air, the water balloon will weigh more because water is denser than air. You know that to have equal weights of sand and Styrofoam packing peanuts the volume of the packing material will be much larger because the packing material is much less dense.

The equation for density is

$$\rho = \frac{m}{V}$$

where the Greek letter ρ (spelled rho and pronounced like row, which rhymes with snow) is the density in kg/m³, m is the mass in kg, and V is the volume in m³. These are the variables and units in our familiar MKS unit system. However, you should be aware that since laboratory work typically involves only small quantities of substances, it is quite common for densities to be expressed in g/cm³. In the examples that follow I will illustrate the use of both of these units. If all you are doing is using the density equation, then either of these sets of units is fine. However, if you are using the density with other equations in a multi-step calculation, such as calculating the weight of an object, then you certainly need to stick with MKS units as always. One final item for you to note is that the density of water at room temperature is

$$\rho_{water} = 998 \ \frac{\text{kg}}{\text{m}^3}$$

This value is useful to know because water comes up in many different applications. So, the density of water is the third and final physical constant listed on the memory list for ASPC (see Appendix A). Now is the time to commit this value to memory.

Determine the density of a block of wood that weighs 18.25 N and has dimensions 4.0 in x 2.5 in x 9.50 in. State your result in g/cm³.

We begin by writing down the given information and converting the length units to centimeters.

$F_w = 18.25$ N

$l = 4.0 \text{ in} \cdot \dfrac{2.54 \text{ cm}}{\text{in}} = 10.16 \text{ cm}$

$w = 2.5 \text{ in} \cdot \dfrac{2.54 \text{ cm}}{\text{in}} = 6.35 \text{ cm}$

$h = 9.0 \text{ in} \cdot \dfrac{2.54 \text{ cm}}{\text{in}} = 22.86 \text{ cm}$

$\rho = ?$

To calculate the density we will need the mass, which we will get from the given weight, and the volume, which we will calculate from the given dimensions. The weight calculation will give us the mass in kilograms and we will convert it to grams.

Calculating the mass we write

$F_w = mg$

$$m = \frac{F_w}{g} = \frac{18.25 \text{ N}}{9.80 \frac{\text{m}}{\text{s}^2}} = 1.86 \text{ kg} \cdot \frac{1000 \text{ g}}{\text{kg}} = 1{,}860 \text{ g}$$

Then we calculate the volume

$$V = l \cdot w \cdot h = 10.16 \text{ cm} \cdot 6.35 \text{ cm} \cdot 22.86 \text{ cm} = 1{,}475 \text{ cm}^3$$

Finally, the density is

$$\rho = \frac{m}{V} = \frac{1{,}860 \text{ g}}{1{,}475 \text{ cm}^3} = 1.26 \frac{\text{g}}{\text{cm}^3}$$

Rounding to two significant figures we have

$$\rho = 1.3 \frac{\text{g}}{\text{cm}^3}$$

The density of germanium is 5.323 g/cm³. A small sample of germanium has a mass of 17.615 g. Determine the volume of this sample.

Begin by writing the givens.

$\rho = 5.323 \frac{\text{g}}{\text{cm}^3}$

$m = 17.615$ g

$V = ?$

Now write the equation and solve.

$$\rho = \frac{m}{V}$$

$$V = \frac{m}{\rho} = \frac{17.615 \text{ g}}{5.323 \frac{\text{g}}{\text{cm}^3}} = 3.309 \text{ cm}^3$$

This value has four significant figures, as it should.

Our final example is to calculate the mass of a cylindrical sample of copper. The sample is 1.5 cm in diameter and 2.5 cm long. The density of copper is 8.96 g/cm³.

We begin by writing the givens and calculating the volume of the cylinder of copper.

$\rho = 8.96 \ \frac{g}{cm^3}$
$d = 1.5 \text{ cm}$
$l = 2.5 \text{ cm}$

$r = \frac{d}{2} = \frac{1.5 \text{ cm}}{2} = 0.75 \text{ cm}$
$V = \pi r^2 \cdot l = 3.14 \cdot (0.75 \text{ cm})^2 \cdot 2.5 \text{ cm} = 1.77 \text{ cm}^3$

Since our final result will need two significant figures it is appropriate for this intermediate result to have three. Now we are ready for the density equation.

$\rho = \frac{m}{V}$

$m = \rho V = 8.96 \ \frac{g}{cm^3} \cdot 1.77 \text{ cm}^3 = 15.9 \text{ g}$

Rounding to two significant figures we have

$m = 16 \text{ g}$

CHAPTER XI EXERCISES

Volume, Mass and Weight Exercises

In addition to volume, we are hitting on mass and weight here to make sure your skills are nice and sharp for the density exercises to follow. Remember to use horizontal bars for every unit fraction, and to show every conversion factor with units in each of your calculations. Also remember that every given quantity represents a measurement of some kind and affects the number of significant digits of your result. Every conversion factor that is not exact also affects the number of significant digits in the result.

1. Convert 98.34 kg/m^3 into g/cm^3.

2. Convert 42 mL into gallons.

3. Given a weight of 18.5 lb, determine the mass.

4. Convert 3.6711×10^4 g/mL into kg/m^3.

5. Convert 1.957×10^4 in^3 into cm^3.

6. Convert 455 mL into m^3.

7. Given a mass of 46,000 kg, determine the weight in both newtons and pounds.

8. Convert 32.11 L into in^3.

9. Given a weight of 14.89 N, determine the mass (kg) and the weight in pounds.

10. Convert 36.00 cm^3 into m^3.

11. Convert 9.11 m^3 into cm^3.

12. Convert 4.11×10^5 m^3 into L.

13. Given a weight of 55,789 lb determine the mass.

14. Convert 5.022 g/cm^3 into kg/m^3.

15. Convert 3.76×10^{-4} g/mL into g/cm^3.

16. Given a weight of 50,000 N determine the mass (kg) and the weight in pounds.

17. Convert 1.75×10^{-6} m^3 into cm^3.

18. Convert 100.5 ft³ into m³.

19. Convert 37 m³ into in³.

20. Convert 750 cm³ into L.

21. Convert 5,755,000 gal into m³.

Answers
1. 0.09834 g/cm³
2. 0.011 gal
3. 8.40 kg
4. 36,711,000 kg/m³
5. 320,700 cm³
6. 0.000455 m³
7. 450,000 N, 101,000 lb
8. 1,959 in³
9. 1.52 kg, 3.35 lb
10. 3.6 x 10⁻⁵ m³
11. 9.11 x 10⁶ m³
12. 4.11 x 10⁸ L
13. 25,300 kg
14. 5,022 kg/m³
15. 3.76 x 10⁻⁴ g/cm³
16. 5,000 kg, 10,000 lb
17. 1.75 cm³
18. 2.846 m³
19. 2,300,000 in³
20. 0.75 L
21. 21,790 m³

Density Exercises

1. What is the density of carbon dioxide gas if 0.196 g of the gas occupies a volume of 100.1 mL?

2. Oil floats because its density is less than that of water. Determine the volume of 550 g of a particular oil with a density of 955 kg/m³. State your answer in mL.

3. A factory orders 15.7 kg of germanium. The density of germanium is 5.32 g/cm³. Calculate the volume of this material, and state your answer both in m³ and cm³.

4. A block of wood 3.00 cm on each side weighs 5.336 x 10⁻² lb. What is the density of this block?

5. A graduated cylinder contains 23.35 mL of water. An irregularly shaped stone is placed into the cylinder, raising the volume to 27.79 mL. If the mass of the stone is 32.1 g, what is the density of the stone?

6. A standard 55-gallon drum is 34.5 inches tall and 24 inches in diameter. Consider a 55-gallon drum filled with kerosene. Using the dimensions in inches to calculate the volume, determine the mass of kerosene that would fill this drum, given that the density of kerosene is 810 kg/m³.

7. Silver has a density of 10.5 g/cm³, and gold has a density of 19.3 g/cm³. Which would have a greater mass, 5 cm³ of silver, or 5 cm³ of gold?

8. Five mL of ethanol has a mass of 3.9 g, while 7.5 liters of benzene has a mass of 6.6 kg. Which one is more dense?

9. Iron has a density of 7,830 kg/m³. An iron block is 2.1 cm by 3.5 cm at the base and has a mass of 94.5 g. How tall is the block?

10. A student measures out 22.5 mL of mercury and finds the mass to be 306 g. Determine the density of mercury, and state your answer in kg/m³.

11. A more precise value for the density of silver is 10.501 g/cm³. A cylinder of pure silver is 4.500 cm long and 2.7500 cm in diameter. How much does this cylinder weigh? State your answer in pounds.

12. The density of water is greatest at 4.0°C. At this temperature the density is 1,000.0 kg/m³. Determine the mass of 5.6 L of water at this temperature.

13. A large contemporary water tower can hold over 3 million gallons of water. How much would 3.0 x 10⁶ gallons of water weigh? State your answer in pounds.

14. Given that the densities of lead and gold are 11.34 g/cm³ and 19.32 g/cm³ respectively, answer the following questions:
    a. Which weighs more, a cubic meter of gold or a cubic meter of lead?
    b. Which is larger, a cube of gold with a mass of 1 gram, or cube of lead with a mass of 1 gram?

15. The density of common steel is 7.85 g/cm³. Determine the weight in pounds of a cylinder of this steel the size of a large coffee can, 6.1875 inches in diameter and 7.0000 inches long.

16. The famous Kon-Tiki was a raft sailed by Norwegian explorers in 1947 from South America to the Polynesian islands in the South Pacific. (Thor Heyerdahl's book about it is really great!) The trip took three and a half months and covered 4,300 miles in the Pacific Ocean. The crew of six men built the main section of the raft out of 9 massive balsa trees, each

2.0 ft in diameter. Assume for simplicity that the balsa trees had a typical balsa wood density of 160 kg/m³, and the logs were each 45 ft in length. Determine the total weight, in pounds, of these logs used to build the raft.

17. A standard Olympic competition pool is 50.0 m long, 25.0 m wide and at least 2.00 m deep. Determine the volume, in gallons, of the water this pool holds, and then use the standard value for the density of water to determine the weight of this water in tons. (A ton is defined as exactly 2,000 lb.)

Answers
1. 1.96×10^{-3} g/mL
2. 580 mL
3. 0.00295 m³, or 2,950 cm³
4. 0.897 g/cm³
5. 7.23 g/cm³
6. 210 kg
7. The answer to this one is top secret.
8. This one is, too.
9. 1.6 cm
10. 13,600 kg/m³
11. 0.618 lb
12. 5.6 kg
13. 24,700,000 lb
14. Another secret!
15. 59.6 lb
16. 12,700 lb
17. 6.60×10^5 gal; 2,750 tons

CHAPTER XII
THE BOHR AND QUANTUM MODELS OF THE ATOM

> **OBJECTIVES**
>
> After studying this chapter and completing the exercises, students will be able to do each of the following tasks, using supporting terms and principles as necessary:
>
> 1. Define these terms: quantum, quanta, quantized, photon, absorption, emission, spectrascope, excitation, atomic spectrum, spectral lines, principle quantum number, atomic energy levels, energy transition and orbital.
> 2. Name, in order, the orbitals available for electrons to fill in the first four energy levels (principal quantum numbers).
> 3. State the number of electrons that can reside in the s, p and d orbitals.
> 4. Describe the Bohr model of the atom.
> 5. Use the Planck relation relating the energy of a photon to its frequency, $E = hf$, to explain why larger electron energy transitions result in light toward the violet in color, and lesser electron energy transitions result in light toward the red in color.
> 6. Given a Periodic Table of the Elements for reference, use electron configuration notation to write the electron configuration for any element from hydrogen (atomic number 1) through krypton (atomic number 36).
> 7. Explain the principle of spectroscopy, or how elements can be identified by the characteristic wavelengths of light emitted by their atoms.
> 8. Describe how the Bohr model of the atom provides a powerful explanation for the phenomenon of atomic spectra.
> 9. Use the Periodic Table of the Elements (PTE) to determine atomic number, atomic mass and mass number.
> 10. Use the atomic mass and atomic number listed in the PTE to determine the numbers of protons, electrons and neutrons an atom of a given element will have.

Danish physicist **Niels Bohr** is one of the major figures in the development of the modern physics of the twentieth century (Figure 12-1). For his work in the development of quantum physics he received the Nobel Prize in Physics in 1922. Bohr's 1913 model of the atom is particularly useful because it provides an explanation for the phenomenon of atomic spectra, which we will discuss in detail shortly.

Bohr theorized that atoms were like little solar systems, with the electrons orbiting the nucleus the way planets orbit the sun. The negatively charged electrons were held in their orbits by electrical attraction to the positively charged nucleus. And a most significant aspect of the model is this: Each electron in an orbit possessed a specific amount of energy. If an electron were to **absorb** additional energy somehow it would move to an orbit that was for electrons with higher energy. If an electron were to **emit** energy somehow, it would move to an orbit that was for electrons with lower energy. Bohr's

Chapter XII   The Bohr and Quantum Models of the Atom

Figure 12-1. Niels Bohr.

"planetary model" is usually depicted as illustrated in Figure 12-2. The model allows a maximum of two electrons in the first energy level, eight electrons in each of the second and third energy levels, and higher numbers in higher levels.

Even though we now know that electrons do not orbit the nucleus like planets, Bohr was completely right about the electrons possessing specific energies. This concept is now well established in atomic theory. When you look at Figure 12-2 it is very difficult to think about the electron energy levels without thinking of them as *spatial*. But a discussion of the spatial arrangement of electrons in atoms is far beyond our objectives for this course. So I advise you to forget thinking about how the electrons are arranged spatially, and simply think in terms of how much energy they have. To facilitate this manner of thinking, let's revise Bohr's model so it looks less spatial, and more oriented toward energy levels, as shown in Figure 12-3. Notice that the higher the energy, the closer the energy levels are to one another.

In this figure I have labeled the energy levels as $n = 1$, $n = 2$, and so on. The lowest energy level is $n = 1$. No electron in an atom can have less energy than this. For an electron to be at level $n = 2$ it must have more energy than an electron in level $n = 1$.

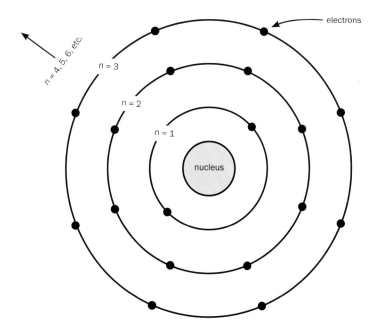

Figure 12-2. Bohr's planetary atomic model.

243

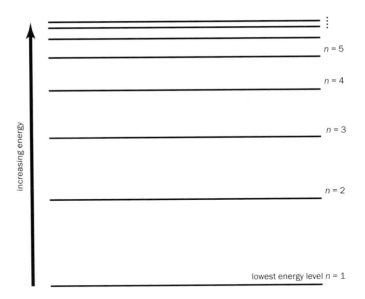

Figure 12-3. Electron energy levels in the atom.

ATOMIC SPECTRA AND THE BOHR MODEL

The Bohr model became popular very rapidly because of the straightforward way it is able to explain the phenomenon of **atomic spectra**. The atoms of every element emit different colors of light when they are excited. (In physics, when we say "excited" we simply mean that the atoms have been energized somehow, such as by heating or being bombarded with electricity or light.) This is how the colors of fireworks are generated. You want green light? Put some copper in the explosion, because when copper atoms get hot they emit a strong green light. You want pale orange? Put some sodium atoms in there. Ever wonder why street lights these days have a pale orange glow? It is because there is sodium vapor inside the lamp fixture and the electricity excites those sodium atoms and causes them to emit their characteristic pale orange color. Why are neon signs red-orange? Because neon atoms give off several specific orange and red colors when they are excited.

Different atoms don't give off just one color when they are excited. Every atom gives off an entire **spectrum** of specific colors. Some of the colors might be emitted very strongly, such as the green light emitted by copper, while other colors are emitted much more weakly. Moreover, the spectrum emitted by the atoms of each element is unique. This means the spectrum of the atoms of a particular element acts as a sort of fingerprint. Every atom of helium, when it is excited, emits the spectrum of colors unique to helium atoms. In fact, this is how we discovered that the sun was composed of a lot of helium, because the light the sun gives off contains the colors of the helium spectrum very prominently. (The Greek word for sun is *helios*, from which we get the element name helium.) **Spectroscopy** is the name for the study of these atomic spectra and the science of identifying the presence of particular elements by looking at light spectra. Astrophysicists use spectroscopy all the time to identify the substances present in distant objects in the galaxy.

The Bohr model of the atom is particularly well suited for explaining atomic spectra. To see this we need to know the relationship between energy and frequency, and how this relates to light and the energy of electrons. As a prelude to this discussion, recall that different colors of light have different wavelengths and frequencies. In the visible spectrum, red light has the lowest frequency and the longest wavelength. Violet light has the highest frequency and the shortest wavelength.

Now, in 1900, German physicist Max Planck (Figure 12-4) put forth two proposals that became the birth moment of quantum mechanics, the new physics that would change the face of physics forever. (He won the 1918 Nobel Prize in Physics for this work.) First, Planck proposed that energy was **quantized**, that is, it was not continuous but was available only in discrete quantities. There is a smallest quantity of energy, one **quantum** of it, and all energies in the universe are multiples of this quantum. Put another way, all energies are actually made up of discrete **quanta**. As it happens, the energy quantum is so small that we never perceive that energies only come in lumps, but they do. This is analogous to the fact that water is lumpy. The smallest bit of water is one molecule of water, but molecules of water are so small that water appears to be smooth and available in any desirable quantity.

Second, Planck put forth an equation relating energy and frequency. This equation is called the **Planck relation**:

$$E = hf$$

Figure 12-4. Max Planck.

where E is energy (J), f is frequency (Hz), and h is a constant called Planck's constant. Planck's constant is incredibly small (6.63 x 10^{-34} J·s, in case you are interested), so even though the frequencies of visible light are on the order of 10^{14} Hz, when you multiply this by Planck's constant you still only have about 10^{-20} J of energy.

The third and final piece of theory we need to present here is that in 1905 Albert Einstein proposed that light itself was quantized, that is, it also came in lumps or **quanta**. (And like everyone else in this chapter, he won the Nobel Prize in Physics in 1921 for this.) This immediately suggests that since light has specific frequencies, the Planck relation gives the amount of energy a single quantum of light has. The term **photon** was coined just a few years later, and is now used to mean a single quantum of light.

We are not going to do any calculations in this course with the Planck relation or Planck's constant. However, to understand how the Bohr model explains atomic spectra you need to notice that in the Planck relation energy varies directly with frequency. This

> **PAUSE FOR A MOMENT AND CONSIDER...**
>
> I want to pause here and simply state that the discoveries that energy and light are quantized are as revolutionary as anything that has ever happened in science. Everything in physics and chemistry changed with these discoveries, and at the dawn of the twentieth century the world became a much different place. Right now I just want you to know this. If you keep studying science, and physics in particular, you will appreciate more down the road just what this meant for the history of the human race. It would be the greatest possible understatement to say that these discoveries were huge.

means that *the higher the frequency of a photon, the higher its energy is*. Now consider the visible light spectrum. From low frequency to high the order of the colors in the spectrum is red – orange – yellow – green – blue – violet. Thus, this is the order of their energies, too. Photons of red light have the lowest frequencies, and thus the lowest energies, in the visible spectrum. Photons of violet light have the highest frequencies, and thus the highest energies, in the visible spectrum. I have summarized these basic conclusions in Figure 12-5.

We are now ready to apply this theoretical background to see how the Bohr model of the atom explains atomic spectra. I have attempted to illustrate these ideas in Figure 12-6. To help you grasp and remember the idea here, I will simply spell this out in the following set of logically ordered propositions. Refer to Figure 12-6 as you study them.

1. When a photon hits an atom, the atom **absorbs** the photon's energy and the photon ceases to exist.
2. The absorbed energy will go to one of the atom's electrons. Since the electrons are in energy levels according to how much energy they possess, an electron that absorbs the energy of an incoming photon will jump up to a higher energy level.

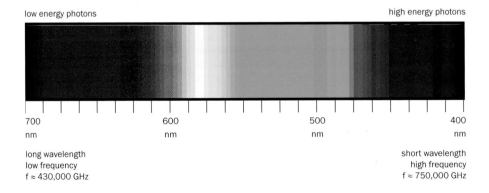

Figure 12-5. Energy and frequency in the visible light spectrum.

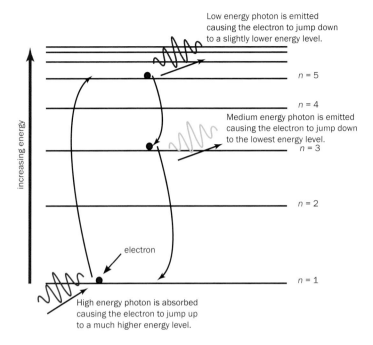

Figure 12-6. Example photon absorption and emission by an electron in an atom.

    This atom, as we say, has been **excited**.
3. The energy of the incoming photon is directly proportional to its frequency, and the more energy an electron absorbs, the higher up it will jump in the energy levels of the atom. Violet photons have high frequencies and thus high energies, and will cause the electrons to make a large energy jump. Red photons have lower frequencies, and thus lower energies, and so cause a smaller energy jump when absorbed. Green photons have intermediate frequencies and energies, and will cause an electron to make an intermediate jump in the energy levels. Photons of ultraviolet light have higher frequency and thus more energy than any visible light, and photons of infrared light have lower frequency and thus lower energy than any visible light photons.
4. When an electron absorbs energy from a photon, it will only stay briefly in the higher energy level. It will work its way back down to a lower energy level by **emitting** one or more new photons, light quanta that fly out of the atom. These emitted photons are the source of the specific colors of light the atom will emit.
5. In every atom there are many different energy levels available. But as we will discuss shortly, a specific energy level can only hold a specific and restricted number of electrons. So when an electron in an excited atom jumps down an energy level and emits a photon of light, the energy it will emit, and thus the frequency of the emitted light, will depend on where there are vacancies in the atom's energy levels. In other words, the energy levels available for all of this electron jumping depend on the atom. Atoms of every element have different numbers of electrons, and

thus different energy levels will be in use or available. Thus, every element has its unique spectrum of light frequencies it will emit. Every element has a unique atomic spectrum.

It is a lot of fun to use a **spectrascope** to look at the atomic spectra for different elements. A spectrascope contains a prism or diffraction grating that spreads out the colors in the light by refraction or diffraction. One looks through the eyepiece to see the spectrum of light entering the spectrascope. If the light entering the spectrascope consists of specific wavelengths of light (instead of containing all of the colors, like the light from a flashlight), then when the light is spread out in the spectrascope the specific colors of light will appear as sharp, separate lines of color. These lines are called **spectral lines**, the lines in the spectrum of a particular element or compound.

In the lab we usually look at atomic spectra by heating a metallic sample in a flame, which is called a *flame test*, or by connecting a glass *spectrum tube* containing the vapor or gas of a particular element to a high-voltage power supply and zapping it with electricity to make it glow. The heat or the electricity provide the energy necessary to excite the electrons in the atoms of the sample. The sample will then begin emitting photons at the specific frequencies in the spectrum for the particular element or elements in the sample. When you look at the light emitted by the sample with a spectrascope you see the individual lines of color in the spectrum. Particular electron energy transitions that happen frequently will produce lots of photons of the corresponding color, resulting in a very bright line in the spectrum. Less common energy transitions resulting in dimmer lines because fewer photons of that energy and frequency are being produced.

Figure 12-7 shows the light given off by three spectrum tubes, neon gas, mercury vapor, and nitrogen gas, when the gases are excited by a high-voltage power supply. These images are actual photographs of energized spectrum tubes, though the light coming from a spectrum tube is so intense that it has to be filtered down quite a bit to enable the digital camera to function properly. Figure 12-8 shows the corresponding spectra for these three elements. These spectra are fascinating to study, and each one is fascinating in its own way!

THE QUANTUM MODEL

Bohr's atomic model was the last developmental stage before the present-day quantum model was put together by the pioneers of quantum mechanics back in the 1920s. Although a lot has been added to the model since then (such as neutrons, discovered in 1932), the quantum model developed then is still our best model today.

In this course we are not going to explore the details of this model too much. Quantum mechanics is a vast and complex subject that raises the hair of every physics student who gets into it. Our goal here is to understand this model well enough for our introduction to chemistry, and chemical reactions are all about electrons. So the only thing we will address here is the way electrons are now regarded in the atom.

One of the major laws of quantum mechanics, proposed by Austrian physicist Wolfgang Pauli in 1925, is called the Pauli exclusion principle. (Do you think Pauli won the Nobel Prize in Physics? Yes he did, in 1945.) The Pauli exclusion principle, which is correct as far as we know, says that no more than one electron in an atom can occupy the same *quantum state*. It takes four different values, called *quantum numbers*,

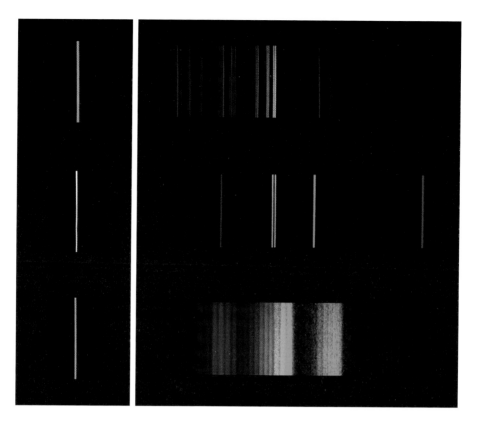

Figure 12-7. The light emitted by spectrum tubes containing neon (top), mercury vapor (center) and nitrogen (bottom).

Figure 12-8. Atomic spectra of neon (top), mercury vapor (center) and nitrogen (bottom).

to specify the quantum state of an electron. Every electron in an atom has to be in a unique quantum state, and to specify this unique quantum state takes four parameters. This is analogous to every mailing address being a unique address. In the U.S. to specify someone's address it takes four parameters – the street number, the street name, the city, and the state. (The zip code doesn't add any new information to the address; it just helps the Postal Service deliver the mail faster.) So just as it takes four parameters to specify your unique mailing address (or five, if you live in an apartment and have an apartment number), the electron has four parameters to specify its unique address – its quantum state – in the atom.

I have been waiting to tell you this: You already know one of these parameters! Remember the energy levels in the atom, numbered as $n = 1$, $n = 2$, and so on? This energy level number n is called the *principal quantum number*. This is the first of the four quantum numbers.

As I have mentioned a couple of times, electrons do not orbit the nucleus in circular pathways. So the term orbit is no longer used. But the electrons do inhabit some kind of region around the nucleus. We now call these regions **orbitals**. There are five different types of orbitals, and the different ones can hold different numbers of electrons. Spelling this out is where we get the other three parameters. As you read, refer to Figure 12-9, which shows the first 20 orbitals arranged according to ascending energy in the energy

ENERGY TRANSITIONS IN THE HYDROGEN ATOM

Are you interested in seeing the *actual* wavelengths emitted by electron energy transitions? The simplest ones are in the hydrogen atom, which only has one electron. This diagram shows the transitions involving the first six energy levels. Johann Balmer discovered the formula for the visible transition wavelengths in 1885, and in 1888 Johannes Rydberg figured out the formula that included all of the hydrogen wavelengths. In 1906 Theodore Lyman observed the ultraviolet transitions that bear his name, and in 1908 Friedrich Paschen first observed the infrared series that bears his. Interesting, isn't it, that the wavelengths follow a *formula*? The mathematical structure of Creation just keeps showing up!

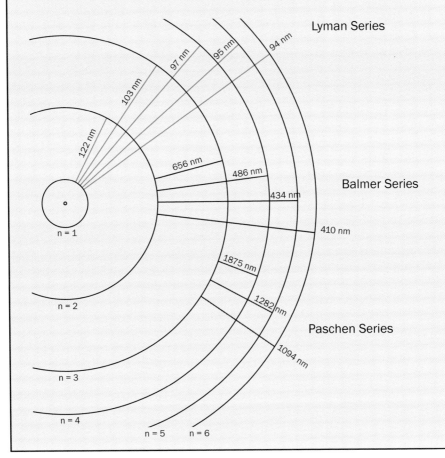

Chapter XII The Bohr and Quantum Models of the Atom

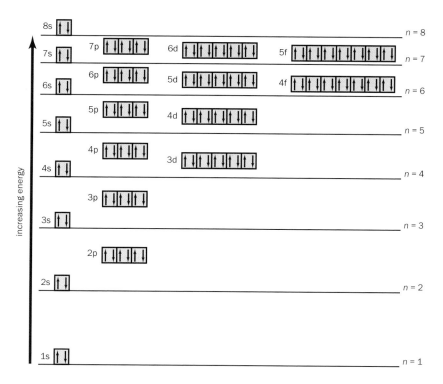

Figure 12-9. The first 20 orbitals in the energy levels in an atom, with arrows representing electrons in the orbitals.

levels of the atom. In this figure the lowest electron energy is at the bottom, and the higher we go up the diagram the higher the energy is. When an electron is not excited it is said to be in the *ground state*. If all of the electrons in an atom are in the ground state, the orbitals in the atom are filled with electrons from the bottom up, that is, from lowest possible energy first and then up to higher energies, in the order shown in Figure 12-9.

I am going to make the presentation of the details about the orbitals short and sweet – just the basics you need to know for now. There is a lot more to all of this than I will present here, but you will pick up this study and take it farther when you get to chemistry.

So there are five different basic kinds of orbitals. They are denoted by the letters *s*, *p*, *d*, *f* and *g*. These letters specify the second of an electron's four quantum numbers.

An *s* orbital can hold two electrons. That's it. But *p*, *d*, *f* and *g* orbitals all have *sub-orbitals*. Every one of these sub-orbitals can hold two electrons. The way these sub-orbitals are structured is very bizarre – we are not going to go there. But putting all of the sub-orbitals together, a set of *p* orbitals can hold six electrons, two in each of three sub-orbitals. A *d* orbital set can hold 10 electrons, an *f* orbital set can hold 14 electrons, and a *g* orbital set can hold 18 electrons. When you specify the sub-orbital an electron is in, you have specified the third quantum number for that electron.

In Figure 12-9 each green square represents a sub-orbital that can hold two electrons. Three green squares are joined together to represent a *p* orbital, because *p* orbitals have three sub-orbitals. Similarly, *d* orbitals have five squares and *f* orbitals have seven

251

squares. As you can see, the *s* orbitals have no sub-orbitals; an *s* orbital just holds two electrons without any sub-orbitals inside of it. In Figure 12-9 I have shown the first 20 orbitals available in every atom. I didn't put any *g* orbitals in the figure, but if I had kept going up to even higher energies beyond the 8*s* orbital the next one up would have been the 5*g* orbital, which would have had 9 squares so it could hold a total of 18 electrons. In Figure 12-9 there are 59 sub-orbitals shown in the first seven energy levels, and together they can hold 118 electrons. The largest atom presently known is element number 118. As we will see shortly, an atom of this element will have 118 electrons. In this atom every orbital in the first seven energy levels will be full. So according to what we know at present, the only way an electron could be in the 8*s* orbital, or in any higher orbital, is for the electron to absorb energy and leave the ground state to jump to a higher energy state.

As for the fourth quantum number, we are not going to study it further. But I know you are dying to know what it is, so I am going to tell you. Electrons have this weird property called spin. From the name you would think this means the electron is spinning on an axis, but forget that. The quantum world of electrons is very strange. Like orbits, spin is just an analogy, not a physical description. Anyway, there are two kinds of spin, up and down. Once you have specified an electron's spin, you have specified the fourth quantum number. So since the Pauli exclusion principle requires every electron to be in a unique quantum state, if two electrons are in the same sub-orbital they must have opposite spins, one up and one down. To represent this I have placed an up arrow and a down arrow in each sub-orbital in the figure. Each arrow represents one electron. In summary, the four quantum numbers to specify the quantum state of an electron are the principle quantum number ($n = 1$, $n = 2$, etc.), the type of orbital (*s*, *p*, *d*, *f* or *g*), the sub-orbital (these are specified by letters like *x*, *y* or *z*), and the spin (up or down).

Let's now use this same diagram to show where the electrons would be in an atom of a specific element. An ordinary atom will have the same number of electrons as it has protons. For this example we will use germanium, which has atomic number 32. This means an atom of germanium has 32 protons, and thus 32 electrons.

Here are the rules for how electron locations are determined. First, in an unexcited atom electrons occupy the lowest available energy levels first. In other words, an atom's electrons will fill in the orbitals in Figure 12-9 from the bottom up. Second, within the sub-orbitals a single electron will reside in each sub-orbital of a particular orbital before any sub-orbital gets a second electron to fill it up. Using these rules, the placement of the electrons in the germanium atom will be as shown in Figure 12-10.

As you see, there are 32 little arrows representing the 32 electrons in the atom. After filling up orbitals up to the 4*s* orbital, the next higher energy is the 3*d* orbital, so this fills up before the 4*p* orbital. Finally, when we arrive at the 4*p* orbital we only have two electrons left. These will go one each in the first two sub-orbitals of the 4*p* orbital.

Students often ask why the numbered orbitals such as 3*d* are not always in the correspondingly numbered energy level. For example, the 3*d* orbital is in the fourth energy level, and its energy is higher than the energy of the 4*s* orbital. The answer to that question goes beyond this text, but you may find at least a little comfort in noticing the patterns the orbitals make in Figure 12-9. In addition to ascending up the energy scale, notice that any sequence of numbered orbitals, such as the 4*s*, 4*p*, 4*d*, 4*f* sequence, form a diagonal in the chart. Atoms are complicated, but when we start seeing patterns it is

Chapter XII
The Bohr and Quantum Models of the Atom

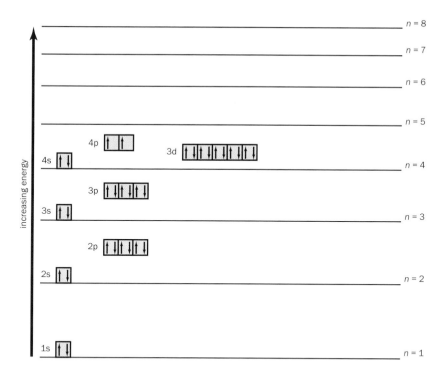

Figure 12-10. Electrons in orbitals in an unexcited germanium atom.

comforting because we know there is much more going on than just random electrons hanging around!

ELECTRON CONFIGURATION NOTATION

Since we just finished discussing the orbitals in the atom where the electrons are located, this is a perfect time to introduce you to a notation that is used to summarize the orbital configuration for the atoms in a given element. This notation is called **electron configuration notation**, and it is basically a list of the orbitals and how many electrons are in each one for a given element. As an example, a chromium atom, as we will discuss in the next section, has 24 electrons. The orbitals fill up the orbitals in Figure 12-9 beginning with the lowest energy until all of an atom's electrons has a place. The electron configuration notation for chromium is written like this:

$1s^2 2s^2 2p^6 3s^2 3p^6 4s^2 3d^4$

As you can see, this is just a list of the orbitals, in order, occupied by the 24 electrons in a chromium atom. The number of electrons in each orbital is written as a superscript after the orbital, so these superscripts need to add up to the number of electrons, which is 24 in the case of chromium. For another example we will write

253

out the electron configuration notation for germanium, which we discussed earlier. Germanium, atomic number 32, has 32 electrons, so the notation is

$$1s^2 2s^2 2p^6 3s^2 3p^6 4s^2 3d^{10} 4p^2$$

ATOMIC MASSES, MASS NUMBERS AND ISOTOPES

There are a few basic pieces of information about elements that we can obtain from the Periodic Table of the Elements (PTE) that we will review here. For this discussion we will use the element chromium as an example. The PTE entry for chromium is shown in Figure 12-11.

Figure 12-11. PTE entry for chromium.

First, the integer with each element that appears at the top of each element's position in the PTE is called the **atomic number**. As I have mentioned a couple of times already, an atom's identity as a particular element depends only on the number of protons an atom has. The atomic number, which is the number the elements are ordered by in the PTE, is the number of protons the element has in each of its atoms. Chromium is element number 24, and thus every chromium atom has 24 protons in its nucleus.

Second, unless an atom is ionized, which we will discuss in the next chapter, the atom has a balance of positive and negative charges. This means that the atomic number also indicates the number of electrons the atom will have. Every non-ionized chromium atom has 24 electrons.

Third, the other number in the element's PTE entry is called the **atomic mass**. This number actually gives the mass in grams of a particular quantity of atoms of the element. However, reading the atomic mass this way is a subject for a full chemistry course, and we will not look into that any further here. But there is another way to interpret the atomic mass, and that is as the average number of particles the atom has in its nucleus. You know that the nucleus contains all of the protons and neutrons in an atom. Together these particles are called **nucleons**. The atomic mass is the average number of nucleons in the atoms of the element. If we round off the atomic mass to the nearest whole number we have the **mass number**, which I will say more about shortly.

You may recall from our discussion of Dalton's atomic model that although every atom of a particular element has the same number of protons, the number of neutrons in an atom can vary. We call these varieties of atoms of the same element **isotopes**. For example, there are three isotopes of hydrogen. The most common form of hydrogen has no neutrons at all (just the single proton in the nucleus). Then there are two other isotopes, one with one neutron and one with two neutrons. (The isotopes with one and two neutrons are called deuterium and tritium, respectively, terms one comes across occasionally in discussions of nuclear energy.) About 99.98% of the hydrogen atoms in the universe have no neutrons. About 0.02% of them have one neutron. And the percentage of hydrogen atoms with two neutrons is minuscule. Most elements, like hydrogen, have one isotope that is very abundant that makes up most of the naturally occurring atoms of the element. All atoms have other isotopes as well, quite a few actually, many of them existing in small percentages. Elements in periods 2 and 3 of the PTE usually have anywhere from a half dozen to 10 or 20 isotopes. Elements in the higher periods tend to have even more because they have a lot more neutrons to play around with.

Since the number of neutrons in an atom can vary, the masses of the atoms of a particular element will vary as well, depending on if the atom has more or fewer neutrons. The atomic mass is the average mass of all of the isotopes of the element, taking into account the relative abundance of the different isotopes in nature. Now since the atomic mass is the average mass of all the isotopes, the mass number I just mentioned above is a very good estimate for the number of nucleons that are present in the most abundant isotope. And once we know the number of nucleons present in a typical atom of some element, we can subtract the atomic number from it (the number of protons) to determine a typical number for how many neutrons the atom will have.

Looking back at the chromium information in Figure 12-11 for an example, an atom of chromium has 24 protons. The atomic mass is 51.996, which rounds to 52. This is the mass number, and tells us the number of nucleons that exist in a typical chromium atom. Subtracting the number of protons from the mass number gives $52 - 24 = 28$ neutrons in a typical chromium atom.

CHAPTER XII EXERCISES

To answer many of the questions below you need to recall the regions in the electromagnetic spectrum (Chapter VII) and the order they are in from low frequency (low energy) to high frequency (high energy).

Spectroscopy and the Bohr Model: Practice Questions Part I

Consider two electron transitions in the hydrogen atom, as shown in the figure below. Each of these results in the emission of a photon from the atom. A particular transition in a hydrogen atom from the fourth energy level, $n = 4$, to the second energy level, $n = 2$, results in the emission of a photon of blue light with a wavelength of 486.1 nm. The color of this light is referred to as *cyan*. Use this information to answer the questions below. You need to answer these questions by reasoning from the principles covered in this chapter, without reference to the box showing all of the hydrogen transitions. You should be able to answer each of these questions without the information given in that box.

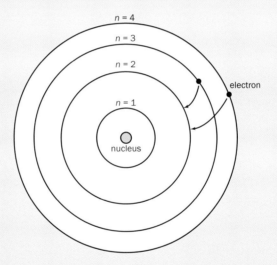

1. The transition from $n = 3$ to $n = 2$ also results in the emission of a photon of visible light. What color might it be? Explain the possibilities.

2. The transition from $n = 4$ to $n = 3$ results in the emission of a photon that is not in the visible spectrum. What kind of radiation might this photon represent?

3. Compare the frequencies, wavelengths and energies of the two transitions shown in the figure.

4. Since the electrons in atoms fill up the lower energy levels first, how could an electron in an atom be in the $n = 4$ energy level with a vacancy in the

$n = 2$ energy level to jump to? How could this happen?

5. What is going on when an electron jumps up to a higher energy level?

6. When an electron jumps to a lower energy level, does it always emit a photon of visible light? If so, why? If not, why not?

Spectroscopy and the Bohr Model: Practice Questions Part II

1. Which has more energy, a gamma ray photon or a microwave photon? Justify your answer.

2. From the point of view of photon energy, why does it make sense that the electromagnetic waves from AM and FM radio stations, which are bombarding us all the time, are not harmful to us?

3. The magnetic field of the earth shields us from the storm of cosmic rays constantly blasting us from the sun. Without this magnetic field life on earth would be destroyed. Explain this from the point of view of the energy in the photons of cosmic rays.

4. Use what you have learned to explain why modest intensities of X-rays (such as are present in a medical procedure) are harmful to humans (necessitating shielding and the wearing of lead aprons during X-ray procedures), but modest intensities of infrared radiation (such as those warming you at a camp fire) are not.

Electron Configuration Notation Exercises

Refer to the Periodic Table of the Elements and write the Electron Configuration Notation for each of the following elements:

1. Carbon

2. Manganese

3. Zinc

4. Bromine

5. Sulfur

PTE Atomic Data Exercises

Use the Periodic Table of the Elements to answer the questions below.

1. State the number of protons in atoms of each of these elements: iodine, iron, magnesium, silver, and potassium.

2. State the number of electrons in non-ionized atoms of each of the following elements: silicon, phosphorus, uranium, gold, and lithium.

3. Estimate the number of neutrons likely to be found in the most common isotope of each of the following elements: fluorine, zinc, calcium, aluminum, and lead.

4. From the mass numbers for neon and sodium, make a list containing as many specific pieces of information as possible about these elements.

5. What are the atomic masses of hydrogen and copper, and what do these values mean?

6. The element copernicium received its official name in 2010 in honor of our old friend Nicolaus Copernicus. It is element 112, symbol Cn. How many neutrons does the most common isotope of copernicium have?

CHAPTER XIII
ATOMIC BONDING

> **OBJECTIVES**
>
> After studying this chapter and completing the exercises, students will be able to do each of the following tasks, using supporting terms and principles as necessary:
>
> 1. Define ion, ionize and polyatomic ion.
> 2. Describe what ionic bonds (crystals) and covalent bonds (molecules) are and how they form.
> 3. Describe what metallic bonds are and how they relate to the formation of crystals and to the "electron sea."
> 4. State the two major goals atoms seek to fulfill when they form bonds with other atoms.
> 5. Given a PTE, state the valence number for atoms of elements in Groups 1-2 and 13-18.
> 6. State the formulas for the polyatomic ions listed in Table 13-5 and give the charge of each. *It is very important that you memorize this list well.*
> 7. Describe what the valence number of an atom is, and how it relates to the position of an element in the PTE.
> 8. Write the formulas for these covalently bonded substances: water (H_2O), carbon dioxide (CO_2), ammonia (NH_3), methane (CH_4) and the diatomic gases H_2, O_2, N_2, F_2 and Cl_2.
> 9. Use electron dot diagrams to represent covalent bonds for the nine substances listed in the previous item.
> 10. Use the valence numbers of elements to identify the number of atoms that will be involved in the formation of ionic compounds.
> 11. Use the charge on a polyatomic ion to identify the number of ions that will be involved in the formation of ionic compounds.
> 12. Write the binary formulas of ionic compounds that will form with given ingredients, including metals, nonmetals, and polyatomic ions.
> 13. Explain in detail why a given element will or will not react with another element, and if they will react, what kind of bond they will form, and how many atoms of each of the elements will be involved in the bond. Explain how all of this relates to the position of an element in the PTE, an element's valence number, and the electrons in an element's valence shell.
> 14. Explain the bonding involved within a polyatomic ion, and between the ion and other elements or ions.

Now that we have covered the history and the essential information about substances and atoms we are ready to get to the real meat of chemistry proper: How chemical bonds work in chemical reactions to form compounds.

Have you studied about the alchemists of old, the men from back in the middle ages practicing alchemy? Their craft was all bound up in mysticism and secret philosophy, but their goal was fairly simple: They were trying to learn how to transmute one metal into another. In particular, they sought ways of turning various metals into gold. They thought this was possible because they saw materials engaging in chemical reactions and thought that changing one metal into another would just be another variety of the same kind of transformation. They didn't know about the chemical bonding processes that are the subject of this chapter.

To discover a process of changing, say, lead into gold would obviously have made one rich (then and now). As we saw back in Chapter X, the characteristic that makes an atom one element or another is the number of protons it has. Gold atoms have 79 protons, period. Lead atoms have 82 protons, period. The only way an atom of lead could be transmuted into an atom of gold would be some kind of nuclear reaction. Nuclear reactions involve changes to the nuclei of atoms. (That is why they are called *nuclear*.) Chemistry doesn't involve nuclei. Chemistry is about *electrons*.

The alchemists didn't know this, and obviously the alchemists didn't have access to any nuclear reactions, either. Even today we have no technology that can simply knock three protons off a lead atom to make a gold atom. So the alchemists' project was doomed to failure. Nevertheless, they were the precursors to the later scientists I wrote about, such as Boyle, Lavoisier, and Priestly, who actually developed the science of chemistry, the science of how atoms bond together to form compounds. And it is to this subject we now turn.

ATOMIC BONDS

There are three basic types of bonds between atoms, **metallic bonds**, **ionic bonds** and **covalent bonds**. I will introduce each of these here, and then treat each one at more length in separate sections below.

Metallic bonds hold the atoms together in metal elements. These are not bonds in compounds, they are just the bonds holding the atoms together in, say, a chunk of iron or an ingot of aluminum. Metallic bonds are crystal structures, like the ionic bonds described next.

Ionic bonds hold the atoms together in bonds formed between metals and nonmetals. Both metallic and ionic bonds result in crystal structures in which the atoms are arranged in a crystal lattice. The arrangement of the atoms in a calcium chloride crystal, an ionically bonded compound, is depicted in Figure 13-1. A crystal lattice, as you recall, is a regular, geometric arrangement of atoms in a three-dimensional structure. (The shaded octahedrons on left side of the figure are there to help you visualize the very amazing geometry of this crystal structure. The green chlorine atoms are visible on the corners of the octahedrons, and there is a calcium atom out of view in the center of each octahedron.)

The third type of bond is the covalent bond. Covalent bonds form between nonmetal elements. Hydrogen also makes a lot of covalent molecules with nonmetal atoms, as we will see. Covalent bonds result in molecules, as in the carbon dioxide molecule, depicted in Figure 13-2. Molecules are individual bundles of atoms bonded together in a group. The atoms are not locked into a lattice the way they are in ionically bonded compounds and metals. (The double bars connecting the atoms in Figure 13-2 represent

Chapter XIII — Atomic Bonding

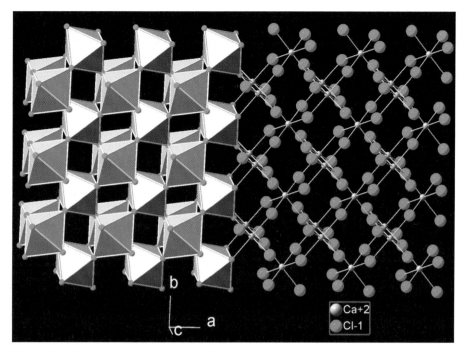

Figure 13-1. Crystal lattice for the ionically-bonded compound calcium chloride.

the double bonds holding the atoms together in the molecule. We will study a few of the possibilities for single, double and triple covalent bonds later in this chapter.)

VALENCE ELECTRONS AND ENERGY LEVELS

Chemical reactions are about electrons. I repeat, chemical reactions are about *electrons*, and where the electrons are in an atom determines how that atom will behave in different reactions. Figure 13-3 shows the energy levels and orbitals again for your reference while you read this description. Remember that each arrow represents where one electron can be stationed in the energy levels of the atom. Each *s* orbital can hold two electrons, each set of *p* orbitals can hold six, and so on. Also remember that an atom has an equal number of protons and electrons, and in an unexcited atom the electrons fill in the energy levels from the bottom up.

There is a very close correspondence between where an atom is in the PTE and where the electrons are in the atom. Take a look at the PTE and notice that there are two elements in the first period. Then notice that the first energy level, n = 1, can hold exactly two electrons. In the PTE there are eight elements in the second period, and the second energy level can hold up to eight electrons. The third period, eight elements; the third energy level, eight electrons. This goes on. Now, as you will read just below, it matters a lot whether the highest occupied energy level in an atom is full or not. In fact, the real action in chemical reactions involves the electrons that happen to be in the highest occupied energy level. The electrons in this highest occupied energy level

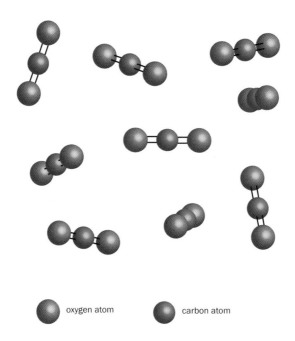

Figure 13-2. Carbon dioxide molecules.

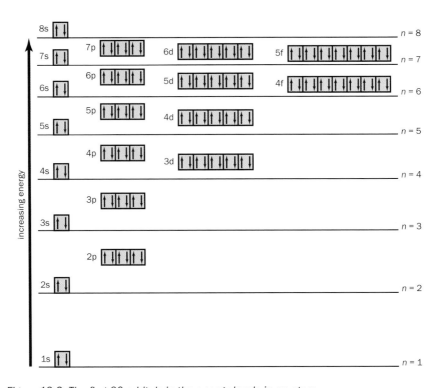

Figure 13-3. The first 20 orbitals in the energy levels in an atom.

are called **valence electrons**. We need to pay attention to how many valence electrons an atom has. The PTE makes this easy.

The energy levels, n = 1, n = 2, etc., depicted in Figure 13-3 are usually called **shells**. This is because the energy levels containing the electrons are often visualized as layers within the atom, like the layers of an onion, one shell inside the other. Remember, this is just a visualization aid; the actual placement and arrangement of electrons in an atom is much more complex (and bizarre) than this (and very difficult to visualize). At any rate, the highest occupied energy level where the valence electrons are is called the **valence shell**. It is the valence shell, and the valence electrons in it, that we have to learn about in order to understand how chemical reactions work.

Let's pick a few elements at random to see how this works. You need to be looking at the PTE inside the back cover of this book while we go through this. The two key factors are how many groups over from the left side of the PTE the element is, and what period the element is in. Specifically notice this: Any atom in Group 1 will have 1 electron in its valence shell, that is, its highest occupied energy level. Any atom in Group 2 will have 2 electrons in its valence shell. Any atom in the third column from the left, which could mean it is in either Group 14 or Group 3, will have three electrons in its valence shell. And so on. Here are some specific examples.

- Magnesium is element number 12. There are 12 electrons in each atom of magnesium. It is in Group 2 and Period 3. Now look at Figure 13-3 to determine how many valence electrons it has. With 12 electrons the first two energy levels will be full, with 2 electrons in level 1, and 8 electrons in level 2. This leaves 2 electrons to go in level 3. Group 2 means 2 valence electrons. Period 3 means the valence electrons are in energy level 3, and energy level 3 is not full, because it can hold 8 electrons. Energy level n = 3 is the valence shell for all of the elements in Period 3.

- Carbon is element number 6. There are 6 electrons in each atom of carbon. It is in Group 14, but more importantly, it is the fourth element from the left in Period 2. Now look at Figure 13-3 to see how many valence electrons it has. With 6 electrons the first energy level is full with 2 electrons. This leaves 4 electrons to go in level 2. Fourth from the left means 4 valence electrons. Period 2 means the 4 valence electrons are in energy level 2, and energy level 2 is not full, because it can hold 8 electrons.

- Chlorine is element number 17 and thus has 17 electrons in an atom. Chlorine is in Group 17, but most important is that it is the seventh element from the left in the PTE in Period 3. Looking at Figure 13-3, we see that 2 electrons will be in energy level 1, 8 in level 2, and 7 in level 3, the valence shell. Seventh element from the left, 7 valence electrons, and the valence shell is not full.

- Neon is element number 10 with 10 electrons. It is in the right hand column of Period 2, the noble gases. Its 10 electrons will completely fill energy levels 1 and 2 with none left over. The highest occupied energy level in a neon atom is full. You may recall that the noble gases like neon are practically nonreactive. This is because the valence shell in a noble gas atom is full.

263

In summary, you can easily spot how many valence electrons an element has by counting what group it is in from the left side of the PTE. The valence electrons will be in the energy level corresponding to the period the element is in in the PTE. If the element is a noble gas the valence shell is full. Otherwise, the valence shell is not full.

GOALS ATOMS SEEK TO FULFILL

Like you, atoms have "goals" for themselves. Unlike you, their goals are not about getting into colleges or living heroically. Atoms do have goals though, two of them. If you will keep these two goals in mind in the sections that follow, the descriptions of atomic bonds will make a lot more sense. I call these Goal Number One and Goal Number Two.

Here are the two goals for an atom, in order of priority:

1. The highest occupied energy level in the atom needs to be full.

   It doesn't matter which energy level this is, as long as it is full. Whatever the easiest way is for this to happen, that's what the atom will do – gaining electrons from some other atom to fill up the valence shell, giving away electrons to another atom to empty out the valence shell, or sharing electrons with some other atom or group of atoms. This is goal number one. Atoms are not content unless the highest occupied energy level is full, and they will interact (i.e., bond) with other atoms to achieve this goal.

2. The atom needs to be in an environment that is electrically neutral.

   In other words, there is no net positive or negative charge from having an imbalance in the number of protons and electrons the atom has. Unless the atom has an equal number of protons and electrons it will have a net electric charge. If it can achieve electrical neutrality by joining together with another atom so that as a team they have an equal number of protons and electrons, that's fine. This makes sense, because if an atom did have a net electric charge on it, it would attract atoms that have the opposite charge and bond with them!

Now that you know about these goals, a couple of comments need to be made about how active the elements are. *Activity* in this context refers to how readily elements will react with other elements. First, the closer an atom is to fulfilling Goal Number 1 the more aggressively it will react. The most aggressive elements in the PTE are those in Groups 1 and 17, because these need only to lose or gain one electron to have the highest occupied energy level full. Alkali metals like lithium and sodium are quite dangerous, because they react quite readily and violently with almost any nonmetal that will enable them to lose their one valence electron. You will see this first hand when we toss a small chunk of sodium into a tub of water. Watch out! It is a very exciting demonstration. Similarly, halogens like chlorine and iodine are dangerous, too. They only need one electron to fill the valence shell and will react violently with many different substances in order to get it. At water treatment facilities trace quantities of chlorine are always added to the water to kill bacteria, which makes our water supply safe. But the piping systems that handle the chlorine, even if made of high grade stainless steel, will often only last a year or two before the chlorine gas corrodes the metal so badly that the pipes and valves have to be replaced. This corrosion is caused by the aggressive chlorine reacting with the metal in the pipes.

The alkaline earth metals of Group 2 are also lively reactors, since they only need to lose two electrons. These elements are not quite as active as the alkali metals, but they are still very active. The same thing goes for Group 16, whose elements only need to gain two electrons.

By contrast to these very aggressive groups, the noble gases of Group 18 don't react much at all under normal conditions. Now you know why. Being in Group 18, their valence shells are full, so these elements are content. With no electrical charge and a full valence shell, there is nothing to motivate these elements to react with others and form compounds. They can be forced to bond with other elements under special conditions in the laboratory or in industrial processes, but the traces of helium and neon atoms we have in our atmosphere are fat and happy and have no plans of reacting at all.

Now you can understand more about the Hindenburg zeppelin disaster of 1937. I am sure you have already learned about this tragedy, which killed 36 people. The Hindenburg airship was full of hydrogen! Hydrogen is a highly flammable gas. Its position in Group 1 indicates how readily it will react. Since the Hindenburg disaster blimps have been filled with helium, a noble gas, instead of hydrogen. From the atomic masses of hydrogen and helium, you can easily see that helium weighs four times as much. But its non-reactivity makes it safe, and this makes it preferable to hydrogen for blimps and balloons.

METALLIC BONDS

Most of our discussion about chemistry will be about ionic and covalent bonds. In fact, almost all of it. But rather than put metallic bonds at the end, where they might be forgotten about, I like to treat them first. They aren't complicated. Then we will move on and spend a lot more time with the other two types of bonds.

As I wrote above, the metallic bond is the bond holding the atoms together in a pure metal. Metals are crystals and their atoms are arranged in a crystal lattice. In Figure 13-4 I have depicted the crystal lattice in a metal. The large circles represent the atoms in the metal. Scattered throughout the lattice are the valence electrons from all of the atoms in the lattice, forming a veritable sea of free electrons. In fact, scientists actually call this sea of electrons the "**electron sea**." These electrons have no idea which atom they belong to. They are free to move around under any electrical influence that might make them move. This is the reason metals conduct electricity so well. Since electric current is electrons moving in a metal wire, all one has to do to get the electrons in a metal moving is apply a voltage difference to the metal and away they go. Electrons in other substances are not free to move this way. Instead, they are locked into place in the bonds holding the atoms together in the substance. As a result, many substances don't conduct electricity very well at all.

Only the valence electrons from the metal atoms are in the electron sea. The electrons in the lower energy levels of each atom remain with the atom in the crystal lattice.

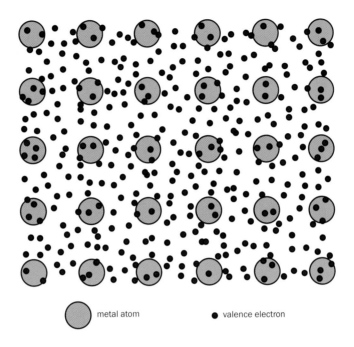

Figure 13-4. Metallic crystal with the "electron sea."

IONIC BONDS

You already know that ionic bonds form between metal and nonmetal atoms, and that the result of this bonding is a crystal structure. Now more specifically, here is what happens. Both the metal atom and the nonmetal atom seek to fulfill Goal Number One. They want their highest occupied energy levels to be full. They will do this the easiest way they can. For a metal, the easiest way to end up having its highest occupied energy level full is usually to give away electrons, emptying out the valence shell. Metals are on the left side of the PTE, which means that they often don't have very many valence electrons, the electrons in the valence shell. So the easiest way to achieve Goal Number One for a metal is to get rid of the valence electrons. Nonmetals are closer to the right-hand side of the PTE, so their valence shells are nearly full. This means that the easiest way for a nonmetal to achieve Goal Number One is to receive enough additional electrons to fill up the valence shell.

When an atom gains or loses one or more electrons it is called an **ion**. The process of gaining or losing electrons is called **ionization**. So ionization enables the atoms to fulfill Goal Number One. They simply ionize and gain or get rid of as many electrons as necessary to have the highest occupied energy level full. However, when an atom ionizes it now obviously does not have an equal number of protons and electrons anymore, so it has an electrical charge. This means the ion is now seeking to fulfill Goal Number Two. But when a reaction is happening there is another ion nearby – if a metal atom ionized by losing one electron, that means there is a nonmetal atom nearby that ionized by gaining one electron. These two ions have opposite charges. One ion is positively charged because it lost an electron and one ion is negatively charged because it gained an electron. The electrical attraction between them because of these opposite charges simply pulls them toward each other and they stick together. Together like this they fulfill Goal Number One by the metal atom transferring one or more electrons to the nonmetal. And they achieve Goal Number Two by sticking together and being electrically neutral as a team.

The key features of ionic bonds are listed in Table 13-1.

Ionic Bonds
1. Ionic bonds form between metal and nonmetal atoms.
2. The metal atom(s) transfers one or more electrons to the nonmetal atom(s) so that the valence shells for both (or all) of them are full. All of the atoms become ions.
3. The ions stick together due to their mutual electrical attraction, so that as a team they are electrically neutral.
4. The ionically bonded atoms form a crystal structure.

Table 13-1. Key features of ionic bonds.

Let's look at a couple of very basic examples of this. Follow along by referencing the PTE as we go. Consider sodium and chlorine. Sodium has one valence electron, because it is in Group 1. Sodium ionizes by losing this valence electron, leaving it with no electrons in energy level 3. Energy level 2 is now the highest occupied energy level and it is full. Chlorine has 7 valence electrons, so it ionizes by gaining 1 more electron to have 8 in energy level 3. So the transfer is 1 electron from the Na atom (energy level 3) to the Cl atom (energy level 3), as shown in Figure 13-5. This chemical reaction forms the compound sodium chloride, NaCl.

Now consider magnesium and oxygen. Magnesium has 2 valence electrons. It ionizes by losing both of them, leaving it with no electrons in energy level 3. Energy level 2 is now the highest occupied energy level and it is full. Oxygen has 6 valence electrons, and it ionizes by gaining 2 more electrons to have 8 in energy level 2. So the transfer is 2 electrons from the Mg atom (energy level 3) to the O atom (energy level 2), as shown in Figure 13-6. This chemical reaction forms the compound magnesium oxide, MgO.

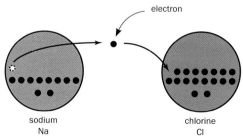

Figure 13-5. Electron transfer in ionic bonding (Na and Cl).

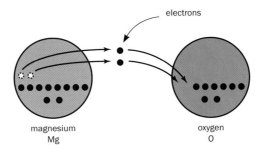

Figure 13-6. Electron transfer in ionic bonding (Mg and O).

So far our examples have matched up metal and nonmetal atoms in one-to-one ratios. But ionic bonds can also form when the number of valence electrons in the metal does not match the number of electrons the nonmetal needs to fill its valence shell. In such cases the ratio of metal to nonmetal atoms will not be one-to-one, but will still be a whole number ratio, just as Dalton said. We will look at two examples.

Calcium, symbol Ca and atomic number 20, has 2 valence electrons since it is in Group 2. Fluorine, symbol F and atomic number 9, needs only one electron to fill energy level 2 since it is in Group 17. Thus, a single calcium atom will donate its 2 electrons to 2 fluorine atoms, 1 electron to each atom of fluorine. I have shown this in Figure 13-7. The compound formed from this reaction is calcium fluoride, CaF_2.

A final example is the reaction between lithium, Li, and sulfur, S, depicted in Figure 13-8. Since lithium is a Group 1 metal, each lithium atom has 1 valence electron to get rid of by ionizing. Sulfur has 6 valence electrons in energy level 3, which holds 8. Thus, 2 lithium atoms will each give up 1 electron, and both of them will go into the third energy level of 1 sulfur atom. Since there are 2 Li atoms and 1 S atom the formula for the compound lithium sulfide will be Li_2S.

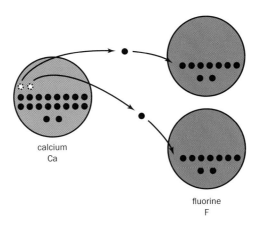

Figure 13-7. Electron transfer in ionic bonding (Ca and F).

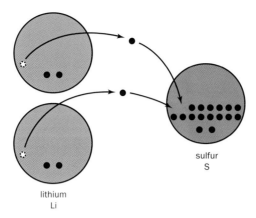

Figure 13-8. Electron transfer in ionic bonding (Li and S).

VALENCE NUMBERS AND IONIC COMPOUND BINARY FORMULAS

When an atom ionizes by losing all of its valence electrons and emptying out its valence shell (metals) or by gaining enough electrons to fill up the valence shell (nonmetals) the atom becomes an ion. We often identify such ions by denoting the charge the ion will have after it ionizes. For example, magnesium ionizes by losing two electrons, and the atom then has 12 protons and only 10 electrons, so it will have a +2 charge. We symbolize this ion by writing Mg^{2+}. Similarly, the lithium ion is Li^+. (Notice that we don't write the number if the number is one, only if it is more than one.) The nonmetals ionize by gaining one or more electrons. So, since they then have more electrons than protons, they become negative ions. Chlorine gains one electron and so becomes the Cl^- ion, and oxygen gains two electrons to become the O^{2-} ion.

These numbers with the + or − sign are called **valence numbers**. The valence number is the *net charge* an atom has after it ionizes. Once we are familiar with how ionic bonding works we can simply write an element's chemical symbol with its valence number and this notation contains all the information we need to put together the formulas for many ionic compounds. The formulas for ionic compounds are called **binary formulas**, because there are two parts to the formula, the symbol for the positive ion and the symbol for the negative ion. I have already written several of these formulas, such as MgO, NaCl, CaF_2 and Li_2S. The rules for writing ionic compound formulas are simple. First, we always write the chemical symbol of the positive ion first. Second, if an element contributes more than one atom to the bond we place a subscript after the element symbol to indicate how many atoms of that element will be involved. Table 13-2 shows several more examples of how binary formulas for ionic compounds are written.

I have two final points to make about ionic bonds. First, the naming convention for the ionic compounds formed is probably obvious to you by now. We simply write down the name of the positive ion (the metal) followed by the name of the negative ion (nonmetal), with the -ide suffix on the nonmetal name. Second, you may have

metal	nonmetal	ionic compound binary formula	compound name
K^+	O^{2-}	K_2O	potassium oxide
Be^{2+}	Br^-	$BeBr_2$	beryllium bromide
Ca^{2+}	Cl^-	$CaCl_2$	calcium chloride
Na^+	S^{2-}	Na_2S	sodium sulfide
Ag^+	P^{3-}	Ag_3P	silver phosphide
Fe^{3+}	O^{2-}	Fe_2O_3	iron oxide

Table 13-2. Writing binary formulas for ionic compounds.

noticed that in my examples the metals always came from Groups 1 and 2. This is because metals in Groups 3 and higher usually have more than one way they ionize when forming compounds, and can even form more than one ionic compound with a particular nonmetal. Dealing with this goes beyond what we will treat in this course. You will learn more about it when you take chemistry. In ASPC when you write formulas for ionic compounds using metals from Groups 1 and 2 you will need to determine the valence numbers for yourself by looking at the PTE. If you are asked to make an ionic compound with a transition metal such as iron or copper you will be told what valence number to use. Further, you only need to know the energy levels and orbitals for elements in Periods 1 through 4, so we will only work with compounds formed from elements in these first four periods.

We will come back to ionic compounds once more near the end of this chapter, but for now we will move on to covalent bonding.

COVALENT BONDS

Covalent bonds only involve nonmetals and hydrogen. Looking at the PTE, there are only 17 nonmetals, six of which are noble gases. The noble gases, as I have said, don't react under normal conditions, so we don't need to worry about them in this course. That leaves only 11 nonmetals plus hydrogen to use for making an enormous number of well-known compounds. Examples of simple molecules you have probably heard of include methane, carbon dioxide, carbon monoxide, propane, nitrous oxide (laughing gas), water, and ammonia. Additionally, there are five common gases that exist naturally as molecules that we will discuss.

As you know, covalent bonds make molecules. In covalent bonds the atoms share electrons in pairs to accomplish Goal Number One. Further, since no atoms are gaining or losing electrons, electrical neutrality never becomes an issue with many compounds, so Goal Number Two is taken care of by default. The main exception to this rule is when polyatomic ions are formed, which we will deal with in a separate section. Our main objective in this section is to understand how covalent bonding works and to look at a few examples. Table 13-3 lists the key features of covalent bonds.

As I have said, covalent bonding involves atoms sharing pairs of electrons. Here is an analogy that may help. Imagine that you and a friend are both into collecting coins. You both have coin collections with those blue coin collecting books (with

Chapter XIII — Atomic Bonding

Covalent Bonds
1. Covalent bonds form between nonmetal atoms, and also hydrogen.
2. The atoms share electrons in pairs.
3. Electrical neutrality is not an issue, except for the case of polyatomic ions.
4. The covalent bond forms a molecule.

Table 13-3. Key features of covalent bonds.

holes punched in the cardboard for you to insert specific coins as you acquire them) and you both have good collections of American quarters. Your friend has two empty places in his book, but you already have these two coins in your book. Likewise, you have two empty places in your quarter collecting book, but your friend already has them both in his book. So here is what you decide to do. Rather than searching around for more coins, you and your friend decide to share. From now on the two quarters he has that you need, and the two that you have that he needs, are going to be considered community property. These four quarters belong to both of you. When you need the quarters for showing off your coin set, you get them. Your friend gets the same privilege. This sharing arrangement has allowed both of you to complete your collections and show them off as complete, and no one has lost anything because the coins are shared.

Covalent bonding works like this. Groups of two or more atoms share electrons – always in pairs – so that all of them are contented and feel like Goal Number One, which is to have the valence shell full, is satisfied. No atoms actually gain or lose electrons. They are always shared. They are always shared in pairs. Sometimes only one pair is shared. Sometimes, like the coin collectors, two pairs are shared. Sometimes even three pairs are shared between two atoms.

We are almost ready for some examples, but first let me introduce you to a kind of sketch called an **electron dot diagram**. In Figure 13-9 I have shown the electron dot diagrams for nitrogen and hydrogen. Electron dot diagrams show an atom, symbolized by its chemical symbol, and its valence electrons. The other electrons in the lower energy levels are not shown. Normally, electron dot diagrams just show the black dots for the valence electrons. I have added in little dotted circles to show where vacancies are in the valence shells of the atoms.

It is customary in electron dot diagrams to place the empty places wherever it is convenient, so don't get too hung up on how I knew where to put the black dots. The important thing is that the atoms are going to fill empty places by sharing electrons in pairs. So the valence electrons need to be drawn around the atom in a way that shows that either a pair is taken care of (like the pair below the nitrogen atom) or there are

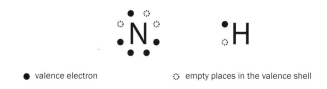

Figure 13-9. Electron dot diagrams for nitrogen and hydrogen.

places where one electron is available to share with an electron from another atom to fill that pair. For elements in Periods 1-3, an atom never has more than 8 electrons in its valence shell, so electron dot diagrams can always be represented with pairs of dots on fours sides of the atom. Or, as with hydrogen, only one pair of dots on one side of the atom.

Looking now at the specifics for these elements, nitrogen, atomic number 7, has 7 electrons. Five of these electrons are in energy level two, which can hold 8 electrons and so is unfilled. These are the 5 valence electrons. Hydrogen, atomic number 1, has one electron in energy level 1, which can hold 2. So to satisfy Goal Number One the nitrogen atom wants 3 more electrons to fill up energy level 2. For its part, the hydrogen atom wants 1 more electron to fill up energy level 1.

Now don't these two diagrams in Figure 13-9 suggest to you a possibility for how nitrogen and hydrogen can share electrons so that all of the atoms feel like they have satisfied Goal Number One? What if that hydrogen atom were just to move over and connect to the nitrogen on the right-hand side? Then the hydrogen and the nitrogen could share a pair of electrons. The hydrogen would be happy. The nitrogen will want to do this two more times on two other sides so that there is a shared pair to fill up each of its three vacancies. This is where a couple of more hydrogen atoms come in. All four of these atoms will get involved in a big group hug that we depict in an electron dot diagram as shown in Figure 13-10. In this figure I have used color to represent the valence electrons belonging to each of the atoms. As you can see, between the nitrogen atom and each of the hydrogen atoms is a shared pair of electrons, one from each atom. This molecule is the ammonia molecule. Its chemical formula is NH_3.

Figure 13-10. Electron dot diagram for ammonia (NH_3).

The electron dot diagram for a covalently bonded molecule shows all of the valence electrons for each of the atoms in the molecule. The element symbols are arranged to show the sharing of electron pairs between each of the atoms in the molecule.

Let's look at a few more examples of molecules with electron dot diagrams. Figure 13-11 shows diagrams for water, H_2O, methane, CH_4, and carbon dioxide, CO_2. In the water molecule, think of the two electrons above the oxygen atom and the two below it and one of the electrons on each side as the valence electrons of oxygen atom (6 of them). Each hydrogen brings 1 valence electron to share in a pair with 1 of the oxygen's valence electrons. You see now the oxygen atom is surrounded by 8 electrons, symbolizing its now full second energy level. Each hydrogen is looking at 2 electrons, symbolizing the full first energy levels of the hydrogen atoms.

In the methane molecule in the center of Figure 13-11, the hydrogen atoms each bring one valence electron as hydrogen always does. The other electron in each of the shared pairs is one of the carbon atom's 4 valence electrons. As the diagram visually

Chapter XIII   Atomic Bonding

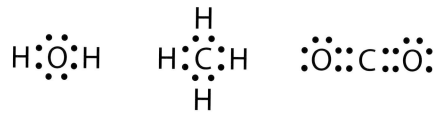

Figure 13-11. Electron dot diagrams for water (H_2O), methane (CH_4), and carbon dioxide (CO_2).

symbolizes, the carbon atom now has 8 electrons in energy level 2, and each hydrogen atom has 2 electrons.

Finally, the carbon dioxide electron dot diagram illustrates a new type of covalent bond. Atoms can share more than one pair of electrons at a time. Each of the oxygen atoms is sharing two pairs of electrons with the carbon atom. The sharing of two pairs is called a **double bond**. But as the diagram depicts, each of the three atoms in the molecule is now surrounded by 8 electrons, indicating that each of them now has a full second energy level. Study this diagram to make sure that the right number of valence electrons is shown. Each oxygen atom has 6 and carbon has 4, for a total of 16 valence electrons. Accordingly, the diagram has 16 dots in it representing these 16 electrons. As you would guess, the bonds we have looked at in the other covalent molecule examples are called **single bonds**, the sharing of a single pair of electrons. In the next section we will encounter a **triple bond**.

DIATOMIC GASES

Back in Chapter XI I mentioned an Italian Scientist, Amedeo Avogadro, who proposed that hydrogen and oxygen gases were composed not of individual atoms, but of molecules, with multiple atoms in each molecule. No one believed him at the time, but Avogadro was right. (I love it when underdogs like Avogadro turn out to be right, so I am a fan of Avogadro. He also has a cool name.) What he actually proposed was the existence of **diatomic gases**, which we are going to learn about here.

In a diatomic gas atoms of the same element covalently bond in pairs to make molecules of the gas. In these gases you will basically never have an atom by itself. They always bond in pairs to fulfill Goal Number One, having the highest occupied energy level full. There are five common elements that do this, hydrogen, nitrogen, oxygen, fluorine and chlorine. Since the atoms in the molecules are in pairs, the formulas for these five gases are H_2, N_2, O_2, F_2, and Cl_2. From their positions in the PTE you can see that hydrogen, fluorine and chlorine each need only one more electron to fill the unfilled energy level. Thus, atoms of each of these elements will form single bonds in the molecule, sharing one pair of electrons with another atom of the same element in the molecule. Oxygen atoms need two electrons, and will thus form double bonds to share two pairs of electrons. Nitrogen atoms need three more electrons, so they will share three pairs of electrons, a triple bond. I have summarized this information in Table 13-4 along with the electron dot diagrams for these molecules, which you need to know.

gas	chemical formula	atomic number	number of valence electrons	number of electrons needed	bond	electron dot diagram
hydrogen	H_2	1	1	1	single	H:H
fluorine	F_2	9	7	1	single	:F̈:F̈:
chlorine	Cl_2	17	7	1	single	:C̈l:C̈l:
oxygen	O_2	8	6	2	double	:Ö::Ö:
nitrogen	N_2	7	5	3	triple	:N:::N:

Table 13-4. Summary information for five diatomic gases.

HYDROGEN

As you have seen in this chapter, hydrogen acts like a Group 17 nonmetal to make single bonds in covalent compounds, even though it appears over on the left side of the PTE with the metals. Like all Group 1 metals, hydrogen ionizes to become H^+, but it also it makes covalent molecules with nonmetal elements. The explanation for this unique behavior should become clear from studying hydrogen's position in the PTE. Because hydrogen is in Period 1, its valence shell only holds 2 electrons, and the hydrogen atom has 1 electron. This means it can ionize by losing its valence electron, just like the other Group 1 metals. But it can form single bonds in a covalently bonded molecule, just like the other Group 17 nonmetals do. As you might guess, hydrogen can also act like a Group 17 nonmetal by forming ionic bonds with metals! The crystal structure for sodium hydride and lithium hydride look just like the crystal structure for sodium chloride. Hydrogen does it all.

As if this weren't enough, hydrogen also acts like a metal, the positive ion, in ionic bonds! These are acids, and though we will not study them this year, you will in chemistry. An example is HCl, known as hydrochloric acid. Many of the acids involve hydrogen bonding with one of the polyatomic ions, which we will discuss next.

POLYATOMIC IONS

Our survey of atomic bonding is nearly finished, but there is one more group of molecules you need to know about. The polyatomic ions, as their name suggests, are covalently bonded molecules made of two or more atoms. ("Poly" means *many*, so polyatomic means "many-atomed"). But unlike ordinary covalently bonded molecules like methane or ammonia, the polyatomic ions are not electrically neutral. When the atoms join up in the molecule, they figure out how to satisfy the goal of filling the unfilled energy level by sharing pairs of electrons, but they do it in a way that does not result in equal numbers of protons and electrons in the molecule, and so the molecule has a net electrical charge.

The atoms in the polyatomic ions are bonded together covalently. Goal Number One is satisfied for this group of atoms, and so they stay together and travel as a group. But since there is the net charge on the molecule, they have not satisfied Goal Number Two, neutral charge. They thus proceed to satisfy this goal by forming ionic bonds with other ions. In other words, a polyatomic ion will act like a single atom and form an ionic bond with another ion, just as we learned in the section on ionic bonds.

There are a lot of different polyatomic ions, but there are 10 you will need to learn for this course, and they are listed in Table 13-5. For each one you need to memorize its name, formula and electrical charge so that you can figure out the ionic compounds it will form.

name	chemical formula	charge	formula with charge
ammonium	NH_4	+	NH_4^+
bicarbonate	HCO_3	–	HCO_3^-
chlorate	ClO_3	–	ClO_3^-
nitrite	NO_2	–	NO_2^-
nitrate	NO_3	–	NO_3^-
hydroxide	OH	–	OH^-
carbonate	CO_3	2–	CO_3^{2-}
sulfite	SO_3	2–	SO_3^{2-}
sulfate	SO_4	2–	SO_4^{2-}
phosphate	PO_4	3–	PO_4^{3-}

Table 13-5. Ten polyatomic ions.

At first, memorizing the information in this table may seem like a major task. However, when you begin studying the table you will start to notice patterns in the formulas and the charges that will expedite the memory work. It is actually not all that bad. If you make flash cards and start practicing you can have it all down in one decent study session. Then be sure to review the ions regularly to keep the material fresh.

When we learned how to write the formulas for ionic compounds, the way I presented it was in terms of electron transfer from the metal to the nonmetal. With polyatomic ions, however, there is a better way to think of it. Instead of trying to follow electrons around simply think in terms of balancing charges. In the ionic compound there will always be a positive ion, which is written first (as usual), joined to a negative ion. You need to have the same number of positive charges as negative charges in the compound. To do this, find the least common multiple between the two charges. Then multiply the charge on the ion or element by applying appropriate subscripts. Note from the examples below that when placing a subscript on one of the polyatomic ions you must put the ion in parentheses to indicate that the subscript applies to the entire molecule.

Table 13-6 lists several examples of ionic compounds formed with polyatomic ions. Note that in the list of polyatomic ions in Table 13-5 there is only one positive ion, ammonium. If the ammonium ion is not being used for the positive ion in the compound then a metal element will be used. When writing formulas for ionic compounds with polyatomic ions in this course we will only use positive ions from elements from Groups 1 and 2 (and ammonium).

As a final check to make sure you understand what the formulas for these compounds represent, consider the number of atoms represented by each of the formulas in Table 13-6. In potassium carbonate there are 2 potassium atoms, 1 carbon atom, and 3 oxygen atoms. This formula therefore represents 6 atoms together in the compound. In the ammonium sulfate, there are 2 ammonium ions, each of which contain 1 nitrogen atom and 4 hydrogen atoms. That's 10 atoms so far. Add in the 1 sulfur atom and the 4 oxygen atoms and we have 15 atoms represented by this formula. You should verify by counting atoms that the calcium phosphate formula represents 13 atoms, and the lithium chlorate formula represents 5 atoms.

positive ion	negative ion	ionic compound formula	compound name
K^+	CO_3^{2-}	K_2CO_3	potassium carbonate
NH_4^+	SO_4^{2-}	$(NH_4)_2SO_4$	ammonium sulfate
Ca^{2+}	PO_4^{3-}	$Ca_3(PO_4)_2$	calcium phosphate
Li^+	ClO_3^-	$LiClO_3$	lithium chlorate

Table 13-6. Writing binary formulas for ionic compounds with polyatomic ions.

Chapter XIII — Atomic Bonding

CHAPTER XIII EXERCISES

Number of Atoms Represented by a Chemical Formula

For each compound listed in this table, determine the number of atoms represented by the chemical formula of the compound. When there are parentheses, the subscript that follows applies to every atom inside the parentheses.

1. $Ca_3(PO_4)_2$	5. $KMnO_4$	9. $CaCl_2$	13. $Al_2(SO_4)_3$	17. KNO_3
2. H_2O_2	6. H_2SO_4	10. NH_4Br	14. $Ba(OH)_2$	18. $Cu(NO_3)_2$
3. $(HH_4)_3PO_4$	7. KCl	11. $Mg(NO_3)_2$	15. $Mg(C_2H_3O_2)_2$	19. Fe_2O_3
4. C_3H_8	8. Hg_2Cl_2	12. $NaC_2H_3O_2$	16. K_2SO_3	20. H_3PO_4

Writing Binary Formulas

Write the binary formulas for the ionic compounds that will form from each of the pairs of ions shown in this table. Valence numbers are given for all metals that are not in Groups 1 or 2.

1. Cr^{2+} S	5. Ni^{2+} O	9. Cr^{3+} S	13. Li Br	17. Ag^+ O
2. Ba F	6. Zn^{2+} S	10. K S	14. Fe^{2+} O	18. Cu^+ Cl
3. Al^{3+} I	7. Mg P	11. Mn^{2+} Br	15. Cu^{2+} Cl	19. Pb^{4+} O
4. Fe^{3+} O	8. Pb^{2+} O	12. Na Cl	16. Mn^{4+} Br	20. Ni^{3+} O

Types of Chemical Bonds

For each compound listed in this table, write the name of the compound and classify the bond in the compound as ionic or covalent. If the compound involves both kinds of bonds write "both."

1. LiCl	5. HCl	9. H_2O	13. Na_2CO_3	17. $CaCl_2$
2. CH_4	6. NO_2	10. SO_3	14. NaF	18. MgO
3. $BaSO_4$	7. NH_4Cl	11. KI	15. $AlPO_4$	19. P_2O_5
4. FeF_3	8. $Ca(OH)_2$	12. CO_2	16. N_2O_3	20. K_2O

Writing Formulas with Polyatomic Ions

Make a table like this one and fill it in with the formula for the compound formed in each cell, putting together the ion at the left end of the row and the top of the column.

	hydroxide	carbonate	phosphate	sulfite	chlorate
aluminum (3+)					
ammonium					
potassium					
hydrogen					
calcium					
lead (4+)					
magnesium					
sodium					

Atoms and Atomic Bonding Study Questions and Exercises

1. Distinguish between neutrons and nucleons.

2. Distinguish between atomic mass and mass number.

3. Calculate the total mass of all the electrons, neutrons and protons (all of them together) in a common boron atom. Forget about significant digits in this calculation. Use every digit you've got.

   The subatomic particle masses are:

   $m_p = 1.673 \times 10^{-27}$ kg
   $m_n = 1.675 \times 10^{-27}$ kg
   $m_e = 9.11 \times 10^{-31}$ kg

4. Define isotope.

5. Describe two important features of the noble gases.

6. Give the names for the elements in the following groups.
   a. 1
   b. 2
   c. 17
   d. 3-12
   e. elements along the diagonal separating metals from nonmetals

Chapter XIII — Atomic Bonding

7. Find tungsten in the Periodic Table of the Elements and list the following information.
   a. atomic number
   b. atomic mass
   c. mass number
   d. number of protons
   e. number of neutrons (common isotope)

8. Define *ductile* and state what type of element has this property.

9. Use electron configuration notation to describe the electron configurations of the following elements.
   a. Al
   b. K

10. Write a few complete sentences which explain the difference between the three types of atomic bonds.

11. Consider the two primary goals atoms have when forming chemical bonds. Write two complete sentences which describe how these two factors are involved in an ionic bond.

12. For the following pairs of elements, indicate the element which is most chemically active, and describe how the PTE informs your answer.
    a. Ar and O
    b. N and O
    c. Na and O

13. Write two complete sentences which explain how and why the atoms in diatomic gases bond together.

14. Think about the arrangements of the electrons in the orbitals of the halogens. Write one or two complete sentences which explain what all of these elements have in common.

15. Will sodium and sulfur react together? Why or why not? If they will, describe the bond they will form (what kind of bond, how many atoms, etc.).

16. In the reaction represented by the chemical equation below, two hydrogen molecules combine with one oxygen molecule to produce two molecules of water. But within the water molecule, two hydrogen atoms will combine with one oxygen atom to form a water molecule. Based on what you know about valence electrons, why does the H_2O bond happen this particular way?

    $$2H_2 + O_2 \rightarrow 2H_2O$$

17. For the following pairs of reactants, state whether they will react, what

279

type of bond they will form if they will, and how many of each atom will be present in the compound formed. Write out the chemical formula for the compound formed by the reaction (if there is one). Finally, write the name of the compound formed (except letter (f), which requires a rule we have not discussed).
   a. Ca and Cl
   b. Ca and Ar
   c. Ca and O
   d. Na and N
   e. Mg and K
   f. S and O
   g. H and O
   h. Mg and NO_3
   i. NH_4 and Na
   j. Be and PO_4

18. Write the electron dot diagrams for methane, ammonia, carbon dioxide, water and the five diatomic gases. For each one write a sentence or two explaining how the shared pairs of electrons in the molecule allow each atom to achieve the goal of filling its valence shell.

19. Write a short but solid explanation for why neon (Ne) doesn't appear in any of the compounds in the examples or exercises in this chapter. (Hint: Consider its position in the PTE.)

Answers
3. $1.8419555 \times 10^{-26}$ kg

CHAPTER XIV
CHEMICAL REACTIONS

> **OBJECTIVES**
>
> After studying this chapter and completing the exercises, students will be able to do each of the following tasks, using supporting terms and principles as necessary:
>
> 1. Describe four different types of chemical reactions and identify examples of each.
> 2. Balance chemical equations.
> 3. Explain how a balanced chemical equation illustrates the law of conservation of mass.
> 4. Distinguish between exothermic and endothermic reactions.
> 5. Use energy diagrams to discuss how activation energy relates to exothermic and endothermic reactions.
> 6. Describe collision theory and how it relates to three factors that govern the rate of chemical reactions.
> 7. Explain what a catalyst is.
> 8. Relate collision theory to the factors affecting the rate of chemical reactions, and use these concepts to explain various common chemical reactions and technologies.

FOUR TYPES OF CHEMICAL REACTIONS

In this last chapter we will take a brief look at chemical reactions and the chemical equations we write to represent them. The elements or compounds that go into a reaction at the beginning are called **reactants**. The elements or compounds that are formed during the reaction and are present at the end are called the **products**. We write **chemical equations** that show the reactants combining to form the products as shown in Figure 14-1. The arrow shows the direction the chemical reaction goes, from reactants to products. The reactants in the equation are always placed on the left side of the arrow,

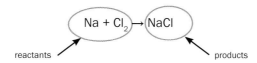

Figure 14-1. Reactants and products in chemical equations.

products on the right. This figure shows the reaction between sodium and chlorine. In this reaction an ionic bond is formed between ionized sodium atoms and the chlorine ions that will come out of the diatomic chlorine molecule. The ions come together to produce, or synthesize, sodium chloride, or table salt. (In the figure the reaction is not "balanced." We will discuss balancing equations in the next section.)

There are four basic types of reactions. These four types do not describe every possibility, and some reactions do not fit neatly into one of the four categories. But these four types are a helpful way to begin thinking about chemical reactions.

In a **synthesis reaction** two elements or compounds are joined together to form a new compound. If we called one of the reactants A and the other one B, this reaction can be symbolized as follows:

$$A + B \rightarrow C$$

The reaction between sodium and chlorine shown in Figure 14-1 is an example of a synthesis reaction. The reactants are separate, and after the reaction they are together in the product compound. Another example is the reaction between hydrogen and oxygen to form water:

$$2H_2 + O_2 \rightarrow 2H_2O$$

Those peaceful symbols belie the fact that to make this reaction happen we need some energy, like a spark, and then the hydrogen explodes. That is, this reaction with oxygen happens very dramatically and rapidly. (Remember my reference to the Hindenburg in the previous chapter?) The hydrogen reaction with oxygen to form water is an explosion. Remarkable, isn't it, that a fiery explosion can produce water?

I am sure you noticed the coefficients in the hydrogen/oxygen reaction. A **coefficient** is a number placed just before one of the elements or compounds in the chemical equation. Mentally cover up those two coefficients for a moment. Without the coefficients you can easily see that on the left we have diatomic hydrogen gas, H_2, and diatomic oxygen gas, O_2. These are the reactants. The product is water, H_2O. The coefficients come in during the process of balancing the equation, which we will address soon.

The second type of reaction is the **decomposition reaction**. The decomposition reaction is the opposite of the synthesis reaction. Instead of elements or compounds joining together, we have a compound that is the reactant, and this compound comes apart to form two or more products. This is usually symbolized like this:

$$A \rightarrow B + C$$

An example of a decomposition reaction is the decomposition of carbonic acid, H_2CO_3, into water and carbon dioxide. The equation for this reaction is

$$H_2CO_3 \rightarrow H_2O + CO_2$$

As you probably know from reading the labels on soda cans, soft drinks often contain carbonic acid. But at room temperature there is enough energy available to get this decomposition reaction going. (We will discuss this energy, called the *activation energy*, later in the chapter.) So the carbonic acid decomposes into CO_2, which contributes to the bubbles that come out of the beverage, and water, which dilutes the beverage. So, after this reaction has run its course, the drink will have no fizz and it will be more watery. Yuck.

The third type of reaction is the **single replacement reaction**. Let's describe this with an analogy. A gentleman and a lady are dancing together in the traditional ballroom style. As I am sure you know, in such dances another gentleman may "cut in" by tapping the dancing man on the shoulder and requesting to dance with the lady in his place. (I have never understood why the second gentleman couldn't just wait until the next dance. If I was dancing with a desirable lady I wouldn't want any other guys cutting in. But that's a different issue.) Before the cutting-in takes place, we have gentleman A alone and gentleman B dancing with the lady, whom we will call C. After the cutting in gentleman A is dancing with the lady and gentleman B is alone. This interaction can be symbolized as follows:

$A + BC \rightarrow AC + B$

You can easily recognize single replacement reactions by noticing that an element is alone among the reactants on the left, but among the products on the right that element is no longer alone. It is now in a compound and a different element is alone.

A very exciting example of a single replacement reaction is what happens when sodium is placed in water. The chemical equation is

$2Na + 2H_2O \rightarrow 2NaOH + H_2$

As you can see, sodium is alone on the left while hydrogen and oxygen are together in the water molecules. After the reaction hydrogen gas is by itself and the sodium is in the compound NaOH. You should be able to recognize this compound from our study of polyatomic ions. It is called sodium hydroxide.

Now by itself, this reaction is amusing to watch but no big deal. The thing is, this reaction produces a great deal of heat. In fact, so much heat that the hydrogen gas that is one of the products will ignite (because there is oxygen handy in the air) and explode. This is the hydrogen and oxygen synthesis reaction I mentioned earlier, and it is quite exciting. I always show this reaction to my students.

The fourth type of reaction is the **double replacement reaction**. How about another dancing analogy? This time we have two couples dancing and they trade partners in the dance. If we call A and C the gentlemen and B and D the ladies, we could symbolize this as follows:

$AB + CD \rightarrow AD + CB$

You might just notice that if we assume that the gentlemen, A and C, represent the positive ions, then they have to be written first in all four compounds, as I have done.

An example of double replacement reaction is

$MgCO_3 + 2HCl \rightarrow MgCl_2 + H_2CO_3$

From our studies on bonding in the previous chapter you can figure out every compound in this reaction. You can also see how the positive ions have switched places with the carbonate and chlorine negative ions.

BALANCING CHEMICAL EQUATIONS

Think back to our survey of the history of chemistry and atomic models. Several scientists were discovering a fundamental principle about chemical reactions. I mentioned that in the 1790s, French chemist Joseph Proust proposed the law of definite proportions, stating that elements combine in definite mass ratios to form compounds. Antoine Lavoisier, the father of chemistry, formulated the **law of conservation of mass** in chemical reactions, which says that the mass of the reactants is equal to the mass of the products. Then a few years later John Dalton proposed his famous atomic theory, which included the basic principle that atoms combine in whole number ratios to form compounds.

All of these statements are correct, and they are all getting at the same basic idea. Chemical reactions occur when groups of atoms combine together or come apart, and this happens because of the dance between atoms' valence electrons that we have studied. In each chemical reaction there is a specific ratio of elements or compounds in the reactants to elements or compounds in the products. Further, all of the atoms present at the beginning of the reaction are there somewhere in the end.

For example, since a water molecule has two hydrogen atoms in it and only one oxygen atom, we will need twice as many hydrogen atoms as oxygen atoms to form water molecules. Actually, since both hydrogen and oxygen are diatomic gases, they are already molecules to start with, so we will need twice as many hydrogen *molecules* as oxygen *molecules* to form a given number of water molecules. All chemical reactions are like this. Definite proportions between reactants will produce specific quantities of products.

When we write chemical equations the equations themselves show this proportionality. Put another way, every chemical equation is an illustration of Lavoisier's law of conservation of mass. All of the atoms that go into the reaction are there somewhere in the products. In balancing chemical equations we are, in effect, identifying the proportions of the reactants necessary to make the reaction work.

When considering the reaction between certain reactants that will produce certain products we start with the chemical equation showing the reactants and products. To balance the equation we have to make sure that every element is represented in equal quantities on both sides of the equation. We do this by placing coefficients in front of the compounds in the equation. We can never change the subscripts on the compounds, because the compounds are *defined* by the subscripts. The subscripts are determined by examining how the elements will bond to produce the compound, and once that is determined the subscripts are a given and cannot be changed.

Sometimes balancing chemical equations takes some trial and error. But by following some general principles you can arrive at a balanced equation quickly. Here are some guidelines to balancing chemical equations.

1. The goal is to have the same number of each type of atom represented on both sides of the equation.

2. You may only add coefficients; you may not change or add any subscripts. The subscripts define the compounds that take part in the reaction. Only coefficients are used to specify proportions.
3. A coefficient in front of a compound like $3CH_4$ applies to every element in the compound. Thus $3CH_4$ indicates 3 atoms of carbon and 12 atoms of hydrogen.
4. If there is an element by itself in the equation save it for last, because you can always add a coefficient in front of it at the end of the exercise without affecting anything else.
5. If an element appears in an even quantity on one side of the equation and an odd quantity on the other side, a good place to start will be by placing coefficients to make it even on both sides.
6. Think of polyatomic ions as single objects in the equation, since any coefficient placed in front of a compound will affect the entire polyatomic ion together.
7. Make sure at the end you have removed common factors from the coefficients. For example, if all of the coefficients are even, then they should all be divided by two.
8. Double check your atom count at the end.

Balancing chemical equations isn't really very hard. Some examples will illustrate the procedure.

Consider the reaction between aluminum oxide and carbon to form elemental aluminum and carbon dioxide gas. As you can see, this is a single replacement reaction in which carbon replaces the aluminum in a bond with the oxygen. We start with the unbalanced equation simply showing the compounds in the reaction.

$$Al_2O_3 + C \rightarrow Al + CO_2$$

The thing to check first is if there are an odd number of atoms of an element on one side of the equation and an even number of the same element's atoms on the other side. In this reaction there are 3 oxygens on the left and 2 on the right. The only way to make them match is to place coefficients on both compounds so that the oxygens go to 6, the least common multiple of 2 and 3. This gives us the equation

$$2Al_2O_3 + C \rightarrow Al + 3CO_2$$

Now carbon and aluminum are both isolated elements in the equation. Always save these isolated elements for last because you can place coefficients on them without affecting anything else. Placing the coefficients to balance these two elements gives us

$$2Al_2O_3 + 3C \rightarrow 4Al + 3CO_2$$

This is the balanced equation. As a final check count elements on both sides to make sure they all match. On each side we have 4 aluminums, 6 oxygens, and 3 carbons.

In our next example hydrogen combines with nitrogen to form ammonia. This is a synthesis reaction, and the basic compounds are all covalent compounds you know by heart:

$$H_2 + N_2 \rightarrow NH_3$$

Both nitrogen and hydrogen have the odd-even issue, so it doesn't really matter which side we start on. So I will start with balancing the nitrogen by placing a coefficient of 2 on the ammonia molecule.

$$H_2 + N_2 \rightarrow 2NH_3$$

Finally, balancing the hydrogen completes the equation.

$$3H_2 + N_2 \rightarrow 2NH_3$$

Our third example is the burning of methane, which produces carbon dioxide and water. Methane is the main constituent of natural gas. The compounds are

$$CH_4 + O_2 \rightarrow CO_2 + H_2O$$

Here, oxygen has the odd-even problem. Placing a coefficient of 2 in front of the water molecule gives us an even number of oxygens on the right and takes care of the hydrogen at the same time. Then a coefficient of 2 on the oxygen molecule completes it.

$$CH_4 + 2O_2 \rightarrow CO_2 + 2H_2O$$

As a final example we will look at the following double replacement reaction with polyatomic ions.

$$Fe_2(SO_4)_3 + Ba(OH)_2 \rightarrow BaSO_4 + Fe(OH)_3$$

Since there are 3 sulfates on the left and only one on the right, I will start by placing a coefficient of 3 on the barium sulfate on the right. Doing so throws the barium out of balance, so I will also go ahead and place a three on the barium hydroxide on the left.

$$Fe_2(SO_4)_3 + 3Ba(OH)_2 \rightarrow 3BaSO_4 + Fe(OH)_3$$

Now we see that with 6 hydroxides on the left, we will need a coefficient of 2 on the iron hydroxide on the right to have 6 hydroxides there as well. This takes care of balancing the iron at the same time, giving

$$Fe_2(SO_4)_3 + 3Ba(OH)_2 \rightarrow 3BaSO_4 + 2Fe(OH)_3$$

ENERGY IN CHEMICAL REACTIONS

Analyzing the way chemical reactions use energy is a very powerful tool in understanding the reactions. In this section we look at two major ways energy considerations apply to chemical reactions.

You know that if I open the gas jet in the lab and natural gas, which is basically methane, sprays into the room, the gas will not automatically ignite. However, if a source of flame is nearby the gas will ignite and remain lit. You also know that you can mix the ingredients together for a cake and pour them in a pan, but unless the batter is placed in an oven the batter will not turn into cake. The heat is necessary to get the reaction going that will convert the batter into cake.

The minimum energy necessary to initiate a particular chemical reaction is called the **activation energy**. The internal energy of the atoms or molecules in the reactants must acquire this amount of energy, the activation energy, for the reaction to occur. When the methane gas comes into the room the methane molecules do not have enough energy to react with the oxygen molecules in the air and burn. A flame gives off enough heat that the methane molecules near the flame will absorb enough energy so that their internal energy acquires the activation energy and the reaction will occur. After that the reaction will continue because as the methane burns it gives off heat energy (as all flames do), providing the activation energy other methane molecules need in order to react with the oxygen.

You will be able to understand this more clearly if we now bring in another energy concept. During chemical reactions the reactants will either absorb energy from their surroundings or release energy into the surroundings. In **exothermic reactions** heat energy is released. In **endothermic reactions** heat energy is absorbed. These two types of reactions are illustrated in Figure 14.2.

The two curves in this figure represent the internal energy in the compounds in a reaction over time as the reaction progresses. The upper figure shows how this internal energy changes over time in the case of an exothermic reaction when heat is released, and the lower curve illustrates the same thing for an endothermic reaction when heat is absorbed. In each case the compounds possess a certain initial energy, E_i, before the reaction begins. This energy is the internal energy of these compounds. As you recall from back in Chapter VI, this internal energy depends on the temperature. Then the energy curve goes up because the compounds are being heated in order to get the reaction going. Once the reactants have been provided with the activation energy, E_a, the reaction can occur. The right end of the curve is at energy E_f, the final energy after the reaction has runs its course. In the exothermic reaction E_f is lower than E_i because energy was released during the reaction. In an endothermic reaction E_f is higher than E_i because the compounds absorbed energy during the reaction.

In some cases reactants at room temperature already possess enough internal energy for a particular chemical reaction to occur. In this case the only thing required to get the reaction to happen is to put the reactants together. Pouring vinegar into baking soda

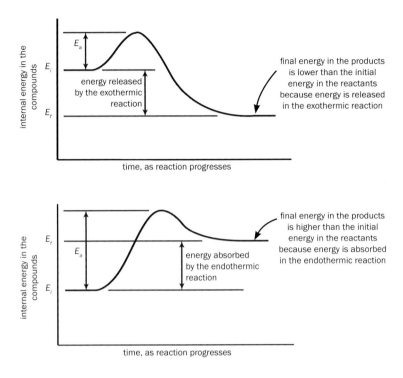

Figure 14-2. Energy in compounds during exothermic and endothermic reactions.

or tossing sodium into water are both examples of reactions that will happen at room temperature without the need for additional heat.

An exothermic reaction releases energy in the form of heat, light, or sound. As you know, fires give off both heat and light. In explosions, which are just rapid combustions, acoustic energy is released as well in the sound of the explosion. If you hold a beaker of chemicals in your hand while an exothermic reaction is going on in the beaker, the beaker will feel hot. Similarly, if you hold a beaker of chemicals in your hand while an endothermic reaction is going on in the beaker, the beaker will feel cold. The compounds in an endothermic reaction are absorbing heat, so heat flows from your warm hand into the cold compounds. When heat flows out of our hands this way the object we are holding feels cold.

Chemical reactions are not the only processes that can be endothermic or exothermic. Any process that absorbs or releases heat energy can be called an endothermic or exothermic process. Just to illustrate, instant hot and cold packs are great examples of endothermic and exothermic processes. A cold pack contains separate compartments of ammonium nitrate and water. When you twist the pack the separation between these compounds is broken and the ammonium nitrate begins dissolving into the water. This process absorbs a lot of heat, so the pack feels cold on your skin as heat flows from your skin into the cold pack. Hot packs contain ionic compounds such as calcium chloride and magnesium sulfate. Both of these dissolve in water exothermically. That is, heat is released. When acids dissolve in water they release heat as well, sometimes a *lot* of heat. This makes mixing an acid in water a potentially dangerous process, so chemistry students all learn how to do it properly. You will learn more about this in your future chemistry course.

REACTION RATES AND COLLISION THEORY

In this final section we will examine four factors that affect how fast a chemical reaction will proceed. For example, as everyone who has ever tried to build a fire knows, a fire will start easier with tiny twigs than with big logs, and once started the fire burns brighter and hotter when you blow on it. We will look at the chemical theory that provides an explanation for these facts.

One factor that affects how fast a reaction will proceed is the presence of a **catalyst**. A catalyst is a substance that makes a reaction go faster by its mere presence, without actually taking part in the reaction itself. In other words, the catalyst is not one of the reactants in the chemical reaction, but is merely present while the reaction is taking place. There are many different catalysts used for many different reactions, but I discovered one interesting use of a catalyst during the years I worked as an engineer in the oil industry. To produce enough gasoline to meet demand, oil companies have to convert heavy molecules found in crude oil into lighter molecules that are used in gasoline. This process is called catalytic cracking, because the larger molecules in the crude oil are "cracked" into smaller molecules. Essentially, the oil is combined with a platinum powder. The platinum powder has been heated to a very high temperature, and when it is mixed with the oil the molecular bonds in the oil are actually broken (or cracked), forming the smaller molecules gasoline is made of. The platinum is not one of the reactants or products in the cracking reaction (it is simply cleaned off and reused), but it is essential to the mechanism that makes the reaction happen.

Catalysts work by lowering the activation energy required for the reaction to occur, as illustrated in Figure 14.3. When a catalyst is present with the reactants the activation energy necessary for the reaction to occur is lower, so the reaction progresses more rapidly. And one more thing about catalysts before we move on. In biology, catalysts are called *enzymes*. Nearly everyone these days has at least heard of enzymes because we hear about them all the time in reports about nutrition or treatments for diseases. Now you have a bit of an idea of what they are. No doubt your future biology teacher will tell you a lot more.

Collision Theory provides us with three more factors that will affect the rate of chemical reactions. According to collision theory, chemical reactions are enabled when atoms or molecules of the reactants are able to collide with one another. Any set of circumstances that increases the frequency of the collisions between the reactants' particles will thus increase the reaction rate. There are three ways to increase the collisions between particles.

The first is to increase the **temperature** of the reactants. As you recall, the temperature of a substance is a measure of the internal energy of the substance, which is, in turn, the sum of the kinetic energies in the particles of the substance. A higher temperature means higher kinetic energy in the particles. If particles have a higher kinetic energy they are moving faster and collisions between them will increase. This is how we use collision theory to explain why reactions happen faster at higher temperatures. I like to use this example to help students visualize this: If we all stood up and roamed around at

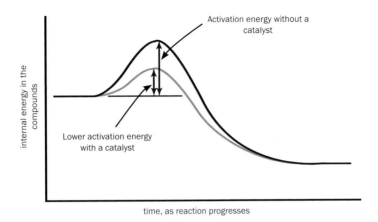

Figure 14-3. A catalyst reduces the activation energy necessary for a reaction to occur.

slow speeds throughout the classroom, we might occasionally bump into one another, but it wouldn't happen frequently because of our slow speed. But if we all began dashing around the room at top speed we would be bumping into one another constantly. In the same way, hot, fast moving atoms or molecules collide more frequently than cooler ones moving at slower speeds, so the reaction proceeds more rapidly.

The second way to increase collisions between the reactants' particles is by increasing the **surface area** of the substances that are reacting together. A log of wood has a certain amount of surface where the carbon atoms in the wood can come in contact with the oxygen molecules in the atmosphere and react (burn). But if the same log is split into smaller sticks of wood, they will burn much faster (and thus hotter), because many more of the carbon atoms are exposed to the air. If we went to an extreme with this and ground the wood all the way to sawdust, we would find that the wood can burn so fast that it can actually explode. Professional woodworkers know that sawdust in a wood shop can be explosive, so they avoid sparks and smoking cigarettes, and they use vacuum systems to keep the sawdust contained to reduce the danger of an explosion from a random spark. (Interestingly, some catalysts work by increasing the surface area available for reactions.)

The third way to increase a reaction rate by exploiting collision theory is to increase the **concentration** of the reactants. An increase in concentration simply means there are more of the particles of the reactants around, so more collisions occur. Oxygen is one of the reactants in every combustion, but only about 20% of the molecules in the air are oxygen molecules. Any technique that increases the concentration of oxygen will increase the reaction rate of a combustion reaction. When you blow on a fire to get it going, you are simply pumping loads of fresh oxygen molecules into the region where the reaction is happening, so more collisions occur between the oxygen molecules and the particles of the fuel. Thus, the reaction rate increases. Now, if just blowing on a fire makes it burn faster and hotter, consider what might happen if one actually pumped some pure oxygen into the fire. Brother, watch out! A few years ago I looked into getting liquid oxygen at our school for some demonstrations. As I was talking to the local distributor of such products about what was involved, he told me that liquid oxygen is so dangerous that if the delivery truck comes in the heat of July and dispenses the liquid oxygen into the customer's container out in an asphalt parking lot, a simple spill of a few drops of the stuff onto the hot asphalt can make the asphalt burst into flames! Yikes!

Think back to my analogy about us bumping into one another in the classroom. Clearly, even if we all move slowly there will be more collisions between us if there are more of us in the room. That's what is going on here. The concentration of oxygen atoms in liquid oxygen is so high that just spilling it onto warm asphalt can start the combustion reaction.

CHAPTER XIV EXERCISES

Balancing Chemical Equations

For each of the equations listed below, balance the equation and identify what type of chemical reaction it represents.

1. $HCl + NaOH \rightarrow NaCl + H_2O$

2. $Mg + HCl \rightarrow MgCl_2 + H_2$

3. $NaCl \rightarrow Na + Cl_2$

4. $KClO_3 \rightarrow KCl + O_2$

5. $N_2 + H_2 \rightarrow NH_3$

6. $Al + NiBr_2 \rightarrow AlBr_3 + Ni$

7. $Cu + Ag(NO_3) \rightarrow Cu(NO_3)_2 + Ag$

8. $Al + O_2 \rightarrow Al_2O_3$

9. $CaCl_2 + F_2 \rightarrow CaF_2 + Cl_2$

10. $(NH_4)_2SO_4 + Ba(NO_3)_2 \rightarrow BaSO_4 + NH_4NO_3$

Energy and Rates in Reactions Study Questions

1. One of the headache medicines from years back was called BC Powder. This medicine was a powder, not a tablet or capsule. Why would BC Powder help a headache faster than a tablet or capsule? Relate your answer to collision theory.

2. Using the language of chemistry, explain what "burning" is. Then explain why wood burns faster if it is surrounded by hot coals than it does in a new fire which doesn't have any coals yet. Relate your answer to collision theory.

3. When I demonstrate the burning of magnesium, I always first show the students that the magnesium will not light easily with a match. However, with a Bunsen burner the metal ignites immediately. Explain why was it that the magnesium would not ignite with a match but it would ignite with a Bunsen burner. Relate your answer to activation energy.

4. To cook a pancake I have to put energy into it by heating it. Where does all this energy go? Answer the question by discussing energy in the reactants

and products during the reaction and after the reaction, as well as energy which escapes as heat out into the room at different times.

5. I did some work for a while in an aluminum processing factory that handled finely-powdered aluminum. This powdered aluminum was extremely explosive. Use collision theory to explain why.

6. When painting with some highly aromatic paints one has to keep the room well ventilated or else people can become dizzy or ill. Even with the room ventilated, one can still smell the paint, so why does the ventilation help prevent dizziness? Relate your answer to collision theory.

7. I once saw a physics professor make a cigar-shaped tube of aluminum foil, closed on one end. He lightly packed the tube with cotton wadding and placed the tube on a table top. He then picked up the tube, poured in a small amount of liquid oxygen, quickly set the tube down and lit the end of the cotton with a burning stick of wood (called a splint). Holy rocketry, Batman! That cotton burned so fast that the aluminum foil tube shot off the table like a bullet. Explain why this happened, according to collision theory. While you are at it, explain the tube's motion using Newton's Third Law of Motion.

8. When baking in an oven, if the temperature is too high the outside of the food may burn before the inside of the food is fully cooked. Explain why this happens using collision theory.

9. When cleaning the outside of a house with a cleaning solution (which has chlorine bleach and detergent in the water) the cleaning will go much faster with larger amounts of cleaner in the water. Explain this using collision theory. (And by the way, you should always follow the directions when using products of this kind.)

10. You have probably noticed signs at gasoline pumps prohibiting smoking while filling a car with gasoline. Explain how these signs relate to activation energy.

11. At the beginning of school one year we noticed that many of the little screws in the equipment in the optics lab had rusted because the building air conditioning system had been turned off over the summer. Use collision theory to explain why this happened. (It is likely that there are at least two different effects going on.)

12. What is collision theory? Explain it clearly in four short sentences or less.

APPENDIX A
CONVERSION FACTORS, PREFIXES AND PHYSICAL CONSTANTS

In ASPC you are required to know the conversion factors, prefixes and physical constants listed in the first table on this page by memory. You will be required to use these values throughout the year in computations on quizzes and assignments. Other conversion factors that you will need, such as the ones given in the second table, will be provided to you on quizzes. The values in the yellow cells are exact and thus will not affect your significant digit count. Values in white cells are rounded, and you must include the precision of the value in your assessment of the significant digits in the problem.

Conversion Factors and Equations	
5280 ft = 1 mi	2.54 cm = 1 in
3 ft = 1 yd	1000 cm³ = 1 L
365 days = 1 year	1000 L = 1 m³
3600 s = 1 hr	1 mL = 1 cm³
$T_C = \frac{5}{9}(T_F - 32°)$	$T_K = T_C + 273.2$
Metric Prefixes	
centi- (c) = 1/100 = 10^{-2} (Thus, there are 100 cm in 1 m.)	kilo- (k) = 1000 = 10^3 (Thus, there are 1,000 m in 1 km.)
milli- (m) = 1/1000 = 10^{-3} (Thus, there are 1,000 mm in 1 m.)	mega- (M) = 1,000,000 = 10^6 (Thus, there are 1,000,000 m in 1 Mm.)
micro- (μ) = 1/1,000,000 = 10^{-6} (Thus, there are 1,000,000 μm in 1 m.)	giga- (G) = 1,000,000,000 = 10^9 (Thus, there are 1,000,000,000 m in 1 Gm.)
nano- (n) = 1/1,000,000,000 = 10^{-9} (Thus, there are 1,000,000,000 nm in 1 m.)	
Physical Constants	
c = 3.00 x 10^8 m/s (speed of light in a vacuum or in air)	ρ_{water} = 998 kg/m³ (density of water at room temperature)
g = 9.80 m/s² (acceleration due to gravity)	

Additional Conversion Factors You do not need to memorize these. They are written here for your convenience.			
0.3048 m = 1 ft	1,609 m = 1 mi	4.45 N = 1 lb	3.786 L = 1 gal

APPENDIX B
CHAPTER OBJECTIVE LISTS

CHAPTER I THE NATURE OF SCIENTIFIC KNOWLEDGE

1. Define science, theory, hypothesis, and scientific fact.
2. Explain the difference between truth and scientific facts, and describe how we obtain knowledge of each.
3. Describe the "Cycle of Scientific Enterprise," including the relationships between facts, theories, hypotheses, and experiments.
4. Explain what a theory is and describe the two main characteristics of a theory.
5. Explain what is meant by the statement, "a theory is a model."
6. Explain the role and importance of theories in scientific research.
7. State and describe the steps of the "scientific method."
8. Define explanatory, response and lurking variables in the context of an experiment.
9. Explain why an experiment must be designed to test only one explanatory variable at a time. Use the procedures the class followed in the Pendulum Lab as a case in point.
10. Explain the purpose of the control setup in an experiment.
11. Describe the possible implications of a negative experimental result. In other words, if the hypothesis is not confirmed, explain what this might imply about the experiment, the hypothesis or the theory itself.

CHAPTER II MOTION AND THE MEDIEVAL MODEL OF THE HEAVENS

1. Define and distinguish between velocity and acceleration.
2. Use scientific notation correctly with a scientific calculator.
3. Calculate distance, velocity and acceleration using the correct equations, MKS units, and correct dimensional analysis.
4. Use from memory the conversion factors, metric prefixes, and physical constants listed in Appendix A.
5. Explain the difference between accuracy and precision, and apply these terms to questions about measurement.
6. Demonstrate correct understanding of precision by using the correct number of significant digits in calculations and rounding.
7. Draw and interpret graphs of distance, velocity, and acceleration vs. time, and describe an object's motion from the graphs.
8. Describe the key features of the Ptolemaic model of the heavens, including all of the spheres and regions in the model.
9. State several additional features of the medieval model of the heavens and relate them to the theological views of the medieval Church.
10. Briefly describe the roles and major discoveries of Copernicus, Tycho, Kepler, and Galileo in the Copernican Revolution. Also, describe the significant later contributions of Isaac Newton and Albert Einstein to our theories of motion and gravity.
11. Describe the theoretical shift that occurred in the Copernican Revolution and how the Christian Church was involved.

12. State Kepler's three Laws of Planetary Motion.
13. Describe how the gravitational theories of Kepler, Newton and Einstein illustrate the way the Cycle of Scientific Enterprise works.

CHAPTER III NEWTON'S LAWS OF MOTION

1. Define and distinguish between matter, inertia, mass, force and weight.
2. State Newton's Laws of Motion.
3. Calculate the weight of an object given its mass, and vice versa.
4. Perform calculations using Newton's Second Law of Motion.
5. Give several examples of applications of the Laws of Motion that illustrate their meaning.
6. Explain why the First Law is called the Law of Inertia.
7. Use Newton's Laws of Motion to explain how a rocket works.
8. Apply Newton's Laws of motion to application questions, explaining the motion of an object in terms of Newton's Laws.

CHAPTER IV VARIATION AND PROPORTION

1. Identify and graph linear and nonlinear variation.
2. Identify the constant of proportionality in a physical equation.
3. Identify direct, inverse, square, inverse square and cubic proportions when seen in equations.
4. Identify direct, square and inverse proportions when seen graphically.
5. Identify dependent and independent variables in physical equations.
6. Graph functional relationships, making proper use of dependent and independent variables.
7. Normalize non-essential constants and variables in a given physical equation, and describe how the remaining two variables vary with respect to each other.
8. Describe the way variations occur between variables in physical equations, including:
    a. area of a triangle
    b. area of a circle
    c. volume of a sphere
    d. gravitational potential energy
    e. kinetic energy
    f. pressure under water
    g. force of gravitational attraction
    h. Charles' Law
    i. Boyle's Law

CHAPTER V ENERGY

1. State the Law of Conservation of Energy.
2. Describe how energy can be changed from one form to another, including:
    a. different forms of mechanical energy (kinetic, gravitational potential, elastic potential)

b. chemical potential energy
   c. electrical energy
   d. elastic potential energy
   e. thermal energy
   f. electromagnetic radiation
   g. nuclear energy
   h. acoustic energy
3. Briefly define each of the types of energy listed above.
4. Describe two processes by which energy can be transferred from one object to another (work and heat), and the conditions that must be present for the transfer to occur.
5. Describe in detail how energy from the sun is converted through various forms to end up as energy in our bodies, or as energy used to run appliances in your home or machines in industry.
6. Define work, gravitational potential energy and kinetic energy.
7. Calculate kinetic energy, gravitational potential energy, work, heights, velocities, and masses from given information using correct dimensional analysis.
8. Define friction.
9. Using the Law of Conservation of Energy and equations for W, E_K and E_G, calculate and graph the energy an object has at various stages in a mechanical system.
10. Describe how gravitational potential energy and kinetic energy vary with respect to height, mass and velocity.
11. Using the pendulum as a case in point, explain the behavior of ideal and actual systems in terms of mechanical energy.
12. Explain how friction affects the total energy present in a mechanical system.

CHAPTER VI HEAT AND TEMPERATURE

1. Define and distinguish between heat, internal energy, thermal energy, thermal equilibrium, specific heat capacity, and thermal conductivity.
2. State the freezing/melting and the boiling/condensing temperatures for water in °C, °F, and K.
3. Convert temperature values between °C, °F, and K.
4. Describe and explain the three processes of heat transfer: radiation, conduction and convection.
5. Describe how temperature relates to the internal energy of a substance and to the kinetic energy of its molecules.
6. Explain the kinetic theory of gases, and use the kinetic theory of gases to explain why the pressure of a gas inside a container is higher when the gas is hotter and lower when the gas is cooler.
7. Apply the concepts of specific heat capacity and thermal conductivity to explain how common materials such as metals, water, and thermal insulators behave.

CHAPTER VII WAVES, SOUND AND LIGHT

1. Define what a wave is.

2. On a graphical representation of a wave, identify the wave parameters and parts: crest, trough, amplitude, wavelength and period.
3. Define the frequency and period of a wave.
4. Describe the following five wave phenomena, giving examples of each:
   a. Reflection
   b. Refraction
   c. Diffraction
   d. Resonance
   e. Interference
      i. Constructive interference
      ii. Destructive interference
5. Give examples of longitudinal, transverse, and circular waves and the media in which they propagate.
6. Define infrasonic and ultrasonic, and give examples of these types of sounds.
7. Define infrared and ultraviolet, and give examples of these types of radiation.
8. Calculate the velocity, frequency, period and wavelength of waves from given information.
9. Given the frequency of a wave, determine the period, and vice versa.
10. List at least five separate regions in the electromagnetic spectrum, in order from low frequency to high frequency.
11. State the frequency range of human hearing, in hertz (Hz), and the wavelength range of visible light, in nanometers (nm).
12. State the six main colors in the visible light spectrum in order from lowest frequency to highest.
13. Identify the relations: frequency and pitch; amplitude and volume.
14. Explain how waves of different frequencies (harmonics) contribute to the timbre of musical instruments.

CHAPTER VIII ELECTRICITY AND DC CIRCUITS

1. Explain what static electricity and static discharges are.
2. Describe three ways for static electricity to form, and apply them to explain the operation of the Van de Graaff generator and the electroscope.
3. Using the analogy of water being pumped through a filter, give definitions by analogy for voltage, current, resistance, and potential difference.
4. Explain what electric current is and what produces it.
5. Explain why electric current flows so easily in metals.
6. Describe the roles of Alessandro Volta and James Clerk Maxwell in the development of our knowledge of electricity.
7. State Kirchhoff's two circuit laws.
8. Calculate the equivalent resistance of resistors connected in series, in parallel, or in combination.
9. Use Ohm's Law and Kirchhoff's Laws to calculate voltages, currents, and powers in DC circuits with up to six resistors.
10. Draw graphs of power vs. current, power vs. voltage, and power vs. resistance, and explain the shape of the graphs in terms of the relationships between variables in the equations derived from Ohm's Law.

CHAPTER IX FIELDS AND MAGNETISM

1. Explain what a field is.
2. Describe three major types of fields, the types of objects that cause each one, and the objects or phenomena that are affected by each one.
3. State Ampere's Law.
4. State Faraday's Law of Induction.
5. Explain the difference between the theories of gravitational attraction of Einstein and Newton.
6. Apply Ampere's Law and Faraday's Law of Magnetic Induction to given physical situations to determine what will happen.
7. Explain the general principles behind the operation of solenoids, generators and transformers.
8. Use the right-hand rule to determine the direction of the magnetic field around a wire or through a solenoid.
9. Explain why transformers work with AC but not with DC.

CHAPTER X SUBSTANCES

1. Define and distinguish between these terms pertaining to substances: substance, pure substance, alloy, mixture, compound, heterogeneous mixture, homogeneous mixture, suspension, element and solution.
2. Make a tree chart that relates together in their proper hierarchy the terms in the previous item. In each category of the chart list two or more examples of representative common substances.
3. Use and define the following terms: solution, solute, solvent, solubility, soluble, saturated solution, supersaturated solution and precipitate.
4. Define and perform calculations pertaining to solubility.
5. Define and distinguish between these additional terms: chemical symbol, chemical formula and atom.
6. Given either the name or the chemical symbol of any of the elements listed in Table 10-1, give the corresponding symbol or name.
7. Identify these basic parts of the Periodic Table of the Elements: periods, groups, alkali metals, alkaline earth metals, noble gases, halogens, transition metals, metalloids, inner transition metals (also called rare earth elements) and nonmetals.
8. Define and distinguish between physical properties, chemical properties, physical changes, and chemical changes. Give several examples of each.
9. Define and give examples of the terms malleable and ductile.
10. Write a paragraph explaining the relation between atoms, elements, molecules, chemical reactions and compounds.

The following two objectives are treated in Chapter II and in Appendix C in the experiment on Solubility:

11. Use the terms meniscus and parallax error when discussing experimental error.
12. Describe how accuracy and precision relate to measurements made with a triple-beam balance or graduated cylinder.

CHAPTER XI ATOMIC MODELS

1. State the key features of the atomic models envisioned by Democritus, Dalton, Thomson and Rutherford. State the key atomic discoveries each of these men made, and identify the aspects of their atomic models that we now regard to be correct or partially correct.
2. Describe the key experiments and contributions of Lavoisier, Mendeleev, Thomson, Millikan and Rutherford.
3. State the law of conservation of mass in chemical reactions.
4. Use the density equation to compute the density, volume or mass of a substance.
5. Calculate the volume of right rectangular solids and right cylinders.

CHAPTER XII THE BOHR AND QUANTUM MODELS OF THE ATOM

1. Define these terms: quantum, quanta, quantized, photon, absorption, emission, spectrascope, excitation, atomic spectrum, spectral lines, principle quantum number, atomic energy levels, energy transition and orbital.
2. Name, in order, the orbitals available for electrons to fill in the first four energy levels (principal quantum numbers).
3. State the number of electrons that can reside in the s, p and d orbitals.
4. Describe the Bohr model of the atom.
5. Use the Planck relation relating the energy of a photon to its frequency, $E = hf$, to explain why larger electron energy transitions result in light toward the violet in color, and lesser electron energy transitions result in light toward the red in color.
6. Given a Periodic Table of the Elements for reference, use electron configuration notation to write the electron configuration for any element from hydrogen (atomic number 1) through krypton (atomic number 36).
7. Explain the principle of spectroscopy, or how elements can be identified by the characteristic wavelengths of light emitted by their atoms.
8. Describe how the Bohr model of the atom provides a powerful explanation for the phenomenon of atomic spectra.
9. Use the Periodic Table of the Elements (PTE) to determine atomic number, atomic mass and mass number.
10. Use the atomic mass and atomic number listed in the PTE to determine the numbers of protons, electrons and neutrons an atom of a given element will have.

CHAPTER XIII ATOMIC BONDING

1. Define ion, ionize and polyatomic ion.
2. Describe what ionic bonds (crystals) and covalent bonds (molecules) are and how they form.
3. Describe what metallic bonds are and how they relate to the formation of crystals and to the "electron sea."
4. State the two major goals atoms seek to fulfill when they form bonds with other atoms.
5. Given a PTE, state the valence number for atoms of elements in Groups 1-2 and 13-18.

6. State the formulas for the polyatomic ions listed in Table 13-5 and give the charge of each.
7. Describe what the valence number of an atom is, and how it relates to the position of an element in the PTE.
8. Write the formulas for these covalently bonded substances: water (H_2O), carbon dioxide (CO_2), ammonia (NH_3), methane (CH_4) and the diatomic gases H_2, O_2, N_2, F_2 and Cl_2.
9. Use electron dot diagrams to represent covalent bonds for the nine substances listed in the previous item.
10. Use the valence numbers of elements to identify the number of atoms that will be involved in the formation of ionic compounds.
11. Use the charge on a polyatomic ion to identify the number of ions that will be involved in the formation of ionic compounds.
12. Write the binary formulas of ionic compounds that will form with given ingredients, including metals, nonmetals, and polyatomic ions.
13. Explain in detail why a given element will or will not react with another element, and if they will react, what kind of bond they will form, and how many atoms of each of the elements will be involved in the bond. Explain how all of this relates to the position of an element in the PTE, an element's valence number, and the electrons in an element's valence shell.
14. Explain the bonding involved within a polyatomic ion, and between the ion and other elements or ions.

CHAPTER XIV CHEMICAL REACTIONS

1. Describe four different types of chemical reactions and identify examples of each.
2. Balance chemical equations.
3. Explain how a balanced chemical equation illustrates the law of conservation of mass.
4. Distinguish between exothermic and endothermic reactions.
5. Use energy diagrams to discuss how activation energy relates to exothermic and endothermic reactions.
6. Describe collision theory and how it relates to three factors that govern the rate of chemical reactions.
7. Explain what a catalyst is.
8. Relate collision theory to the factors affecting the rate of chemical reactions, and use these concepts to explain various common chemical reactions and technologies.

APPENDIX C
LABORATORY EXPERIMENTS

The following pages contain your guidelines for the six laboratory experiments we will do in ASPC during the year. For each of these experiments you will submit an individual written report. It is your responsibility to study *The Student Lab Report Handbook* thoroughly so that you can meet the expectations for lab reports in this course.

The instructions written here are given to help you complete your experiment successfully. However, your report must be written in your own words. This applies to all sections of the report. Do not copy the descriptions in this appendix into your report in place of writing your report for yourself in your own words.

LAB JOURNALS

You must maintain a proper lab journal throughout the year in ASPC. Your lab journal will contribute to your lab grade along with your lab reports. In Chapter 1 of *The Student Lab Report Handbook* you will find a detailed description of the kind of information you should carefully include in your lab journal entries. I will not repeat that entire description here, but I will list the important highlights.

A good lab journal will include all of the following features:

1. The pages in the journal are quadrille ruled (graph paper) and the journal entries are in pencil.
2. The journal is maintained very neatly, and is free of sloppy marks, doodling, and messiness.
3. Every entry includes the date and the names of the team members working together on the experiment.
4. Every experiment and every demonstration that involves taking data is documented in the journal.
5. Entries for each experiment or demonstration include:
    - the team's hypothesis
    - an accurate list of materials and equipment, including make and model of any electronic equipment or test equipment used
    - tables documenting all of the data taken during the experiment, including the units of measure and identifying labels for all data
    - all support calculations used during the experiment or in preparation of the lab report
    - special notes documenting any unusual events or circumstances, such as bad data that required doing any part of the experiment over, unexpected occurrences or failures, or changes to your experimental approach
    - little details about the experiment that need to be written in the report that you may forget about later
    - important observations or discoveries made during the experiment.

EXPERIMENT 1 | THE PENDULUM LAB
Experimental Methods and Procedures

This experiment is an opportunity for you to learn about conducting an effective experiment. We are going to conduct an investigation involving a simple pendulum, and in this investigation you will learn about manipulating variables, collecting careful data, and organizing data in tables in your lab journal.

In this fun experiment your goal is to determine the factors (variables) that influence the period of a simple pendulum. A pendulum is an example of a mechanical system that is *oscillating*, that is, repeatedly "going back and forth" in some regular fashion. In the study of any oscillating system an important parameter is the *period* of the oscillation. The period is the length of time (in seconds) required for the system to complete one full cycle of its oscillation. After thinking about the possibilities and forming your team hypothesis, you will construct your own simple pendulum from string and some weights and conduct tests on it to determine which variables actually do affect its period and which ones do not.

In class you will explore the possibilities for variables that may affect the pendulum's period. Within the pendulum system itself there are three candidates, and your instructor will lead the discussion until the class has identified them. (We will ignore factors such as air friction and the earth's rotation in this experiment. Just stick to the obvious variables that clearly apply to the problem at hand.)

Then as a team continue the work by discussing the problem for a few minutes with your team mates. In this team discussion you will form your own team hypothesis stating which factor(s) you think will affect the period. To form this hypothesis you will not actually do any new research or tests. Just use what you know from your own experience to make your best guess.

The central challenge for this experiment will be to devise an experimental method that tests only one variable at a time. Your instructor will help you work this out, but the basic idea is to set up the pendulum so that two variables are held constant while you test the system with large and small values of the third variable to see if this change affects the period. You will have to test all combinations of holding two variables constant while manipulating the third one. All experimental results will be entered in tables in your lab journal. Recording the data for the different trials will require several separate tables. For each experimental set-up you should time the pendulum three times and record the result in your lab journal. Repeating the trials this way will enable you to verify that you have good, consistent data. To make sure you can tell definitively that a given variable is affecting the period, you should *make the large value of the variable at least three times the small value* in your trials.

Let me give you a bit of advice about how to measure the period of your pendulum. The period of your pendulum is likely to be quite short, only one or two seconds, so measuring it directly with accuracy would be difficult. Here is an easy solution: Assign one team member to hold the pendulum and release it on a signal. Assign another team member to count the number of swings the pendulum has completed, and another member as a timer to watch the second hand on a clock. When the timer announces "GO" the person holding the pendulum releases it, and the swing counter starts counting.

After exactly 10.0 seconds the timer announces "STOP" and the swing counter states the number of swings that have been completed. Record this value in a table in your lab journal. If you have four team members, the fourth person can be responsible for recording the data during the experiment. After the experiment the data writer can read off the data to the other team members so they can enter the data in their journals.

This method of counting the number of swings in 10 seconds does not give a direct measurement of the period, but you can see that your swing count will work just as well for solving the problem posed by this experiment, and is a lot easier to measure than the period itself.

One more thing on measuring your swing count: Your swing counter should state the number of swings completed to the nearest 1/4 swing. When the pendulum is straight down, it has either completed 1/4 swing or 3/4 swing. When it stops to reverse course on the side opposite from where it was released, it has completed 1/2 swing.

When you have finished taking data, review the data together as a team. If you did the experiment carefully your data should clearly indicate which variables affected the period of the pendulum and which ones did not. If your swing counts for different trials of the same set-up are not consistent, then something was wrong with your method. Your team should repeat the experiment with greater care so that your swing counts for each different experimental set-up are consistent.

You will have some time in class to discuss your results with your team members and reach a consensus about the meaning of your data. You should expect to spend about four hours writing, editing and formatting your report. Lab reports will count a significant percentage of your science course grades throughout high school, so you should invest the time now to learn how to prepare a quality report.

Your goal for this report is to begin learning how to write lab reports that meet all of the requirements outlined in *The Student Lab Report Handbook*. One of our major goals for this year is to learn what these requirements are and become proficient at generating solid reports. Nearly all scientific reports involve reporting data, and a key part of this first report is your data tables, which should all be properly labeled and titled.

After completing the experiment all of the information you will need to write the report should be in your lab journal. If you properly journaled the lab exercise you will have all of the data, your hypothesis, the materials list, your team members' names, the procedural details, and everything else you need to write the report. Your report must be typed, and will probably be around 2 to 2 ½ pages long. You should format the report as shown in the examples in *The Student Lab Report Handbook*, including major section headings and section content.

Here are a few guidelines to help you get started with your report:

1. There is only a small bit of theory to cover in the Background section, namely, to describe what a pendulum and its period are. You should also explain why we are using the number of swings completed in 10 seconds in our work in place of the actual period. As stated in *The Student Lab Report Handbook*, the Background section must include a brief overview of your experimental method and your team's hypothesis.
2. In this experiment we are not making quantitative predictions, so there will be no calculation of experimental error. We are simply seeking to discover which variables affect the period of a pendulum, and which do not.

3. Begin your Discussion section by discussing your data and considering how they relate to your hypothesis. Since there was no quantitative prediction for this experiment, there will be no calculation of experimental error. Your goal in the Discussion section is to identify what your data say and relate that to your reader.
    a. What variables did you manipulate to determine whether they had any effect on the period of the pendulum?
    b. What did you find? Which ones did affect the period?
    c. Were you surprised by what you found?
4. If you would like to produce a really outstanding report, consider exploring the following questions in your discussion:
    a. Many clocks use pendulums to regulate their speed. What is it about pendulums that makes them good for this?
    b. How would this work in an actual clock?

EXPERIMENT 2	THE SOUL OF MOTION LAB, OR THE PICK-UP TRUCK LAB (or your own clever title!)
	Newton's Second Law of Motion

Note: The report for this experiment requires the student to set up a graph showing predicted and experimental curves on the same set of axes. Procedures for creating such a graph in two different versions of Microsoft Excel (one for PCs and one for Macs) are described in detail in *The Student Lab Report Handbook*. Procedures for creating graphs in other applications, such as on a Mac using Pages, are available as free downloads from novarescienceandmath.com.

You will have a great time with this experiment. We will meet out in the parking lot as a class. We are going to push a vehicle from the rear using scales that measure how hard the pushers are pushing. We will time the vehicle as it accelerates from rest through a ten-meter timing zone and use the time data to calculate the experimental values of the vehicle's acceleration. We will use the mass of the vehicle and Newton's Second Law to predict what the acceleration should be for each amount of pushing force used. Our goal will be to compare our predicted accelerations to the experimental values for four different force values. We will graph the results and calculate the experimental error to help us see how they compare.

This experiment is an excellent example of how experiments in physics actually work. The scientists have a theory that enables them to predict, in quantitative terms, what the outcome of an experiment should be. Then the scientists carefully design the experiment to measure the values of these variables and compare them to the predictions, seeking to account for all factors that could affect the results. If the theory is sound and the experiment is well done, the experimental results should agree well with the theoretical predictions and the experimental error should be low.

In our case, when a force is applied to a vehicle at rest, it should accelerate in accordance with Newton's Second Law of Motion, $a = \dfrac{F}{m}$, which predicts that the acceleration depends on the force applied. So Newton's Second Law is our theoretical model for the motion of an accelerating object. Now, we know that a motor vehicle has a fair amount of friction in the brakes and wheel bearings, which means that not all of the force applied by our pushers will serve to accelerate the car. Some of it will simply overcome the friction. Also, the ground will probably not be perfectly level either, so this will affect the acceleration as well. So to make our model as useful as possible we will want to use the actual *net* force in our predictions so that they will be as accurate as possible. More on this below.

For our data collection we need a way to measure what the vehicle's acceleration actually is, so that we can compare it to our predictions. You already know an equation that gives the acceleration based on velocities and time. However, we have no convenient way of measuring the vehicle's velocity. (The vehicle will be moving too slowly for the speedometer to be of any use.) Fortunately, there is another equation we can use if we time the vehicle with a stop watch as it starts from rest and moves through a known distance. If we know the distance and the time, we can calculate the vehicle's acceleration with the equation $a = \frac{2d}{t^2}$. This is the equation we will use to determine the experimental acceleration value for each force, using the average time for each set of trials.

Here are some crucial details to make our experiment as successful as possible:

1. We will always have two students pushing on the vehicle, so for each force value that our pushers use the total applied force will be twice that amount. We will use four different force values in the experiment.
2. We need to measure friction so we can subtract it from the force the pushers are applying to get the net force applied for our predictions. To measure the friction we will simply get one pusher and estimate the absolute minimum amount of force needed to keep the vehicle barely moving at a constant speed. As you know from our studies of the Laws of Motion, vehicles move at a constant speed when there is no net force. So if the vehicle is moving at a constant speed it means that the friction and the applied force are exactly balanced. This allows us to infer what the friction force is.
3. We will use four different values of pushing force. For each force value used for the pushing, we need to time the vehicle over the ten-meter timing zone at least three times. The force the pushers apply to the vehicle will vary quite a bit, so if we get three valid trials at each force we will have three good data points for the time. You can then calculate the average of these times and use it to calculate the experimental value of the acceleration of the vehicle for that force.
4. The major factor introducing error into this experiment is the forces applied by our pushers. Pushing at a constant force while the vehicle is accelerating is basically impossible. (The dial on the force scale will be jumping all over the place.) But if our pushers are careful they can push with an *average* force that is pretty consistent. We need some kind of standard to judge whether or not we have had a successful run with good consistent pushing. Here is the criterion we will use: When we get three trials that have times that are all within a range of one second from highest to

lowest, we will accept these values as valid. If our times are not this close together, we will assume that the pushing forces are not consistent enough and we will keep running new trials until we get better data.

5. The instructor will take the vehicle, with a full gas tank, to get it weighed and will report this weight to the class. We need to make sure to measure the weight of the driver and the weight of the scale support rack (if there is one). These weights will need to be added to the weight of the vehicle, and the mass determined for this total weight. (Of course, the instructor will also make sure the gas tank is full on the day of the experiment, since the fuel in the tank could amount to 1-2% of the vehicle weight.)

Considerations for Your Report

In the Background section of your report, be sure to give adequate treatment to the theory we are using for this experiment. Describe why a graph of *acceleration* vs. *force* should be linear and why we expect our experimental acceleration values to vary in direct proportion to the force. Explain the equations we are using to get the predicted and experimental acceleration values. Since we are using two different equations your Background section should include explanations of both of them and what they are needed for. The force we are using to make our predictions takes friction into account. You need to explain this also, why we are doing it and how it relates to the equations.

In the Procedure section don't forget the important details, such as how we measured the friction force, weighed the driver, judged the validity of our time data and other details.

In the Results section, all time data should be presented in a single table, along with the average times for the trials at each force value applied by the pushers. All of the predicted values, experimental values and experimental errors should be presented in another table or two. Do not forget to state all of the other values used in the experiment, such as the vehicle weight, the weights of the driver and support rack, the distance, the total mass you calculated, and the friction force.

In the Discussion the main feature will be a graph of *acceleration* vs. *force*, showing both the predicted and experimental values on the same graph for all four force values. Carefully study Chapter VII on graphs in *The Student Lab Report Handbook* and make sure your graph meets all of the requirements listed.

For your predicted values of acceleration, use the total mass of the vehicle, driver and support rack. The instructor will tell you the weight of the vehicle, which you should record in your lab journal. The weights of the driver and support rack determined during the experiment should also be recorded in your journal. Convert the total weight from pounds to newtons, then determine the mass in kilograms by using the weight equation.

For the force values in your predictions, use the nominal amount of force applied (the two pushers' forces combined) less the amount of force necessary to overcome the friction (which will be determined during the experiment).

Table C-1 summarizes the calculations you need to perform for each set of trials.

The heart of your discussion will be a comparison of the two curves representing acceleration vs. force (displayed on the same graph), and a discussion of how well the actual values of acceleration match up with the predicted values. In addition to this graphical comparison you must compare the four predictions to the four experimental acceleration values by calculating the experimental error for each one, presenting these values in a table and discussing them.

variable	equation	comments
force	net force = (2 x force for each pusher) − friction force estimate	There are four values of net force, one for each set of trials.
predicted acceleration	predicted accel = (net force)/(total mass)	Net force is as calculated above. Mass is determined from the total weight. There is a predicted acceleration for each value of net force.
experimental acceleration	experimental accel = (2 x distance)/(avg time)^2	Distance is the length of the timing zone. Average time is the average of the three valid times for a given trial. There is an experimental acceleration for each value of net force.

Table C-1. Summary of equations for the calculations.

To compare the curves, think about the questions below. Do not write your discussion section by simply going down this list and answering each question. (Please spare your instructor the pain of reading such a report!) Instead, use the questions as a guide to the kinds of things you should discuss and then write your own discussion section in your own language.

Thought Questions and Considerations for Discussion

1. Are both of the curves linear? What does that mean?
2. Do they both look like direct proportions? What does that imply?
3. Do the curves have similar slopes? What does that imply?
4. How good are the error figures? An experimental error of less than 5% for an experiment as crude as this would be remarkably good. If the error is greater than 5%, you must identify and discuss the factors that could have contributed to the experimental error. In this experiment there are several, including wind that may have been blowing on the vehicle.
5. Do not make the mistake of merely assuming that the fluctuations in the pushers' forces explains everything, without taking into account the precautions we took to eliminate this factor from being a problem (our time data validity requirement).
6. Also do not make the mistake of assuming that friction explains the error. Friction can only affect the data one way (slowing the vehicle down), so if it was a factor, the data have to make sense in light of what friction would do. But further, since measuring friction and taking it into account in our predictions was part of our procedure, a generic appeal to friction will not do.
7. Finally, do not make the mistake of asserting that errors in the timing or the timing zone distance measurement explain the error. You should consider just what kind of percentage of error could realistically be in these measurements, and whether that kind of percentage helps at all in explaining the experimental error you have. For example, the timing zone was 10.00 meters long. If it was carefully laid out on the pavement, it is unlikely that the distance was in error by more than a centimeter or so. Even including the slight misalignments of the vehicle that cropped up, the

distance could probably not have been off by more than, say, 10 or 20 cm. But this is only 1 – 2% of 10 m, and if you are trying to explain an error of 5 or 10% or more this won't do it. Similar considerations apply to the time values. Given the slow speed the vehicle was moving, how far off could the timing have been? What kind of percentage error would this produce?

Alternate Experimental Method

If your class is using digital devices such as the Pasco Xplorer GLX to read forces, you can use a slightly different experimental method that will dramatically improve results and lower the experimental error. One of the major sources of error in this experiment is the difficulty the pushers have in accurately applying the correct amount of force to the vehicle. If you are using bathroom scales to measure the force, there is nothing that can be done about this problem and the pushers will simply have to do the best they can.

However, with the digital devices you can eliminate the problem of force accuracy by using the actual average values of the forces applied by the two pushers to calculate the predicted values. The Xplorer GLX can record a data file of the applied force during a given trial, and when reviewing the data file back at your computer you can view the mean value of the force during the trial. This mean value can then be used to calculate the prediction of the acceleration from Newton's Second Law. Using this method to form your predictions will eliminate much of the uncertainty surrounding the forces that are being applied to the car.

Here are a few details to consider if you will be using this alternative approach to collecting data:

1. You will not need to select four different force values in advance and push the vehicle repeatedly at each force value. Instead, each force need be applied for only a single trial.
2. Instead of selecting four values of force to use and running three or more trials with each one, select 15 different target force values and run a single trial with each. The force targets should range from low values that will barely get the vehicle to accelerate, all the way up to the highest values the pushers can deliver. For each trial tell the pushers the target force, and tell them to do their best to stay on it during the trial. But it won't matter how accurate the pushers are because you will be using the average of the data from the digital file to make the predictions, rather than relying on the pushers to maintain the target force accurately.
3. The method for determining values of net force for the predictions will be similar to that shown in Table C-1. The difference is that instead of using two times the target force for each pusher, you will add together the actual mean forces obtained from the data files for each pusher and subtract out the friction force.
4. The time of each trial will be used to determine the experimental value of the acceleration.
5. Calculate the experimental error for each trial and report these values in the report. Also calculate the average of the experimental error values and use this figure in your discussion of the results.

EXPERIMENT 3	THE HOT WHEELS LAB

Conservation of Energy

    We will perform this experiment together as a class. We will use the principles of conservation of energy to predict the velocity a toy car will have when it rolls to the bottom of a hill, starting from rest, and compare this prediction to our experimental result. The concept is simple. We have a Hot Wheels car on a short ramp of Hot Wheels track. We let the car roll down the ramp and use a digital timing system with a pair of photogates to time the car as it passes through a short timing zone on a horizontal stretch of track at the bottom of the hill. You will use the time data and the distance between the photogates to determine how fast the car is moving at the bottom of the hill, and compare this experimental measurement of velocity to the velocity you predict from performing a conservation of energy calculation based on the height of the hill.

    Data collection involves running the car down the hill several times, recording the time for each trial. You will use the average time from these trials in determining the final velocity. You will also need to measure the initial height, final height, and mass of the Hot Wheels car. And you need to measure the distance between the photogates. Each of these measurements should be taken by four or five different students in the class to improve the accuracy. Everyone can write their measurements in a table on the marker board to make sharing all the data easy. The measurements for each variable will be averaged and these averages used in your calculations. If any single measurement seems to be quite different from the others, it may mean the different measurement is inaccurate. A different student should repeat the measurement and the inaccurate measurement discarded.

    We will not be taking friction into account in this experiment. There are a couple of reasons for this. First, the Hot Wheels cars are so light weight that the friction force is very low and is extremely difficult to measure. Second, since the friction is so low, we can neglect it and still get very good results. With a digital timing system to aid in determining the car's final velocity, the experimental error should be low.

    This experiment is simple enough that not much needs to be said about your report preparation. The report will not require any graphs. There are a couple dozen measurements and their averages, plus the times and their average, that you need to present in a table or two. There will be only one experimental error value, and it should be quite low. If it is low, your discussion section should focus more on why it is so low than on explaining the experimental error. If the error is greater than you expect, then of course you must analyze your data and results to find reasonable possibilities for the source(s) of the error.

    In your Background section, do make sure you explain the theory behind this experiment thoroughly, including the Law of Conservation of Energy, how the law applies to our set up, and the equations you used for the two forms of energy involved. The final velocity prediction is calculated using the conservation of energy principles discussed in class, explained in this text, and rigorously explored in the problem sets. The experimental velocity measurement uses the simple equation relating distance, time and velocity that you mastered back in Chapter II.

EXPERIMENT 4	DC CIRCUITS
DC Circuits and the use of electronic test equipment	

We have two main goals in this laboratory exercise. The first is to use the theory you have learned to make predictions and compare these predictions to experimental results, as you have done a couple of times now. In this case, we want to use the circuit calculation techniques we have learned to determine what the voltages and currents in a circuit should be (your predictions), and then compare these predictions to actual measurements. Our second goal is for you to fool around with the test equipment and learn how to use it. One of the learning objectives for a good science class is to gain experience using unfamiliar equipment. This is a valuable skill for anyone entering a technical field of study, as many of you will. However, for the purpose of writing your report, consider yourself a competent researcher and focus on the first of these two goals.

You will begin by designing a DC circuit with a power supply and the following four resistor values:

$1.1 \text{ k}\Omega$ $\qquad$ $1.5 \text{ k}\Omega$ $\qquad$ $2.0 \text{ k}\Omega$ $\qquad$ $4.7 \text{ k}\Omega$

Your circuit must include both parallel and series resistance combinations, but the particular arrangement of the four resistors is up to you. Next, you must select a voltage to use to power your circuit. You will be using an electronic DC power supply as your voltage source. Your instructor will tell you what the voltage possibilities are for the power supply you will be using. Calculate the equivalent resistance for your circuit in advance of the experiment. But before calculating the voltages and currents for your circuit you should measure the voltage that your power supply produces when it is actually connected to your resistor network. Once you have a good measurement for your power supply voltage, sit down and calculate what the voltages and currents should be for each of the four resistors in your circuit. Use four decimal places in all of your calculations, just as you did in the Chapter VIII exercises. Compare your results with those of your team members to assure that each of you agrees on the numbers. These voltage and current values are your predicted values. Since there are six or eight measured values to compare to your predictions you will have six or eight experimental error values to present and discuss in your report.

On the day of the experiment you will hook up your circuit and measure the experimental values for the voltage and the current for each resistor. You may also wish to measure the equivalent resistance of your resistor network and the current going from the power supply into the network. These values may help in the discussion of your results. You may wish to compute the experimental error for the equivalent resistance as an aid to explaining the performance of your circuit.

Technical Notes

1. Resistors come in different "tolerances." Most common resistors have a 10% tolerance, which means that the resistor value can vary by +/− 10% from the value indicated by the colored bands on the resistor. You will be using "precision

resistors" that have a 1% tolerance. You should make note of this so that you can take it into consideration when you write the discussion section of your report and calculate the experimental errors.

2. To determine which of your resistors is which, you can measure the resistance using your digital multi-meter (DMM). You can also learn how to read the resistor code in the colored bands on the resistors. This is not hard and some of you will find it very interesting, so I will explain it here. Let's begin with the way the code works for ordinary 10% tolerance resistors. (This is one of those things that science geeks *all* know! Most of them learned it in ninth grade and have known it ever since.) For 10% resistors, the resistance values are in colored-coded scientific notation, with 2 significant digits. Since the value is in scientific notation, there is an implied decimal between the first and second digits. The first colored band corresponds to the first significant digit, the second band to the second digit, the third band to the power of 10, and the fourth band (if there is one) to the resistor tolerance. No fourth band usually indicates a 20% tolerance resistor. Tolerances of 10%, 5% and 1% are represented by silver, gold and brown bands, respectively.

Table C-2 shows the resistor color code for the digits (first two bands) and the power of 10 (third band).

	black	brown	red	orange	yellow	green	blue	violet	gray	white
sig digs	0	1	2	3	4	5	6	7	8	9
power of 10	$\times 10^0$	$\times 10^1$	$\times 10^2$	$\times 10^3$	$\times 10^4$	$\times 10^5$	$\times 10^6$	$\times 10^7$	$\times 10^8$	$\times 10^9$

Table C-2. Resistor color code.

Here is an example of how the color code would be used for standard resistors. (Note, however, that we are using precision resistors, not standard resistors, in our experiment.) If the first three bands are yellow, violet and orange, the resistor value is determined as follows:

yellow = 4 violet = 7

These two bands give you the value 4.7.

orange = $\times 10^3$

Putting these together gives 4.7×10^3, or 4.7 kΩ.

For precision resistors with 1% tolerance like those used in this experiment the code works slightly differently. These have three colored bands for significant digits with no implied decimal. The fourth band indicates the power of ten needed to multiply the first three digits by to bring the total to the correct value. Thus, if the first four bands were yellow, violet, green and brown, the resistor value would be

yellow = 4 violet = 7 green = 5 brown = $\times 10^1$

475 x 10^1 = 4750, or 4.75 k.

3. Your DMM will measure voltages, currents, and resistances. To make a measurement you must attend to these three things: connect the test leads to the correct terminals on the DMM, set the DMM selector switch to the appropriate setting, and connect the DMM properly to the circuit you are testing. Depending on the type of meter you have, you may also have to set a range switch. The range switch is set to a low range for low measurement values (such as voltages in the mV range or currents in the mA range), or to a higher range for higher values (such as voltages close to 1 V or currents close to 1 A). However, many DMMs these days will automatically set the range to that the meter can measure any value without fooling with a range switch.

   The DMM has a "common" or "negative" terminal, colored black, where the black test lead connects. The red test lead will connect to one of two different red terminals, depending on what measurement you are making. The two red terminals on the DMM are labeled to indicate which one is to be used for voltage and resistance measurements, and which one is to be used for current measurements.

   a. Voltage and resistance measurements are made by connecting the DMM *across* a resistor or power source. In other words, the DMM is connected in *parallel* to the resistor or power supply. To connect the DMM this way for a voltage measurement, you simply clip the red test lead to the higher voltage side of the device and clip the black test lead to the lower voltage side of the device. If you are measuring a resistance, be sure the resistance you are measuring is completely disconnected from everything else, and simply connect the two test leads to the ends of the resistor or resistor network. It does not matter which lead connects to which end of the resistor.

   Caution: Never place the DMM selector switch to the resistance *setting when the DMM is connected to devices with the power supply on.* The DMM does not want to see current flowing through it when the selector switch is in the *resistance* position. If you connect it like this you will either blow a fuse in the DMM or burn up the DMM, depending on what the voltage is at the time. Whenever your DMM is switched to the *resistance* position the power supply to the circuit should be off.

   b. To make a current measurement you must break the circuit and insert the DMM *into* the circuit so the current in the circuit flows through the DMM. In other words, the DMM is connected in *series*, like a resistor. The current should flow from the circuit into the red test lead and out of the black test lead back into the circuit.

4. To make it easy for you to connect your resistors together you will use a small mounting board called a *breadboard*. The breadboard has small rows of holes in which you insert the resistors or connecting wires. Certain rows of holes in the breadboard are connected together inside the breadboard to make it easy to connect

different devices together. The hole patterns in the breadboard and the ways they are internally connected are illustrated in Figure C-1.

In Figure C-1 the horizontal rows of holes shown that are the same color (other than black) are all connected together inside the breadboard. Thus, each of the holes in the four long rows of holes along the two long edges of the breadboard are connected together within a given row. These long rows of holes are usually connected to the power supply, such as connecting the positive power supply terminal to the light green row and the negative power supply terminal to the darker green row. This way anything on the breadboard needing power can get it by just connecting somewhere to those long rows. The short columns of five holes each in the center section of the breadboard (black or gray in the figure) are each connected together within each column. For example, each of the separate columns of five holes shown in gray are connected together within the column and are not connected to anything else. The same thing holds true for each of the columns of black holes shown in the figure on both sides of the horizontal center line. (You may also be interested to know that the spacings of the holes are designed so that many different types of integrated circuits, computer chips, and other devices can be plugged straight into the bread board. Obviously, we aren't using any of those devices in our experiment.)

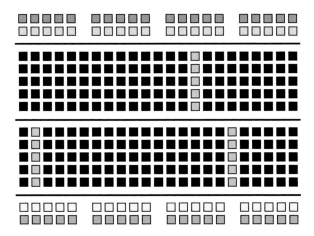

Figure C-1. Hole arrangements in an electronic breadboard.

The breadboard also has a set of terminals you can use to connect the circuit to the power supply. These terminals are called five-way binding posts. The five-way binding posts have a hole in the top that is sized to accept a type of plug called a banana plug. Sometimes these terminals or binding posts are spaced apart to accommodate a dual banana plug connector. These different connectors are shown in Figure C-2. To attach the connecting wire for the breadboard circuit to the binding posts, unscrew the red or back knob on the binding post, insert the wire into the hole through the metal post, and tighten the knob back down.

Appendix C Laboratory Experiments

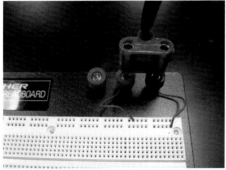

The five-way binding posts on this breadboard are spaced to accommodate a dual banana plug. In this photo there is a dual banana plug inserted into the binding posts, and connecting wires from the binding posts to the holes on the breadboard.

The binding posts on this breadboard won't accept a dual banana plug (they are too far apart), so single banana plugs must be used. If the test leads that come with your DMM have alligator clips, single banana plugs can be inserted into the binding posts to give the alligator clips something to clamp to, as shown in this photo. Connecting wires are also shown from the binding posts to the holes on the breadboard.

Figure C-2. Connecting the power supply to the breadboard.

Using a test lead with the appropriate connectors on both ends, connect the power supply to the five-way binding posts on the breadboard. Then use short lengths of connecting wire to connect the terminals to the rows that you will use for your $+/-$ power supply points.

To illustrate all of this, Figure C-3 shows two simple circuits, one with two resistors in parallel and one with two resistors in series. (Your circuit will have four resistors.) As you can see from these photos, you don't need much connecting wire to put a circuit together. The resistors themselves can simply be the means of connecting one device to another. However, note that if your resistors have been used before by other students and the legs are all bent, they may not insert into the breadboard sockets properly. So take a moment to straighten the resistor legs out so that they insert completely and securely down into the sockets. Figure

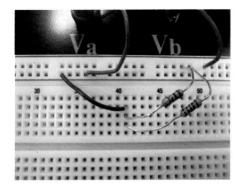

Figure C-3. Parallel (left) and series (right) resistor connections installed on the breadboard.

315

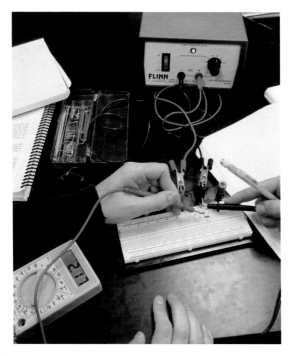

Figure C-4. Making measurements on a circuit.

C-4 shows students making measurements with a complete circuit connected to a power supply.

Procedural Highlights for this Experiment

This is a complex experiment with a lot of details to consider. But now that you are familiar with the background, let's summarize the main elements of the procedure that you need to keep in mind:

1. Design your circuit before the lab session and calculate the R_{EQ} value in advance to save time. (But don't calculate anything else yet.)
2. On lab day your first step is to identify which resistor is which. If you have time you may enjoy doing this by working through the color code. But if time is a factor it will be much faster simply to measure the resistances with your DMM.
3. Build your circuit on your breadboard and get your instructor to check that it is connected correctly.
4. Turn on the power supply and measure the power supply voltage at the terminals on the breadboard.
5. Using the actual power supply voltage from the measurement you just made in the previous step, calculate all of the currents and voltages in your circuit. Enter these all in your lab journal in a table. In this table you can list your variable names in the first column. In the second column enter the calculated values for each variable. (These are your predictions.) Use a third column for the measured values.

6. Now measure each of the voltages and currents. Remember, you measure a voltage by simply touching the DMM leads to each side of the resistor. To measure a current, you must remove one of the resistor legs from the breadboard and connect the DMM in series with the resistor so that the current flows in series through both the resistor and the DMM and then back into the rest of the circuit. Compare your measurements to your calculated values as you go. They should match closely. If they don't you've got a problem somewhere that you need to figure out before you continue. The problem could be in your circuit, your calculations, your measurement techniques, or your equipment. If you don't solve the problem your measurements may be a waste of time and may have to be done all over again.

Additional Notes

1. Be sure to use an accurately measured value of the power supply voltage for all of your calculations of the predicted values of voltage and current. Unless you are using an expensive voltage source with an accurate output voltage display, simply depending on the labels or indicators of the power supply will almost certainly result in large errors.
2. Your report must include a schematic diagram of your circuit. The diagram should have all of the resistances, voltages and currents labeled on it. These labels should match the labels in your data tables, which must also be in the report in the Results section. It will be best to place your schematic in the Procedure section so you can refer to it there.
3. Your schematic diagram can be prepared in a computer application or may be drawn by hand. However, if it is drawn by hand you must prepare it very neatly using a straightedge, and you must scan it so it can be digitally placed in your report.
4. Your report must include all of the experimental error figures for all of the voltages and currents in the circuit.

EXPERIMENT 5	SOLUBILITY
Determining solubility and the effect of temperature on solubility	

Note: The report for this experiment requires the student to set up a side-by-side bar chart showing solubilities of two different solutes in water. In addition to the reference procedures below for Microsoft Excel, specific procedures for creating bar charts in other applications, such as on a Mac using Pages, are available as free downloads from novarescienceandmath.com.

There are two main questions we will investigate in this experiment. First, we will examine how the solubility of a given solute in a given solvent can be experimentally determined. Second, we will investigate whether or not temperature affects solubility. On this second question, consider that when making iced tea, some people prefer to mix the sugar in while the tea is hot, just after it is brewed. They believe that when the tea is hot the sugar dissolves in it better. In technical language, they believe that the solubility of sucrose (sugar) in water is higher at higher temperatures. Is this a myth, or is it correct? If it is correct, does the same principle apply to other solutions, such as salt in water? These are the questions you seek to answer in this experiment.

You will investigate the relationship between temperature and solubility for two different solutes, salt and sugar, in one solvent, water. You will do this by determining the solubilities for four different solutions. That is, you will find out how much salt and sugar will dissolve into water at two different temperatures. Your team's hypothesis, which you need to formulate and write in your lab journal before you begin the experiment, will contain two parts. First, identify the solution (salt or sugar) you predict will have the higher solubility. Second, predict for both solutions whether or not the solubilities will be affected by temperature. Note that your hypothesis is qualitative; you are not predicting any actual solubility values. Thus, there will be no experimental error calculation in your report.

Safety Issues

1. Be very careful with all glassware and thermometers in this activity! There are three ways to break glass – carelessness, silliness, and improper procedures. These are all bad.
2. Use tongs and much caution when handling beakers containing hot liquids.
3. Always wear eye protection when there is a burner flame going at your table.

Using the Apparatus Properly

1. When placed in a graduated cylinder most liquids form a bowl-shaped curve on the top of the liquid. This curved shaped is called a **meniscus**. The proper place to read the liquid level in the graduated cylinder is at the bottom of the meniscus. To see the lower rim of the meniscus clearly it may help to place a dark colored background behind the cylinder while you read it.
2. Avoid **parallax error** in your measurements by positioning your head and eyes directly in line with the marks on the instrument you are trying to read. When reading measurements on a graduated cylinder if your head is slightly too high or too low your reading will suffer from parallax error and will be inaccurate. The same thing applies to readings made on the scale of a triple-beam balance.
3. Read your measurement with the appropriate number of significant digits. Recall from Chapter II that this means recording all of the digits known with certainty, plus one digit you estimate between the marks inscribed on the instrument.
4. Triple-beam balances have an adjustment knob to calibrate the balance so that it reads properly. Always check the calibration before you begin making measurements and adjust it if necessary. When no masses are on the pan and the weights are all set to zero, the alignment marks on the balance should line up perfectly.
5. The proper way to add solute to a solution is with a tool called a *scoop*.
6. If you are using electric/magnetic mixers for stirring, control the speed of the rotating magnet so that none of the solution splashes out. If any liquid splashes out your accuracy is compromised.
7. If you are using a Bunsen burner to heat your solutions, note that the flame from a Bunsen burner has two blue cones, one inside the other. The tip of the inner blue

cone is the hottest part of the flame. Position your burner ring so that the tip of the inner blue cone is right at the underside of the burner pad.
8. Get a fresh weighing tray each time you obtain more solute from the dispensing station. Used trays are contaminated and should be discarded. Make a rough estimate of the weight using the balance at the dispensing station. Then take the tray with the solute to your own lab table and there make your careful and precise measurement of the mass.
9. In the Procedure section of your lab report it is good to note each of these details so that the reader knows that you used all of the apparatus properly.

Procedure

1. Obtain some paper towels to use to keep your work area clean and dry.
2. Add approximately 50 mL of tap water to a 100-mL graduated cylinder. (Do not try to get precisely 50.0 mL of water. Your instructor will not believe your data if you show 50.0 mL.) Measure the volume of the water properly, reading at the proper place on the liquid meniscus and avoiding parallax error. Record the actual amount of water used as accurately as possible, using the correct precision for the apparatus. After measuring, transfer the water to a 250-mL beaker.
3. Go to where the salt is, get a weighing tray, and place it on the triple-beam balance. Measure out approximately 100 g of sodium chloride (table salt) and then carry the weighing tray of salt to your lab table. Use your own balance to obtain an accurate and precise measurement of how much sodium chloride you have and record this measurement.
4. Use a scoop to begin adding small amounts of salt to the water. After each amount is added, stir the solution with a spatula or a magnetic stirrer until the salt is completely dissolved. When the salt begins taking a long time to dissolve, proceed by adding increasingly small amounts each time. If you can still get the salt to dissolve by stirring, keep adding. The water will get cloudy, and it will no doubt have contaminants floating in it that you can see, but ignore these. What you are looking for is crystals of salt collecting on the bottom of the beaker.
5. At the first sign that crystals of salt are accumulating at the bottom of the beaker and are no longer dissolving into the water, stop adding salt. The solution is now saturated.
6. Measure and record the mass of the salt that remains in the weighing tray. The difference between the original and new masses of the salt is the mass of salt that dissolved into the water. (Since the mass of the weighing tray was in both measurements, it has been subtracted out, and thus doesn't matter.) Also record the temperature of the saturated solution.
7. Calculate the solubility of the salt in water as solubility = (mass of solute at saturation)/(volume of solvent). If you use grams of salt and mL of water, your solubility units for salt in water will be g/mL.
8. Repeat steps 2–7 with hot water, approximately 90°C. Use a graduated cylinder to make a precise measurement of a quantity of water, in the range of 50 mL. Then transfer the water to a 250-mL beaker and bring it to a boil. When you finish dissolving the solute into the solvent (i.e., when your solution is saturated), check the temperature. Your final temperature must be at least 70°C. If it isn't, you will need

319

to reheat it. After doing this you may need to top off your solute to make sure that the solution is still saturated. You may even need to reheat and add solute several times until you have a solution that is simultaneously at least 70°C and saturated.
9. Repeat steps 2–8 with sucrose (table sugar).

Analysis

Include in your report a side-by-side bar chart showing the two solubilities for each of the two solutions. Using your team's data, discuss whether your hypothesis was confirmed or not by this experiment. There are no experimental error calculations to perform for this report. You may wish to know this reference information:

Table salt is sodium chloride, NaCl
Table sugar is sucrose, $C_{12}H_{22}O_{11}$

Displaying Your Results

Figure C-5 shows a side-by-side bar chart, similar to the one you will prepare, but with different solutes. The side-by-side bar chart is the classiest and most efficient way to display your results.

There is one tricky thing about setting up a side-by-side bar chart in Microsoft Excel, and that is placing the labels under the groups of bars. In a spreadsheet, type the two solubility values for your first solute in the first row of your spreadsheet (columns A and B), and the two values for the second solute in the second row (columns A and B). Then type the labels you want to use for the groups of bars into a third row, such as row 3, like this:

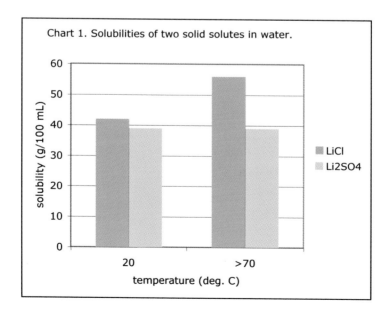

Figure C-5. Example side-by-side bar chart.

	A	B
1	42	56
2	39	39
3	20	>70

When setting up the graph, first highlight only your data (not the row with the labels), click on the Chart Wizard, and select the first graph choice of "Column." This will give you the chart setup you want. In order to label the columns, click on the Series menu in the second step of the Chart Wizard (or from the Source Data menu) and in the field labeled Category (X) axis labels type in

=Sheet1!$A$3:$B$3

This string of characters tells the software where to get your labels from (cells A3 and B3), and uses the same formatting that you can see in the "Values" field just above on the same page of the setup wizard. When you are finished, your chart will look like the one in Figure C-5.

EXPERIMENT 6 | DENSITY

Accurately determining density with correct lab technique and computer-based analysis

Note: The report for this experiment requires the student to determine the slope of the line of best fit through a data set. In addition to the reference procedures below for Microsoft Excel, specific procedures for finding the equation for the line of best fit in other applications, such as on a Mac using Pages, are available as free downloads from novarescienceandmath.com.

The purpose of this experiment is to make accurate measurements of the densities of some standard commercial materials and compare these values to the published values. This experiment is an exercise in careful measurements, the goal being to have the lowest possible error. We will also learn an important computer-based technique for analyzing data.

As you know, density is defined as $\rho = m/V$. Common metric units for density are kg/m^3 (MKS) and g/cm^3. The g/cm^3 units are common for laboratory work, which typically deals with small quantities. We will use g/cm^3 in this exercise.

We can rewrite the density equation as $m = \rho V$. In this linear equation, mass (the dependent variable) varies directly with volume (the independent variable), with the density acting as the constant of proportionality (the slope of the line). If we measured the masses and volumes of several samples of the same material, large volumes would have large masses, small volumes would have small masses, and in a graph of *mass* vs. *volume* the points should all fall on the same line. The slope of that line would be the density of the material. So this is what you will do, using your computer to determine the slope of the "best" line for your data, which gives you the experimental density value.

In this experiment you will work with two materials, aluminum metal and PVC plastic. You will have four samples of each material, each of a different size. You will use a triple-beam balance to determine each mass, and a graduated cylinder to determine each volume using the displacement method. For each of the two materials you will create a plot of all the masses versus their corresponding volumes. Then you will use your spreadsheet to calculate the slope of the "line of best fit" for your data. This slope is your experimental density value. Finally, you will calculate the experimental error of your value compared to the standard values from a technical reference source.

Using a Spreadsheet to Determine the Density (the slope of the "line of best fit")

The density, ρ, of a substance is *defined* in terms of the ratio of mass and volume. Because of this, and because of the fact that mass and volume are always proportional for a given substance (as long as the density of the substance is uniform), your data points for mass and volume for a particular substance are measurements that should fall very close to the line $m = \rho V$. No measurement can ever be exact; there is always error in every measurement. This is certainly the case when measuring volumes with a graduated cylinder as in this experiment, so there will be some error in your measurements. This means your data points will not fall perfectly in a line, but they should be close if your measurements were accurate. Each of your data points will fall just a bit above or below the true line $m = \rho V$ for the substance at hand.

Mathematically, there is a calculation used in statistics that computes the equation for the line that makes the total of all of these errors for a given data set as small as possible. This line is called the *line of best fit*. (The line of best fit actually makes the sum of the *squares* of the errors as small as possible, but that need not concern us here.) Because of the way the line of best fit calculation works, the slope of this line is actually a much better estimate for the density than you would get from merely calculating the mean of the four densities for your four samples. Thus, you will use a computer as a tool to calculate the slope of the line of best fit, and you will use this value as the experimental value for the density of a particular substance.

To use Microsoft Excel to calculate the slope of the line of best fit, first enter the data into the Excel spreadsheet as you normally do by typing the values of the independent variable (the volume) in column A and the values of the dependent variable (the mass) in column B. Then in an empty cell type in the following command:

=SLOPE(B1:B4,A1:A4)

This formula assumes your data begin in Row 1 and end in Row 4, with independent variable values in column A and dependent variable values in Column B. After you type this formula in and hit enter, the slope of the line of best fit will appear in the cell. This slope is the experimental value of the density. Easy!

You will go through this entire procedure twice, once for each of the two materials you are using.

Standard Values

The materials we are using are aluminum and polyvinyl chloride (PVC). The standard density values for these materials are:

aluminum   2.70 g/cm^3
polyvinyl chloride 1.4 g/cm^3

Note: The aluminum value is more precise (3 significant digits) than the PVC value (2 significant digits). This is because in the manufacturing of PVC its density can vary from 1.35 to 1.45 g/cm^3. If the measurements your class makes are very consistent, you might be able to determine more precisely the density of the PVC samples your class is using (to 3 significant digits). Then you could use that more precise value for your error calculations. After everyone has their density values your instructor will analyze them to see if this is possible. This will depend on how consistent the different values are from the student groups.

Important Additional Notes

1. You will be using large, 250-mL glass graduated cylinders for your volume measurements. Avoid breaking one of these costly pieces of apparatus by using proper technique to slide your samples down into the cylinder and to get them out. As I have written before, there are three ways to break glass – carelessness, silliness, and improper procedures. These are all bad in a laboratory!
2. Make sure every single measurement you make is documented in your report.
3. In your report you must include two graphs, one showing *mass* vs. *volume* for aluminum, another for PVC. Do not connect the data points in these graphs. Instead, set up your spreadsheet to show the line of best fit on the graph. To do this in Excel, right click on one of the data points, select Add Trendline, and select Linear. It is the slope of this line that is your experimental density value. In your Discussion section don't forget to describe what that line represents and why it is there in your graphs.
4. Use correct precision and significant digits throughout your measurements, calculations, and report.
5. Demonstrate in your report that you used correct procedures regarding calibration of the triple-beam balance, the way volumes are determined from the liquid meniscus, avoidance of parallax error, and so on.
6. An important feature of your report will be the clever way we are obtaining the experimental density value from slope of the line of best fit through the data. You certainly need to explain this in the Background section of your report when you present the theory behind the experiment. Obviously, you also need to address this in your Discussion section, describing what you are doing, how the experimental values were obtained, what they were, and what the experimental error is when these values are compared to the standard density values for these materials given above.

APPENDIX D
SCIENTISTS REQUIRED FOR CUMULATIVE REVIEW

During the school year in ASPC we will examine several historical episodes in the history of physics and chemistry. With each of them there are a number of well-known scientists we will encounter. Many of these contributors are of such key importance in the history of science that they are worth knowing about, so I have selected some of these for you to remember. To help you organize your study of these important scientists, this table itemizes what you need to know.

Scientists	What You Need to Know
Tycho Brahe Johannes Kepler Galileo Isaac Newton John Dalton Antoine Lavoisier (10 to 20 point quiz items)	• His name and nationality • Key dates related to his scientific work • Where he worked • What his major areas of research were • Summary of his major contribution, including the year (For Kepler, this includes stating the Laws of Planetary Motion; for Newton, stating the Laws of Motion; for Dalton, stating the five principles of his atomic model; for Lavoisier, describing the two tin experiments and the significant discoveries resulting from each one.)
Nicolaus Copernicus J. J. Thomson Robert Millikan Ernest Rutherford (10 point quiz items)	• His name and nationality • Summary of his major contribution, including the year (For Thomson, Millikan and Rutherford this includes describing their famous experiments and the significant discoveries resulting from each one.)
Albert Einstein Alessandro Volta James Clerk Maxwell Democritus Dmitri Mendeleev (5 point quiz items)	• His name and nationality • A single statement describing his major contribution, including the year (Einstein is famous for many things, but in this course it is the general theory of relativity that is of interest.)

APPENDIX E
UNIT CONVERSIONS TUTORIAL

One of the most basic skills scientists and engineers use is re-expressing quantities into equivalent quantities with different units of measure. These calculations are called unit conversions, and mastery of this skill is essential for any student in high school science.

Let's begin with the basic principle of how this works. First, we all know that multiplying any value by unity (one) leaves its value unchanged. Second, we also know that in any fraction if the numerator and denominator are equivalent, the value of the fraction is unity. A "conversion factor" is simply a fractional expression in which the numerator and denominator are equivalent ways of writing the same quantity. This means a conversion factor is just a special way of writing unity (one). Third, we know that when multiplying fractions, factors that appear in both the numerator and denominator may be "cancelled out." So when performing common unit conversions what we are doing is repeatedly multiplying our given quantity by unity until cancellations alter the units of measure so they are expressed the way we wish. Since all we are doing is multiplying by one, the value of our original quantity is unchanged; it simply looks different because it is expressed with different units of measure.

Let me elaborate on this idea of unity just a bit. School kids all learn that there are 5,280 feet in one mile, which means 5,280 ft = 1 mi. One mile and 5,280 feet are equivalent ways of writing the same length. If we place these two expressions into a fraction, the numerator and denominator are equivalent, so the value of the fraction is unity, regardless of the way we write it:

$$\frac{5,280 \text{ ft}}{1 \text{ mi}} = \frac{1 \text{ mi}}{5,280 \text{ ft}} = 1$$

So if you have a measurement such as 43,000 feet that you wish to re-express in miles, the conversion calculation would be written this way:

$$43,000 \text{ ft} \cdot \frac{1 \text{ mi}}{5,280 \text{ ft}} = 8.1 \text{ mi}$$

There are two important comments to make here. First, since any conversion factor can be written two ways (depending on which quantity is placed in the numerator), how do we know which way to write the conversion factor? Well, we know from algebra that when we have quantities in the numerator of a fraction that are multiplied, and quantities in the denominator of the fraction that are multiplied, any quantities that appear in both the numerator and denominator can be cancelled out. In the example above, we desire that the "feet" in the given quantity (which is in the numerator) will cancel out, so the conversion factor needs to be written with feet in the denominator and miles in the numerator.

Second, if you perform the calculation above, the result written on your calculator screen will be 8.143939394. So why didn't I write down all of those digits? Why did I round my answer off to simply 8.1 miles? The answer to that question has to do with the significant digits in the value 43,000 ft that we started with. The concept of significant digits and how to deal with them is explained in Chapter II. In the examples that follow below, I will always write the results with the correct number of significant digits for the values involved in the problem.

STRATEGIES FOR DOING UNIT CONVERSIONS CORRECTLY

There are several important techniques you need to follow when performing unit conversions. I will illustrate them below with examples. You should rework each of the examples on your own paper as practice to make sure you are doing them correctly. The conversion factors used in the examples below are all listed in Appendix A. You should study Appendix A to see which ones you must know by memory and which ones will be provided to you on quizzes.

1. Never use slant bars in your unit fractions. Use only horizontal bars.

In printed materials one often sees a value written with a slant fraction bar in the units, as in the value 35 m/s. Although writing the units this way is fine for a printed document, you should not write values this way when you are working a problem. This is because students often get confused and do not realize that one of the units is in the denominator in such an expression (s, or seconds, in my example), and the conversion factors used must take this into account.

Example: Convert 57.66 mi/hr into m/s.

Writing the given quantity with a horizontal bar makes it clear that the "hours" is in the denominator. This will help you to write the hours-to-seconds factor correctly.

$$57.66 \, \frac{mi}{hr} \cdot \frac{1609 \, m}{mi} \cdot \frac{1 \, hr}{3600 \, s} = 25.77 \, \frac{m}{s}$$

Now that you have your result, you may write it as 25.77 m/s if you wish, but do not use slant fraction bars in the units when you are working out the unit conversion.

2. The term "per" implies a fraction.

Some units of measure are commonly written with a "p" for "per," such as mph for miles per hour, or gps for gallons per second. Change these expressions to fractions with horizontal bars when you work out the unit conversion.

Example: Convert 472.15 gps to L/hr.

$$472.15 \, \frac{gal}{s} \cdot \frac{3.786 \, L}{1 \, gal} \cdot \frac{3600 \, s}{1 \, hr} = 6,435,000 \, \frac{L}{hr}$$

3. Use the ⊠ and ⊡ keys correctly when entering values into your calculator.

When dealing with several numerator terms and several denominator terms, multiply all the numerator terms together first, hitting the ⊠ key between each, then hit the ⊡

key and enter all of the denominator terms, hitting the ⌸ key between each. This way you will not need to write down intermediate results, and you will not need to use any parentheses.

Example: Convert 43.17 mm/hr into km/yr.

The set-up with all of the conversion factors is as follows:

$$43.17 \, \frac{mm}{hr} \cdot \frac{1 \, m}{1,000 \, mm} \cdot \frac{1 \, km}{1,000 \, m} \cdot \frac{24 \, hr}{1 \, day} \cdot \frac{365 \, day}{1 \, yr} = 0.378 \, \frac{km}{yr}$$

To execute this calculation in your calculator you should enter the values and operations in this sequence:

$43.17 \times 24 \times 365 \div 1000 \div 1000 =$

4. When converting units for area and volume such as cm^2 or m^3, you must use the appropriate length conversion factor twice for areas or three times for volumes.

The units "cm^2" for an area mean the same thing as "cm x cm." Likewise, "m^3" means "m x m x m." So when you use a length conversion factor such as 100 cm = 1 m or 1 in = 2.54 cm, you will need to use it twice to get squared units (areas) or three times to get cubed units (volumes).

Example: Convert 3,550 cm^3 to m^3.

$$3,550 \, cm^3 \cdot \frac{1 \, m}{100 \, cm} \cdot \frac{1 \, m}{100 \, cm} \cdot \frac{1 \, m}{100 \, cm} = 0.00355 \, m^3$$

Notice in this example that the unit cm occurs three times in the denominator, giving us cm^3 when they are all multiplied together. This cm^3 term in the denominator will cancel with the cm^3 term in the numerator. And since the m unit occurs three times in the numerator, they multiply together to give us m^3 for the units in our result.

This issue only arises when you have a unit raised to a power, such as when using a length unit to represent an area or a volume. When using a conversion factor such as 3.786 L = 1 gal, the units of measure are written using units that are strictly volumetric (liters and gallons), and are not obtained from lengths the way in^2, ft^2, cm^3, and m^3 are. Another common unit that uses a power is acceleration, which has units of m/s^2 in the MKS unit system.

Example: Convert 5.85 mi/hr^2 into MKS units.

$$5.85 \, \frac{mi}{hr^2} \cdot \frac{1,609 \, m}{1 \, mi} \cdot \frac{1 \, hr}{3600 \, s} \cdot \frac{1 \, hr}{3600 \, s} = 0.000726 \, \frac{m}{s^2}$$

With this example you see that since the "hours" unit is squared in the given quantity, the conversion factor converting the hours to seconds must appear twice in the conversion calculation.

5. Use scientific notation correctly in your calculator.

This item, of course, applies to all types of problems and not just to unit conversions. All scientific calculators have a key for entering values in scientific notation. When using this key to enter a value in scientific notation, you do not press the ⊠ key, nor do you enter the 10. You only enter the digits in the stem of the number, then hit the correct key (labeled EE or EXP on many calculators), then enter the power on the ten. There are many different calculators and many different styles, so your instructor will need to give you more specific advice. But you do need to learn to use this key correctly. You should *not* enter values by typing in the "x 10" and using the power key to enter the power on the 10. The scientific calculator is designed to reduce all of this key entry, and the potential for error, by use of the scientific notation key.

APPENDIX F
REFERENCES

CHAPTER II

The definition for significant digits is quoted from *Trigonometry*, Charles McKeague and Mark Turner, 6th ed.

All the images of scientists are from Wikimedia Commons.

The quote in the box about Copernicus is from *On the Revolutions of the Heavenly Spheres*, Nicolaus Copernicus, Prometheus Books 1995.

The quote in the box about Kepler is from *Harmonies of the World*, Johannes Kepler, Prometheus Books 1995.

Galileo's recantation speech is found in *The Great Physicists from Galileo to Einstein*, George Gamow, Dover 1988.

Some of the historical information about Brahe and Kepler is from *Kepler*, Max Caspar, Dover 1993.

CHAPTER III

Statements about Newton's wording of the Laws of Motion are based on the revised Motte translation, *Sir Isaac Newton's Mathematical Principles*, University of California Press, 1947.

CHAPTER V

The nuclear decay problem was adapted from *College Physics*, 5e, Raymond Serway and Jerry Faughn, Saunders, 1999.

CHAPTER VI

Most of the data in the box about the speed of air molecules are from Penn State College of Earth and Mineral Sciences website, http://www.ems.psu.edu/~bannon/moledyn.html.

Specific heat capacity and thermal conductivity data were obtained from engineeringtoolbox.com.

CHAPTER VII

The harmonic spectrum images were captured on an iPad with an app called "n-track tuner."

CHAPTER VIII

Much of the historical information about the history of electricity was taken from *The Great Physicists from Galileo to Einstein*, George Gamow, Dover 1988.

All the images of scientists are from Wikimedia Commons.

CHAPTER X

In the box on the hydrogen energy transitions the wavelength values and design concept for the diagram were taken from wikipedia.org in the article on Hydrogen spectral series.

The images of oxides are from Wikimedia Commons.

CHAPTER XI

My history of chemistry and atomic models owes a lot to *Conceptual Chemistry*, John Suchocki, Benjamin Cummings/Addison Wesley 2001.

Rutherford's comment about the artillery shell bouncing back from a piece of tissue paper is found on wikipedia.org, among other places.

All the images of scientists are from Wikimedia Commons.

Details about the voyage of the Kon-Tiki were obtained from wikipedia.org.

CHAPTER XII

The design ideas for the energy level diagrams in this chapter and Chapter XIII, as well as the diagram illustrating photons being absorbed and emitted by electrons in different energy levels, are adapted from *Conceptual Chemistry*, John Suchocki, Benjamin Cummings/Addison Wesley 2001.

All the images of scientists are from Wikimedia Commons.

CHAPTER XIII

The diagram of the calcium chloride crystal structure is from Wikimedia Commons.

INDEX

A 440, 130
absolute scale, 102
absolute zero, 103, 105, 106
absorption, 242, 246, 247
AC, 154, 159, 191, 200, 203, 204, 208
acceleration, 11, 16, 17, 19–27, 42–44, 48, 52, 55, 56, 58–60, 64, 79, 305–309, 327
acceleration (of gravity), 52, 58, 64, 69, 70, 100, 294
accumulation (of charge), 146, 147–149
accuracy, 11, 13, 209, 303, 318–322
acids, 274, 288
acoustic energy, 78, 80, 82, 288
action, 48, 54–57, 60
action-reaction pair, 57
activation energy, 281, 282, 287, 289, 290, 292
activity (of elements), 264
air, 42, 72–74, 80, 100, 105, 107, 108, 112, 114, 116, 119, 123–125, 130, 138, 147, 218, 223, 225–228, 232, 234, 287, 294
alchemists, 225, 260
algebra, ii, 85, 103, 104, 113, 325
alkali metal, 209, 211, 213, 264, 2265
alkaline earth metal, 209, 211, 213, 265
alloy, 209, 217
alpha decay, 94
alpha particle, 94, 95, 232, 233
alternating current. *See* AC
aluminum, 103, 110, 111, 115, 116, 206, 217, 218, 221, 285, 293, 322, 323
aluminum foil, 114, 115, 146, 293
amber, 141
ammonia, 259, 272, 280, 286
Ampère, André-Marie, 144
Ampère's Law, 144, 145, 191, 194, 195, 200, 203, 204, 208
amplitude, 117, 118, 132, 135
AM Radio, 133, 134, 138, 257
angle of incidence, 124
angle of reflection, 124
anode, 230
antifreeze, 73
appliances, 183

arc, 147, 148
area, ii, 33, 34, 61, 67, 68, 327
Aristotle, 26, 30, 32, 36
astronaut, 60
Atlantic Ocean, 115
atmosphere, 110, 111, 290
atmospheric pressure, 76, 77, 119, 130
atom, 46, 80, 82, 94, 104–110, 144, 146, 206, 209–211, 215, 216, 225–234, 242, 243, 245–280, 284, 285, 287, 289, 290
atomic bonding, 259–280
atomic energy level. *See* energy level
atomic mass, 229, 242, 254, 255, 258, 278, 279
atomic model, 225, 228, 231, 233, 243, 254, 284, 324
atomic number, 242, 254, 273, 279
atomic spectrum, 242, 244–246, 248, 249
atomic structure, 110
atomic theory, 2, 229, 243, 284
atomic weight, 229
Australia, 41
autobahn, 42
Avogadro, Amedeo, 229, 273
balancing chemical equations, 281, 282, 284–287, 292
Balmer, Johann, 250
Balmer Series, 250
balsa tree, 240
balsa wood, 241
banana plug, 314, 315
bar chart, 320
bass guitar, 130
battery, 154–156, 161–166, 172–184, 187, 204, 207
beta decay, 94
beta particle, 94
Beyond, 29
Bible, 30, 31
binary formula, 259, 269, 270, 276, 277
binding post, 314, 315
blimp, 265
Bohemia, 32
Bohr model, 242–246, 248, 255, 257
Bohr, Niels, 234, 242, 243
boiling temperature, 102, 103
Bologna, 143

bond. *See* chemical bond
Boyle, Robert, 226, 260
Boyle's Law, 61, 76, 226
Brahe, Tycho, 11, 31, 32, 141, 324
brass, 115, 217, 221
breadboard, 313–317
brine, 143
bronze, 217
brushes, 201, 202
Bunsen burner, 318
buret, 74
burner pad, 319
burner ring, 319
burning, 221
calculator, ii, 183, 325–328
calculus, 36, 145
Cambridge, England, 230
Cambridge University, 36, 232, 234
capacitance, 154
carbon, 101, 201, 211, 214, 217, 272, 273, 285, 290
carbon dioxide, 138, 259, 272, 273, 280, 286
catalyst, 281, 289, 290
catalytic cracking, 289
cathode, 230
cathode ray, 230
cathode ray tube, 231
Cavendish, Henry, 228
Cavendish Laboratory, 230
Celsius scale, 102
Chadwick, James, 234
charge, 141–151, 156, 191–194, 210, 232, 234, 242, 254, 259, 264, 266, 267, 275
charge-to-mass ratio, 230, 231
Charles' Law, 61, 72, 73, 75, 76
Charles' Law demonstration, iv, 72–75
chemical bond, 80, 215, 221, 259, 260, 265–280
chemical change, 209, 220, 222, 223
chemical equation, 279, 281, 284
chemical formula, 209, 214, 215, 272, 274, 277, 280
chemical property, 209, 220–222
chemical reaction, 81, 209–211, 214, 215, 217, 221, 223, 226, 228, 248, 259–261, 263, 279, 281–289, 292

331

chemical potential energy, 78–84, 89
chemical symbol, iii, 209, 214, 254, 271
chemical theory, 289
chemistry, i, iii–v, 155, 210, 225, 226, 228, 246, 248, 251, 254, 259, 260, 265, 274, 284, 287, 292, 324
Christianity, 1
Christians, 1
Church, 11, 26, 30–32, 36
circle, 67, 68
circuit, 151–184, 311, 313–317
circuit diagram, 160, 175
circular orbit, 28, 31, 32, 36
circular wave, 117, 120, 121
clarinet, 134, 137
coal, 82, 83
coefficient, 282, 284, 285
coil, 195–200, 203–208
collisions (of particles), 105, 106
collision theory, 281, 289–293
color, 117, 125, 126, 133, 220, 229, 230, 244, 248
comb, 146, 150
combustible, 221
combustion, 57, 216, 221, 288, 290, 291
commutator, 201, 202
compass, 144, 206, 207
compound, 209, 210, 214–217, 226, 228, 248, 259, 260, 267–270, 275, 277–288, 290
compression, 120, 130
computer, 309, 321, 322
computer chip, 115, 314
concentration, 290, 291
condensing temperature, 102, 103
conduction (of electricity), 138, 148–150, 211, 265
conduction (of heat), 102, 106–112, 115
conjunction, 32
conservation of energy, 79, 88, 90–92, 95–101, 162, 163, 310. *See also* Law of Conservation of Energy
constant of proportionality, 61, 63, 64, 79, 142
constants, 61, 63, 65, 66, 68
constant speed, 47, 53, 56, 60, 306
constellation, 28
constructive interference, 117, 128
control, 6

control group, 1, 7, 8, 9
control system, 198
convection, 102, 106–108, 112
conversion factor, iii, vii, 11, 12, 16, 18, 40, 98, 238, 294, 325–328. *See also* unit conversion factor
Copenhagen, 31
Copernican model, 36
Copernican Revolution, 11, 26
Copernicus, Nicolaus, 11, 31, 258, 324
copper, 110, 138, 143, 201, 206, 217, 221, 223, 236, 242
core, 197, 198, 203, 204
corpuscles, 230
cosmic rays, 134, 257
Coulomb, Charles-Augustin de, 142
Coulomb's Law, 142
covalent bond, 259, 260, 265, 270–275, 277
covalent compound, 274, 286
Cram-Pass-Forget cycle, i
Creation, 250
Creator, 2, 30, 31, 35, 36, 63, 142
crest, 117, 118, 128, 129
crude oil, 82, 83, 289
crystal, 216, 219, 259, 265, 266
crystal lattice, 216, 260, 261, 265
crystal structure, 110, 260, 266, 267, 274
current, 140, 143, 144, 146, 151, 154–167, 172–181, 183–190, 194, 195, 198–204, 207, 208, 265, 311, 313, 316, 317
Cycle of Scientific Enterprise, 1, 2, 3, 6, 10, 11
cylinder, 225, 236, 237
Dalton, John, 225, 228, 229, 254, 284, 324
David, 35
Davy, Humphry, 144
data, iii, 4, 6, 9, 15, 35, 67, 72–75, 112, 141, 302–307, 309, 310, 317, 320–322
DC, 154, 191, 204, 208
DC circuit, 140, 154, 155, 161, 165, 175, 184
DC Circuits experiment, 159, 311–317
decay. *See* nuclear decay
deceleration, 20
decomposition reaction, 282
degrees, 102
Democritus, 225, 324

Denmark, 144
denominator, 325–327
density, iii, 70, 71, 220, 225, 229, 230, 234–240, 294, 321–323
Density experiment, 321–323
dependent variable, 21, 22, 48, 52, 61–72, 76, 321, 322
depth, 67, 70, 71
destructive interference, 117, 129
deuterium, 255
diatomic gases, 259, 273, 274, 280–282
die, 221
diffraction, 117, 126, 129, 248
diffraction grating, 126, 248
digital multi-meter, 312, 313, 316, 317
digital radio, 134
digits. *See* significant digits
dimensional analysis, 11, 12, 78
direct current. *See* DC
direct variation, 62
discharge (pump), 151, 152
discharge (static). *See* static discharge
displacement method, 322
dissipate, 162
dissolve, 217, 219, 288, 319
distance, 11, 16, 22–26, 41, 43, 47, 71, 81, 87, 89, 105, 118, 142, 306, 308, 310
DNA, 105
double bond, 261, 273, 274
double replacement reaction, 283, 286
draw, 221
ductile, 209, 221, 279
earth, 27, 28, 30, 33, 40, 41, 43, 47, 60, 79, 82, 83, 108, 110, 111, 141, 160, 191, 192, 225, 226, 257, 303
eclipse, 38, 192
Eddington, Arthur, 38, 192
Einstein, Albert, 11, 38, 39, 59, 79, 93, 95, 145, 191, 192, 245, 324
ekasilicon, 229, 230
elastic potential energy, 78, 80, 83
electrical conductivity, 110
electrical energy, 78, 80, 83
electrical neutrality, 264, 266, 267, 270, 271, 275
electric field, 132, 191–194, 206
electricity, 140–144, 147, 148, 151, 153, 154, 211, 244, 248

electrolyte, 143
electromagnet, 197, 207
electromagnetic field theory, 2
electromagnetic radiation, 78, 80–83, 108, 132–134, 192, 194
electromagnetic spectrum, 117, 123, 132–134, 255
electromagnetic theory, 145
electromagnetic waves, 80, 83, 107, 108, 119, 123, 132, 138, 145, 257
electromagnetism, 144
electromotive force, 160
electron, 43, 86, 94, 146–151, 155, 156, 206, 210, 229–232, 234, 242–275, 278
electron sea, 259, 265, 266
electron configuration notation, 242, 253, 257, 279
electron dot diagram, 259, 271–274, 280
electronic, 115, 138, 154, 158, 211, 302, 311
electron sharing, 270–275
electron transfer, 267–269, 275
electroscope, 140, 141, 146–150, 191
element, iii, 94, 209–217, 225, 226, 228, 229, 244, 248, 252–260, 263, 264, 274, 276–285
ellipse, 33
elliptical orbit, 33, 36
emission, 242, 244, 247–249, 256
Empyrean, 29
endothermic reaction, 281, 287, 288
energy, 12, 51, 78–101, 104–108, 113, 116, 118, 119, 128, 135, 147, 160–162, 242–244, 246–248, 251–256, 262, 282, 286, 288, 292
energy balance, 90, 101
energy diagram, 281
energy level, 242–244, 247–252, 256, 257, 261–264, 266, 270–275
energy trail, 82
energy transfer, 78
energy transition, 242, 250, 255
engineering, iii, 75, 205, 289, 325
England, 141, 144
enzymes, 189
epicycle, 27, 28, 31
equation for a line, 62

equivalent resistance, 140, 165–172, 175, 184, 185, 189, 190, 311, 316
Eratosthenes, 27
error, 13
ethylene glycol, 73
evidence, 2
excitation, 242, 244, 247, 248, 251–253
exothermic reaction, 281, 287, 288
experiment, iv, 1–10, 15, 141, 148, 218, 225, 226, 231, 302–323
experimental error, 209, 304, 305, 307–310, 312, 317, 318, 320, 322
experimental group, 7
experimental result, 310–311
experimental value, 305–310, 321–323
experimental variable. *See* variable
explanatory variable, 1, 7, 9, 10, 60
explosion, v, 80–82, 130, 221, 223, 242, 282, 283, 288, 290, 293
exponent, 65
eye protection, 318
fact, 1–6, 9
Fahrenheit scale, 102
Fall of man, 30
Faraday, Michael, 144
Faraday's Law of Magnetic Induction, 144, 145, 191, 195, 199, 203, 204, 207, 208
Favorite Experiments for Physics and Physical Science, iv
feathers, 141
fibreglass insulation, 116
field coils, 200, 201
field lines, 191, 194–197, 203
fields, 144, 191–194, 204, 206, 207
fingerprint, 244
fire, 225, 226, 289
firecracker, 81, 82, 221, 223
Firmament, 28, 29
first mover, 29
fission, 80
flame test, 248
flammable, 221, 228, 265
Florence, 36
fluid, 95, 106, 107, 112
fluid dynamics, 65

flute, 134–137
flux. *See* magnetic flux
FM Radio, 134, 138, 257
force, 12, 36, 38, 46–49, 52–60, 67, 71, 81, 87, 88, 98, 142, 160, 191, 192, 194, 305–309
formula, ii, iii, 259, 269, 270, 276–278, 322
fossil fuel, 82, 83
fraction, 325, 326
France, 144
free fall, 36, 52
freezing temperature, 102, 103
French Revolution, 226
frequency, 120–123, 132–134, 136, 138, 242, 244, 246–248, 255
frequency range, 130, 132
friction, 50, 51, 56, 78, 79, 87, 88, 90, 95, 98–101, 146–148, 232, 306–308, 310
frog legs, 143
function (algebraic), 23
fundamental, 135, 136
fur, 141
fusion, 80
fusion reaction, 82
galaxy, 245
Galileo, Galilei, 11, 27, 31, 36, 52, 61, 63, 324
Galvani, Luigi, 143
gamma decay, 94
gamma rays, 134, 139, 257
gas, 72, 76, 102, 105, 106, 109, 210, 216, 218, 226, 229, 239, 248, 265, 283
gasoline, 79, 289, 293
gas turbine generator, 83
gauge, 152, 153
Gay-Lussac, Joseph, 229
gene, 7
General Revelation, 2, 36
general theory of relativity, 2, 38, 39, 192, 324
generator, 145, 191, 197–200, 205, 208
geocentric system, 28
geometry, ii, 32, 197, 260
germanium, 229, 230, 234, 236, 239, 252–254
Gilbert, William, 140
glass, 110, 115, 116, 119, 124
glassware, 318
goals (of atoms), 259, 264, 266, 267, 270–273, 275, 279, 280
God, v, 1, 2, 29, 35, 36, 63, 79, 105, 146
gold, 104

333

gold foil experiment, 232, 233
graduated cylinder, 15, 209, 240, 318, 319, 322, 323
graph (of motion), 11, 22–26, 43–45
gravitational attraction, 47, 61, 67, 71, 142, 191
gravitational field, 60, 79, 191, 192, 194, 206
gravitational potential energy, 61, 67, 69, 78, 79, 83, 85, 87–93, 96–99, 160
gravity, 11, 32, 36–38, 40, 52, 56, 60, 64, 192, 206
Greeks, 141, 148, 225
ground (electrical), 160, 161
ground state, 251
group, 209, 211, 263, 264, 274, 276, 278
Gulf of Mexico, 115
Gulf Stream, 115
guitar, 130
halogen, 209, 211, 213, 264
harmonics, 117, 134–136
harmonic spectrum, 136, 137
Harmonies of the World, 35
hearing protection, 131
heat, 78 – 83, 95, 102, 104, 106–116, 162, 206, 248, 283, 287–289, 292, 293, 318
heat sink, 115
heat transfer, 102, 106–108, 113, 114, 148
heavenly bodies, 27–29, 36
height, 68, 69, 78, 84, 88–90, 92, 93, 96, 118, 160
heliocentric model, 31, 36
helium, 82, 94, 104, 109, 232, 265
heresy, 31, 36
heterogeneous mixture, 209, 217
Hindenburg disaster, 265, 282
Holy Spirit, 46, 206
homework, iii
homogeneous mixture, 209, 217, 218
horizontal, 326
horizontal axis, 62
Hot Wheels Lab, 89, 310
human body, 41
human hearing, 117, 130, 138
human race, 246
human voice, 130
humidity, 130
hydrogen, 82, 104, 228, 229, 250, 254, 255, 256, 265, 270–274, 279, 282–286

hypothesis, 1, 3–6, 9, 10, 38, 302–305, 318, 320
ideal gas, 72, 76
ideal pendulum, 95, 96
incident ray, 124
independent variable, 21, 22, 61, 62, 64, 65, 67–72, 76, 321, 322
induction (of electric current), 195, 199, 204
induction (of static electricity), 149, 150
inert gas, 211
inertia, 46, 47, 53, 59, 60
inference, 2
inflammable, 221
infrared, 107, 108, 117, 133, 134, 138, 257
infrasonic, 117, 130
inner transition metal, 209, 213, 214
instrument (measurement), 5, 13, 14
instrument (musical), v, 135, 136
insulation, 116
integrated circuit, 314
interference, 117, 126–129
interference pattern, 126, 129
internal energy, 102, 105, 106, 109, 287–290
inverse square law, 64, 142
ion, 259, 266, 267, 269, 275–277, 281, 284
ionic bond, 259–261, 265–269, 275, 277
ionic compound, 270, 275, 276, 288
ionization, 259, 266, 267, 269, 270
iPad, 136
iron, 114, 143, 193, 203, 206, 217, 221, 223, 240
isotope, 229, 254, 255, 258, 279
Jesus, 2
junction, 164, 172, 175, 178
Jupiter, 28, 29, 32, 36
Kelvin scale, 102
Kepler, Johannes, 11, 32–36, 38, 63, 141, 324
Kepler's First Law, 33
Kepler's Laws of Planetary Motion, 11, 32, 33, 35, 38, 39, 324
Kepler's Second Law, 33
Kepler's Third Law, 34, 35, 142, 145
keyboard, 130

kinetic energy, 61, 67, 70, 78–80, 82–99, 102, 105–107, 109, 289
kinetic theory of gases, 102, 109, 113, 114, 145
Kirchhoff's Junction Law, 172, 175, 178–181
Kirchhoff's laws, 140, 172, 175
Kirchhoff's Voltage Drop Law, 172–176, 179, 181
knowledge, 1, 3
Kon-Tiki, 240
koozie, 115
lab journal, 15, 302–304, 316, 318
lab report, iii–v, 15, 302, 319
laboratory experiments, 302–323
laser, v, 41, 122, 124–126, 129, 138
Lavoisier, Antoine, 225–228, 260, 284, 324
Lavoisier, Marie, Anne, 226
Law of Conservation of Energy, 78, 79, 82, 88, 89, 90, 95. *See also* conservation of energy
law of conservation of mass in chemical reactions, 225, 226, 281, 284
law of definite proportions, 228
Law of Inertia, 46, 47, 53, 59, 60
law of reflection, 124, 125
Law of Universal Gravitation, 36–39, 64, 71, 142, 192
Laws of Motion. *See* Newton's Laws of Motion
Laws of Planetary Motion. *See* Kepler's Laws of Planetary Motion
least common multiple, 276
leaves, 146
Les Miserables, 60
Leyden jar, 141
light, 36, 38, 41–42, 79–82, 94, 105, 108, 117–120, 123–129, 132, 138, 141, 145, 147, 148, 244, 245, 248, 249, 288
lightning, 148
light (speed of), 40–43, 58, 79, 86, 95, 122–124, 294
light waves, 127–129
light-year, 41
line, 21, 22, 62
linear function, 62, 308, 321
line of best fit, 321–323
liquid, 106, 114, 210, 217, 218, 223
load, 165
logarithmic scale, 136

London, 116, 144
longitudinal wave, 117, 119
loop, 172–174
loudness, 131, 132, 136
loudspeaker, 119, 120
lurking variable, 1, 5, 7, 9, 10, 60
Lyman Series, 250
Lyman, Theodore, 250
Mach, 105
magnet, 141, 144, 191–195, 197, 199, 200
magnetic field, 132, 144, 191–196, 201–207, 230, 257
magnetic flux, 194, 195, 198, 200, 203, 204
magnetic levitation, 100
magnetic stirrer, 318, 319
magnetism, 132, 141, 144, 191, 193, 194, 207, 208
malleable, 209, 221
Mars, 27–29, 60
mass, iii, 38, 46, 47, 52, 57–59, 64, 69–71, 78–80, 84–87, 91–98, 100, 101, 114, 142, 192, 194, 209, 210, 217, 225, 226, 228, 231, 235, 238, 240, 265, 284, 305, 307, 308, 318, 319, 321, 322
mass-energy equivalence, 79, 93
mass number, 242, 254, 255, 258, 278, 279
mastery, i–v
mathematical order, 63
mathematical structure, 35, 142, 250
mathematical symmetry, 166
mathematics, i, ii, v, vii
math skills, iv
matter, 46, 47, 59, 118, 130, 210, 234
Maxwell, James Clerk, 140, 144–146, 324
Maxwell's Equations, 145
measured value, 13
measurement, 5, 11–15, 21, 209, 310, 313, 316–319, 321–323, 325
mechanical energy, 78, 87
mechanical power, i
mechanical system, 78, 83, 87, 88, 95, 303
mechanical wave, 119, 130
mechanical work, 79
medication, 7
medieval model, 11, 26
medieval period, 26, 30, 225

medium, 117, 119, 120, 121, 125, 127, 130, 135, 139
melting temperature, 102, 103
Mendeleev, Dmitri, 225, 229, 234, 324
meniscus, 15, 209, 318, 319, 323
mental model, 2–4
Mercury (planet), 28, 29
metal, 102, 106, 109–112, 115, 116, 140, 146, 149, 194, 198, 211, 213, 217, 218, 221, 225, 259, 260, 265–269, 275, 276, 278, 322
metal foil, 115
metallic bond, 259, 260, 265
metalloid, 209, 211, 213
methane, 259, 272, 273, 280, 286, 287
metric prefix, ii, iii, vii, 11, 294
metric system, iii, 12
microwaves, 80, 108, 123, 133, 134, 138, 257
middle C, 130, 134, 136
Millikan, Robert, 225, 231, 232, 324
missile, 59
Mission Impossible, 222
Mississippi River, 40
mixture, 209, 210, 217
MKS system of units, 11, 12, 17–20, 46, 58, 70, 72, 84, 90, 120, 157, 168, 235, 327
model, 1, 2, 4, 7, 9, 26–28, 31, 32, 38, 61, 145, 146, 225, 306
molecular bond, 81
molecule, 63, 72, 79, 80, 82, 102–109, 153, 209, 215, 229, 259–262, 270–276, 279–287, 289, 290
momentum, i
monopole, 193
moon, 27, 28, 29, 30, 36, 52, 58, 192
motion, 11, 16, 23, 25, 26, 36, 37, 38, 42–44, 46–48, 57, 80, 105, 146, 306
motor, 145, 162, 198–201, 205
motorcycle, 40
musical instrument, v, 117
naming convention, 269
natural gas, 82, 83, 286, 287
negative, 155, 156, 160, 161, 192, 210, 230, 234, 242, 254, 264, 267, 269, 276, 284
net charge, 269, 275
net force, 47, 48, 53, 55, 306

neutron, 94, 146, 206, 210, 229, 232, 234, 242, 248, 254, 255, 258, 277, 279
New Orleans, 40
Newton, Isaac, 11, 33, 36, 38, 46, 48, 49, 59, 127, 145, 191, 192, 324
Newton's First Law of Motion, 47, 53, 55, 56, 59, 60
Newton's Laws of Motion, 36, 38, 39, 46–49, 53–56, 59, 60, 306, 324
Newton's Second Law of Motion, 36, 48–50, 52–59, 72, 305–309
Newton's Third Law of Motion, 48, 49, 54–57, 60, 293
nitrogen, 206, 210, 214, 248, 271, 272
Nobel Prize, 230, 232, 234, 242, 245, 248
noble gas, 209, 211, 213, 263–265, 270, 278
node, 164, 172, 179–181
nonmetal, 111, 209, 211, 213, 259, 260, 264, 266–271, 274, 275, 278
normalization, 61, 64–66, 68
normalized curve, 67
normalized equation, 65, 68–72
normal line, 124
nova, 32
n-track tuner, 136
nuclear bomb, 82
nuclear decay, 93, 94
nuclear energy, 78, 80, 82, 83, 93, 255
nuclear physics, 94
nuclear process, 79, 80, 93, 94
nuclear radiation, 94
nuclear reaction, 79, 82, 260
nuclear weapon, 79
nucleon, 94, 210, 254, 255, 278
nucleus, 80, 82, 94, 95, 146, 210, 229, 232, 234, 242, 243, 250, 254, 255, 260
numerator, 325–327
Objectives List, iii, vi
observation, 2
ocean, 110, 111
octopus, v
Ørsted, Hans, 144
Ohm's Law, 140, 157, 158, 161, 162, 172, 175–181
oil drop experiment, 231, 232
oil industry, 289
Olympic pool, 241

335

On the Revolutions of Heavenly Spheres, 31
optics, 32, 36
orbital, 242, 250–253, 261, 262, 270
oscillation, 119, 120, 122, 123, 132, 203, 204, 303
OSHA, 131
oxidation, 221
oxygen, 104, 193, 210, 211, 216, 221, 228, 229, 273, 279, 282, 283, 286, 287, 290, 293
Pacific Ocean, 240
parallax error, 209, 318, 319, 323
parallel resistors, 140, 164–174, 177, 180, 181, 311, 313, 315
Paschen, Friedrich, 250
Paschen Series, 250
Pauli exclusion principle, 248, 252
Pauli, Wolfgang, 248
peak, 118, 127
pendulum, 60, 78, 95, 96, 99, 100, 303–305
Pendulum Lab, 9, 10, 118, 303–305
period, 34, 117, 122–124, 138, 209, 211, 255, 261, 263, 264, 303–305
Periodic Table of the Elements, iii, 209, 211–214, 229, 242, 254, 255, 257, 259, 261, 263, 264, 267, 270, 279, 280
perpetual motion machine, 60, 96
phase change, 103
phase (of waves), 128, 129
phase (of matter), 210, 218, 220, 223
photon, 94, 242, 245–248, 256, 257
physical change, 209, 220, 222, 223
physical constant, iii, vii, 11, 235, 294
physical property, 209, 218, 220–222
physics, i–v, 246, 248, 293, 305, 324
piano, 130
Pick-up Truck Lab. *See* Soul of Motion Lab
pipe, 151–154, 159, 164, 264, 265
pitch, 117, 121, 131
Planck, Max, 245
Planck relation, 242, 245

Planck's constant, 245
planet, 26–28, 33, 34, 36–38, 46, 52, 59, 63, 111, 210, 242, 243
planetary model, 243
plants, 82, 83
plasma, 206
plastic, 110
Plato, 26, 30
Platonic solids, 35
plumb pudding model, 231
polarization, 132
pole (magnetic), 191–193, 199
polyatomic ion, 259, 270, 271, 274–278, 285, 286
Polynesian islands, 240
Pope John Paul II, 31
Pope Paul III, 31
positive, 154–156, 175, 192, 210, 230, 234, 242, 254, 264, 266, 269, 275, 276, 284, 314
postage stamp, 40
Postal Service, 249
potential difference, 140, 160
power, 12, 51, 140, 157, 158, 161–163, 167, 175–190, 203, 204, 311
power distribution system, 154, 158, 160, 200, 201
power lines, 204
power station, 80, 83, 183, 198, 202
power supply, 183, 313–317
Prague, 32, 141
pre-algebra, ii
precipitate, 209, 219
precision, 11, 13, 14, 17, 209, 294, 319, 323
prediction, 3–6, 38, 89, 305–311, 316, 318
pressure, 61, 67, 70, 71, 76, 77, 102, 109, 113, 114, 119, 130, 151, 153, 226
pressure difference, 151
pressure gauge, 151
pressure profile, 152, 153
Priestly, Joseph, 142, 228, 260
primary coil, 203, 204, 208
Primum Mobile, 29
Principia Mathematica, 37, 38
principle quantum number, 242, 249, 252
prism, 124, 125, 248
problem solving method, 16, 17, 19, 21, 51
problem solving strategy, 12
procedure, 7
product, 226, 281–284, 288, 293

propagation, 117, 119, 120, 123, 124, 125, 127, 130, 132, 139
properties, 216, 225, 229. *See also* chemical properties, physical properties
proportion, 61, 65, 72, 85
proportional, 48, 133, 322
proportionality, 61, 62, 65, 284
proton, 43, 46, 58, 94, 146, 151, 156, 210, 211, 229, 232, 242, 252, 254, 255, 257, 260, 261, 264, 266, 275, 278, 279
Proust, Joseph, 228, 284
Proxima Centari, 41
Ptolemaic model, 11, 26–31
Ptolemy, 26, 30
pump, 151, 153, 160
pure substance, 209, 210, 215, 217
PVC, 322, 323
quadratic, 24, 25
quanta, 242, 245, 247
quantized, 242, 245
quantum, 242, 245, 252
quantum mechanics, 245, 248
quantum model, 242, 248
quantum number, 248–252
quantum physics, 242
quantum state, 248, 249
quantum theory, 2
quark, 63
Queen Elizabeth, 141
radioactive, 94, 214
radiation, 102, 107, 108, 110, 117, 256, 257
radio waves, 80, 108, 119, 123, 145
radium, 94, 95
radius, 68, 69, 77
radon, 94, 95
rare earth elements, 209, 213, 214
rarefaction, 120, 130
ratio, 139
ray, 124
reactant, 226, 279–284, 287–289, 292
reaction, 48, 54–57, 60
reaction rate, 288–292
rectangular solids, 225
reflected ray, 124
reflection, 117, 124
refraction, 117, 124–126, 248
Reign of Terror, 226
relay, 198, 205
resistance, 140, 153–161, 167, 178, 183, 203, 311–313, 316

resistor, 140, 154–190, 311–313, 315, 316
resistor color code, 312, 316
resistor network, 163–172, 180, 184, 311
resolution, 13
resonance, 117, 127, 134, 135
resonant frequency, 134
response variable, 1, 7, 9, 10, 60
retention, i
retrograde motion, 27, 28
retro-rocket, 42, 44, 57
right-hand rule, 191, 195, 196, 207
rocket, v, 42–44, 56–60
roller coaster, 100
room temperature, 105, 106, 114, 235, 282, 288
rotor, 201, 202
Rudolph II, 32
rusting, 221
Rutherford, Ernest, 225, 232–234, 324
Rydberg, Johannes, 250
safety, 318
salt, 217, 318, 319
saturated solution, 209, 218, 223, 319, 320
saturation, 209, 219
Saturn, 28, 29, 32
saxophone, 136, 137
scale, 67, 99, 305, 306, 318
schematic diagram, 154, 317
science, i, ii, v, vii, 1, 2, 13, 75, 139, 141, 221, 245, 325
scientific fact. *See* fact
scientific knowledge, 3, 4
scientific method, 1, 6, 9
scientific notation, ii, 11, 15, 16, 40–42, 312, 328
scientists, 324, 325
scoop, 318, 319
Scripture, 1, 2, 36
secondary coil, 203, 204
Second Law of Thermodynamics, 106
semiconductor, 211
series resistors, 140, 163–172, 175, 311, 315
Shakespeare, 35
shape, 220
shell, 263
sibilance, 130
significant digits, ii, 11, 13–21, 40, 41, 103. 157, 167, 183, 238, 294, 312, 318, 323, 325, 326

significant figures. *See* significant digits
simple machines, i
sine wave, 118, 138
single bond, 261, 273, 274
single replacement reaction, 283, 285
sinusoidal, 118, 200, 203, 204
SI system, 12
size, 220
skills, i, ii
sky, 36
slope, 21, 22, 62, 321–323
slope-intercept form, 21, 62
sodium demonstration, 264
soft drinks, 282
solar flare, 43
solar system, 35, 242
solenoid, 191, 197, 198, 203, 207
solid, 105–107, 210, 217, 218, 223, 320
solubility, 209, 219, 220, 223, 317, 320
Solubility experiment, 317–321
soluble, 209
solute, 209, 217, 219, 223, 317–320
solution, 209, 217–219, 318–320
solvent, 209, 217, 219, 223, 317, 319
Soul of Motion Lab, 60, 305–309
sound, 117–119, 130, 139, 147, 288
sound energy. *See* acoustic energy
Sound Pressure Level, 131
sound wave, 80, 119, 121, 126–130, 135, 138, 139, 206
South America, 240
space, 118, 119
space ship, 43, 44, 58–60
space station, 60
space-time, 38, 192
spark, 147, 282
spatula, 319
Special Revelation, 2
specific heat capacity, 102, 109–112, 114–116
spectral lines, 242, 248
spectrascope, 242, 248
spectroscopy, 242, 244, 255, 257
spectrum tube, 248, 249
speed, 47, 48, 95, 124, 290
speed of light. *See* light (speed of)
speedometer, 306

speed of sound, 105, 123, 130, 139
sphere, 67–69, 76, 146–148, 150
spheres (planetary), 11, 28, 29, 30, 31
spin, 252
standing wave, 127, 128, 135
starlight, 192
stars, 27, 28, 30, 32, 41, 56, 63, 79
state of motion, 46, 48, 55
static charge, 147, 151
static discharge, 140, 147, 148
static electricity, 140–151, 191, 232
statistics, 322
stator, 201
Statue of Liberty, 221
steam, 210
steel, 138, 203–207, 217, 240
string, 119, 127, 135
Student Lab Report Handbook, iv, 302, 304, 305, 307
study strategy, vi, vii
Styrofoam, 148–150
subatomic particle, 146
sub-orbital, 251, 252
subsonic, 130
substance, 79, 80, 104–106, 109, 111, 112, 209, 210, 217–223, 225, 226, 228, 234, 245, 259, 290, 322
suction, 151, 152, 160
sugar, 317, 318, 320
sun, 27–30, 33, 34, 38, 40, 41, 43, 78, 82, 83, 108, 110, 113, 114, 119, 192, 210, 242, 244
supernova, 32
supersaturated solution, 209, 218
supersonic, 130
surface area, 290
suspension, 209, 217, 218
synthesis reaction, 282, 286
Systèm internationale d'unités, 12
table of values, 69–72, 76
Teaching Science so that Students Learn Science, i, iv
technology, 146
telescope, 32, 35, 36
temperature, 41, 67, 72–76, 81, 102–111, 113–116, 130, 287, 289, 293, 317–319
temperature conversions, 103, 104, 113
temperature scales, 102, 105
test, 4, 5, 9
texture, 220

337

theory, 1–6, 9–11, 26, 38, 141, 192, 304, 305, 307, 310, 311, 323
therapy, 7, 9
thermal conductivity, 102, 110–112, 114–116
thermal energy, 78, 80, 83, 102, 104, 105, 111
thermal equilibrium, 102, 106, 114, 115
thermal insulator, 102, 110
thermocouple, 73
thermodynamics, 106
thermometer, 318
threshold of hearing, 131
threshold of pain, 131
Thomson, J. J., 225, 230, 231, 324
thrust, 57–59
timbre, 117, 134, 135
time, 11, 20–26, 43, 64, 118, 288, 289, 303, 305–310
time interval, 25
timing zone, 308
tin experiment, 226–228, 324
tolerance, 311, 312
tools, i
Toronto, 116
train, 42
transformer, 144, 191, 197, 201, 203–205, 208
transition metal, 209, 213, 214
translation, 105
transmission lines, 202
transverse wave, 117, 119
treatment, 7, 9
triangle, 67, 68, 125
trigonometry, 118
triple-beam balance, 209, 318, 319, 322, 323
triple bond, 261, 273, 274
tritium, 255
trough, 117, 118, 127–129
true value, 13
truth, 1, 2, 4, 9
turns, 194, 195, 199, 203
Tycho. *See* Brahe, Tycho
ultrasonic, 117, 130, 138
ultraviolet, 108, 117, 133, 134, 138
uncertainty, 21
uniform acceleration, 19, 20, 22–25, 45
uniform motion, 47
unit conversion, ii, 12, 15–19, 40, 101, 325–328

unit conversion factor, 12, 17, 19, 40. *See also* conversion factor
United Kingdom, 116
unit fraction, 238, 326
units (of measure), vii, 12, 17, 18, 51, 102, 156, 157, 162, 178, 235, 238, 302, 325, 327
University of Copenhagen, 144
U.S. Customary System, iii, 12
vacuum, 42, 101, 113, 114, 116, 119, 123, 130, 134, 206, 294
vacuum tube, 230
valence electron, 261, 263–274, 279, 284
valence number, 259, 269, 270, 277
valence shell, 259, 263–274, 280
Van de Graaff generator, 140, 150, 206
vapor, 210, 218, 223
variable, vii, 1, 6, 7, 9, 10, 17, 60–63, 65–72, 76, 99, 118, 154, 156, 157, 184, 235, 303–305, 310, 316
variation, 61–66, 85, 245
Variation and Proportion Study Packet, 62, 67–77
velocity, 11, 16, 19–26, 42–48, 52, 55, 56, 59, 67, 70, 78, 85, 89, 90, 92, 95–99, 101, 117, 120, 122, 123, 130, 306, 310
Venus, 28, 29
vertical axis, 62
vibration (of a string), 127, 135
vibration (of atoms or molecules), 105, 106, 107
violin, 135
viscosity, 232
visible light, 117, 132–134, 247, 257
visible spectrum, 133, 138, 245, 246
Volta, Alessandro, 140, 143, 324
voltage, 140, 155–167, 172–181, 183–190, 202, 204, 230–232, 248, 265, 311, 313, 316, 317
Voltaic Pile, 143, 144, 155
volume, ii, 61, 67–69, 72–77, 225, 226, 235, 237–240, 319, 322, 323, 327
volume (of sound), 117, 132
water, 97, 100, 102–104, 108–112, 114–116, 119, 124, 130, 153, 154, 160, 164, 172, 210, 215–218, 221–228, 234, 235, 239–241, 259, 272, 273,

279, 280, 282–284, 286, 288, 293, 294, 317, 319, 320
water analogy (for electricity), 140, 151, 164, 172
water circuit, 151, 153, 155
water filter, 151–153, 159
water supply, 264
water wave, 120, 121
wave, 79, 107, 108, 117–129, 135, 138, 139
wavelength, 40, 108, 117, 120–122, 125, 132–135, 138, 139, 242, 245, 248, 250, 256
wavelength range, 133
wave motion, 120
wave phenomena, 117, 124
wave train, 118, 119, 127, 130, 132
weekly quiz, iv, vi
Weekly Review Guide, iv, vi, vii
weighing tray, 319
weight, 46, 52, 58, 64, 87–89, 98, 100, 101, 191, 238, 307, 310, 318, 319
white light, 125
windings, 205
wire, 153–156, 158, 159, 164, 191, 195, 198, 200, 202, 205, 206, 265, 313, 314
wire drawing, 221
work, 78, 79, 81, 87–90, 97–100, 162
World War I, 234
X-rays, 108, 123, 134, 139, 257
y-intercept, 62
zeppelin, 265
zero reference, 84, 160
zip code, 249
zodiac, 28